Springer Proceedings in Materials

Volume 51

Springer Proceedings in Materials publishes the latest research in Materials Science and Engineering presented at high standard academic conferences and scientific meetings. It provides a platform for researchers, professionals and students to present their scientific findings and stay up-to-date with the development in Materials Science and Engineering. The scope is multidisciplinary and ranges from fundamental to applied research, including, but not limited to:

- Structural Materials
- Metallic Materials
- Magnetic, Optical and Electronic Materials
- Ceramics, Glass, Composites, Natural Materials
- Biomaterials
- Nanotechnology
- Characterization and Evaluation of Materials
- Energy Materials
- Materials Processing

To submit a proposal or request further information, please contact one of the following Springer Publishing Editors according to your affiliation location:

European countries: **Mayra Castro** (mayra.castro@springer.com)
India, South Asia and Middle East: **Swati Meherishi** (swati.meherishi@springer.com)
South Korea: **Smith Chae** (smith.chae@springer.com)
Southeast Asia, Australia and New Zealand: **Ramesh Nath Premnath** (ramesh.premnath@springer.com)
The Americas: **Michael Luby** (michael.luby@springer.com)
China and all the other countries or regions, as well as topics in materials chemistry: **Maggie Guo** (maggie.guo@cn.springernature.com)

This book series is indexed in **SCOPUS** and **EI Compendex** database.

Sunita Rattan · Bhuvanesh Gupta ·
Christine Jeyaseelan · Anita Gupta
Editors

Emerging Trends in Synthesis and Catalysis in Chemistry

Proceedings of ETSC 2023

Springer

Editors
Sunita Rattan
Department of Chemistry
Amity Institute of Applied Sciences
Amity University
Noida, Uttar Pradesh, India

Bhuvanesh Gupta
Department of Textile and Fiber Engineering
Indian Institute of Technology Delhi
New Delhi, Delhi, India

Christine Jeyaseelan
Department of Chemistry
Amity Institute of Applied Sciences
Amity University
Noida, Uttar Pradesh, India

Anita Gupta
Department of Chemistry
Amity Institute of Applied Sciences
Amity University
Noida, Uttar Pradesh, India

ISSN 2662-3161 ISSN 2662-317X (electronic)
Springer Proceedings in Materials
ISBN 978-981-97-5171-6 ISBN 978-981-97-5169-3 (eBook)
https://doi.org/10.1007/978-981-97-5169-3

This Springer imprint is published by the registered company Springer Nature Singapore Pte Ltd.
The registered company address is: 152 Beach Road, #21-01/04 Gateway East, Singapore 189721, Singapore

Preface

Synthesis and catalysis are fundamental areas of chemical research, with broad implications for a range of industries, including pharmaceuticals, materials, energy, and many more. In recent years, significant progress has been made in this field, with new methods, techniques, and materials being developed that are revolutionizing the way we think about synthesis and catalysis. Asymmetric synthesis, natural product synthesis, photochemistry, and homogeneous catalysis using organometallic complexes are some of the signature areas along with the inclusion of artificial intelligence and machine learning in synthesis and catalysis. AI is being used to design new catalysts with improved properties and selectivity. Machine learning algorithms are being applied to identify the best catalyst candidates based on vast amounts of data. An in-depth understanding of these areas is extremely useful in developing various applications for the same.

The prime objective of the 2nd International Symposium on Emerging Trends in Synthesis and Catalysis (ETSC 2023) was to bring together scientists and researchers to discuss novel synthesis strategies and highly efficient heterogeneous/homogeneous catalytic methods. It presented a unique opportunity for leading academicians, scientists, researchers, and scholars, to exchange their knowledge base and share their experiences and research outcomes on various aspects of green synthesis, reaction designs, catalyst synthesis, and performance optimization. This two-day event included plenary sessions, invited talks, and oral and poster presentations on the most recent innovations, trends, and concerns as well as practical challenges encountered and solutions adopted in the fields of synthesis and catalysis.

This book includes selected papers from the presentations made at the symposium held on April 12–13, 2023, at Amity University, Noida. It was attended by participants from different parts of the country. It was organized under the auspices of Amity University, Uttar Pradesh, and sponsored by Agnitio Pharma and Springer Nature itself as publishing partner.

This edition includes papers based on organic and inorganic materials like transition metal complexes along with their anticancer activity. Synthesis of various organic compounds based on oxadiazole, thiazole derivatives, etc. has been studied. The synthesis of nano-based materials with their application in catalysis has been described. Recent advances in luminescent materials for dosimetry applications are another area of focus. Application of zeolite synthesized from industrial waste and other modified biopolymers for wastewater treatment is elaborated. The growing importance of synthesis and catalysis makes the interdisciplinary research an upcoming and new area of innovative ideas.

We are sure that the results and findings reported here are the latest in their fields and will be useful for academicians and scientists as well as students of undergraduate and postgraduate level along with research scholars who are specializing in the field of organic and inorganic chemistry studying the applications of the materials in various fields.

We appreciate the efforts of all the participants who have contributed to this book and look forward to enhanced research in the respective areas and increase future collaborations. The knowledge exchanged during the symposium and this proceedings is the dedicated efforts of all the members of the chemistry department who contributed in various aspects to make it a success. We extend our thanks to each one of them for all their efforts.

New Delhi, India

Prof. Sunita Rattan
Prof. Bhuvanesh Gupta
Prof. Christine Jeyaseelan
Prof. Anita Gupta

Contents

About the Editors

Prof. Sunita Rattan is Dean Science and Technology and Director, Amity Institute of Applied Sciences, Amity University, Noida. Her research areas include polymer grafting, ion beam modifications, stimulus responsive polymeric membranes, chemiresistive sensors, nanomaterials, nanocomposites, and membranes for selective separation. She has published more than 75 research papers in reputed international journals and has given more than 25 invited talks at various international conferences. She has filed 15 patents out of which five are granted. She has edited a number of books and has contributed chapters in the books with publishers like Royal Society of Chemistry, Springer, and Nova publishers.

Prof. Bhuvanesh Gupta is working as Professor in the Department of Textile Technology at the Indian Institute of Technology (IIT) Delhi. He completed his M.Phil. and Ph.D. at the Vikram University, Ujjain, India, and IIT Delhi, respectively. He worked as Scientist at various Institutions in Switzerland, France, and Sweden, before returning to IIT Delhi as faculty member. His research interests include modification of polymers, membranes, medical textiles, biomaterials, and tissue engineering. He has authored eight books and more than 170 research papers, holds 25 patents, and serves on the editorial boards of several international journals.

Prof. Christine Jeyaseelan is Professor and Head of Department Chemistry at Amity Institute of Applied Sciences, Amity University, Noida. She has been awarded with doctoral degree by Nagpur University. Topics of research interest are polarographic and voltammetric analysis of metal ions and pharmaceutical compounds and the use of natural and modified adsorbents for the removal of toxic pollutants from water sources, biodiesel production, and analysis. She has also written 2 books and has published several papers in national and international journals. She has 3 patents to her credit of which 1 has been granted. She has about 20 years of teaching experience at both undergraduate and postgraduate level.

Prof. Anita Gupta received her Ph.D. from PEC University of Technology, Chandigarh (India), and has been teaching at Amity University, Noida (India) since 2011, besides her 12 years of teaching experience in State government Institute. She has been granted 3 patents for her novel work in Organometallic Chemistry and is Life Member of Indian Carbon Society.

Impurity Profiling of Antihyperglycemic Drug-Sitagliptin

Aashima and Deepshikha Gupta

1 Introduction

Antihyperglycemic medications are a group of drugs used to lower high blood sugar levels in individuals with diabetes mellitus, a medical condition characterized by elevated levels of glucose (sugar) in the blood. There are several classes of anti-hyperglycemic medications, each with its own mechanism of action like insulin [1], biguanides, sulfonylureas, meglitinides, thiazolidinediones, alpha-glucosidase inhibitors, dipeptidyl peptidase-4 (DPP-4) inhibitors, SGLT-2 inhibitors, GLP-1 receptor agonists, and combination therapies [2]. Dipeptidyl peptidase-4 (DPP-4) inhibitors-like sitagliptin, saxagliptin, and linagliptin increase the levels of incretin hormones, which stimulate insulin release and reduce glucagon secretion and are the current focus class of antihyperglycemic drug [3]. The choice of antihyperglycemic medication(s) depends on various factors, including the type of diabetes, the patient's overall health, their blood sugar levels, and other individualized considerations [4].

Sitagliptin belongs to the gliptin family of dipeptidyl peptidase-4 (DPP-4) inhibitors [5]. Sitagliptin works by blocking DPP-4, into which it degrades, from acting proteolytically. Sitagliptin was developed by Merck & Co. and approved for medical use in the United States in 2006 [6]. The incretins, gut hormones, play a critical role in glucose regulation by decreasing glucagon release and increasing insulin formation. Incretin (Glucagon-like peptide-1) levels are increased by the DPP-4 inhibitor [7]. Glucagon release is inhibited, and insulin secretion is increased by gastric inhibitory polypeptide and its mode of action, which lowers blood glucose levels. Glycemic management using DPP-4 inhibitors has been found to be successful in treating diabetes [8].

Aashima · D. Gupta (✉)
Amity Institute of Applied Sciences, Amity University Uttar Pradesh, Sector 125, Noida, India
e-mail: dgupta2@amity.edu

S. Rattan et al. (eds.), *Emerging Trends in Synthesis and Catalysis in Chemistry*,
Springer Proceedings in Materials 51, https://doi.org/10.1007/978-981-97-5169-3_1

The active pharmaceutical ingredients (API) contaminated with impurities can have a substantial impact on the quality, safety, and efficacy of therapeutic treatments, making impurity profile analyses an essential part of the development of APIs. Because of the strict quality requirements [9], the API used to make a drug product must have its impurity profile studied. According to the most recent International Conference on Harmonisation (ICH) rules, any impurity with a concentration of $\geq$ 0.10% should be identified and characterized [10]. Quantifying the contaminants in the drug material and the drug product is extremely important to regulatory bodies. Standards for impurities are thus needed in their purest form. These contaminants are necessary at the conceptual stage of the analytical process. These impurities are needed to evaluate various performance parameters, such as specificity, accuracy and precision, linearity, range, robustness, system suitability testing, limit of detection (LOD), limit of quantification (LOQ), and relative response factor (RRF), during the design and development phase of analytical methods. Small quantities of contaminants are also required for analysis of standard production batches [9].

2 Synthesis of API

Six impurities were found to be present in sitagliptin in the impurity profiling literature search. For the synthesis of the active pharmaceutical ingredient, Sitagliptin (**1**), sketched in Fig. 1, the 2,4,5-tri fluorophenyl acetic acid (**9**) was used as the starting material that reacted with Meldrum's acid (**8**) in the presence of a base to produce (**7**). The essential intermediate, β-ketoamide (**5**), is created by combining this intermediate (**7**) with triazole. HCl (**6**) in the presence of ethyl acetate at 35 °C for six hours.

The intermediate was produced in the next stage by heating the (R)-(+)-phenylethylamine (**4**) in toluene to 110 °C and reacting it with the β-ketoamide (**3**). The diastereomer compound was created by further reducing the β-enamino amide (**3**) using the $NaBH_4$/MsOH reagent. It was subsequently purified and separated as HCl salt (**2**). The necessary diastereomer was simply transformed into sitagliptin (**1**) by utilizing 10% Pd/C in 50% isopropanol at 65–70 °C and 105 psi pressure (H_2) for 10 h [11].

3 Impurities of Sitagliptin:

Our next step after identifying the various sitagliptin impurities was to illustrate the synthetic pathway for production. Acid (**Impurity III**), triazole (**Impurity II**), alkene (**Impurity IV**), cyclohexyl (**Impurity V**), enantiomer (**Impurity VI**), and phenylethylamine (**Impurity I**) are among the impurities found in the sample of API. The structure of each impurity is given in Fig. 2. The manufacture of acid impurities, alkene impurities, and enantiomer impurities is described in earlier literature.

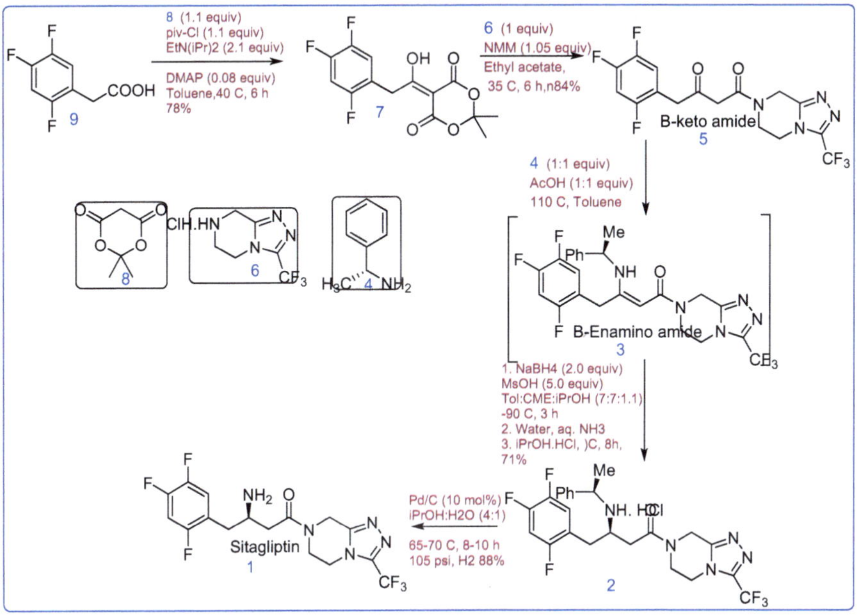

Fig. 1 Synthetic scheme for sitagliptin (API)

Numerous techniques have been documented for the manufacture of acid impurities to the date till now [12–15].

4 Synthesis of Acid Impurity

Precedent Approach

To carry out the reduction of methyl 3-oxo-4-(2,4,5-trifluorophenyl) butanoate stereo selectively, several of these techniques have used the expensive chiral ligand-based catalyst (S)-BINAPRuCl_2, which is followed by the Mitsunobu reaction to create acid impurity [16]. Use of [Rh (COD)Cl]$_2$, i.e., cyclooctadiene rhodium chloride dimer in addition to the presence of a chiral ferrocene ligand to achieve asymmetric hydrogenation is also reported. Most of these methods require the use of expensive chiral ligands in costly enzymes that break down or costly chiral catalysts like 2R(+) Diisopino campheyl chloroborane. One of the known synthetic methods for acid impurities involves a more involved synthetic process and enzymatic resolution in addition to a number of other processes [17–20].

Modified Synthesis

Therefore, a slightly modified technique was used to enable the production of this acid impurity as given in Fig. 3. The synthesis is split into 2 steps: first, Meldrum's

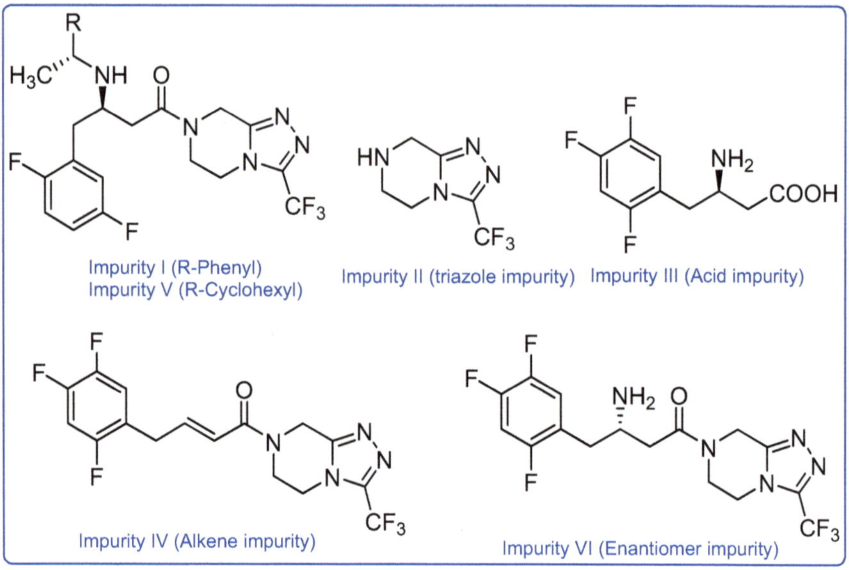

Fig. 2 Structures of six identified impurities associated with sitagliptin

adduct (**7**) was converted to the appropriate β-keto ester by reaction with ethyl alcohol at boiling temperature for 3 h; and second, enamine was synthesized using chiral auxiliary (R)-(+)-phenylethylamine in presence of ethanoic acid in toluene at 110 °C. The resultant **β-enamino ester** was reduced using the $NaBH_4$/MsOH reagent and treated with HCl to be isolated as HCl salt. Following this hydrolysis of **β-enamino ester** with 2.2 equiv of NaOH is done. The hydrolyzed product is hydrogenating using 10% Pd/C (50% wet) in methanol at 65–70 °C and 105 psi H_2 pressure for 12 h, to give **acid impurity III** [21].

4.1 Alkene Impurity

Precedent Approach

Impurity IV can be synthesized when KBH_2 and $BF_3.Et_2O$ are present, a more sophisticated method is used to convert 2,4,5-trifluorophenylaceticacid to alcohol. Alkene impurity is produced by combining triazole, NEt_3, EtOAc, and Et_2O with tetrahydrofuran, Et_2O, and TEMPO to oxidize aldehyde, which is subsequently followed by a Doebner alteration of the Knoevenagel condensing process [22].

Modified Synthesis

Alkene impurity for formed by using an acid-catalyzed β-elimination reaction; the acetate salt (in Fig. 4) heated to reflux for 3 h in n-BuOH rapidly produces an alkene

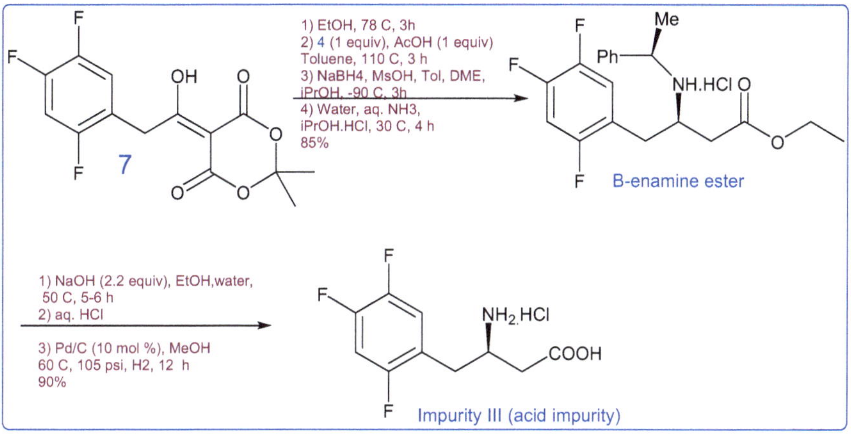

Fig. 3 Synthetic scheme for the impurity III

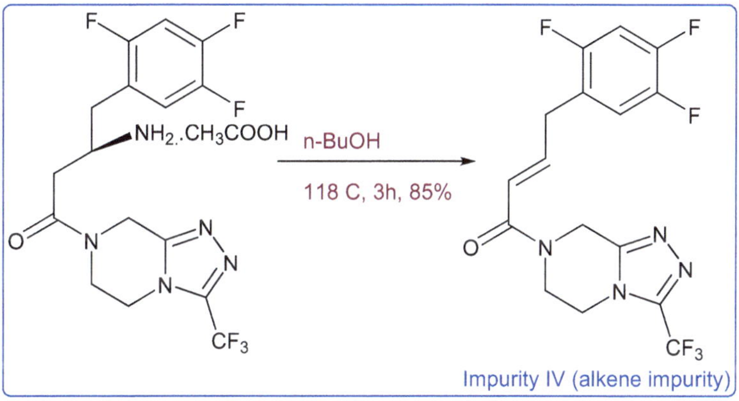

Fig. 4 Synthetic scheme for the impurity IV

impurity IV that is separated from toluene. Sitagliptin can be co-injected to validate its presence in the sample during the HPLC analysis.

4.2 *Cyclohexyl Impurity:*

Precedent Synthesis

According to the reported method, enantioselective catalytic allylic amination of olefin use chiral ligand 1,2-dichloroethane and a Pd(TFA)$_2$ catalyst, i.e., palladium (II) trifluoroacetate takes place over the course of seven days. This is

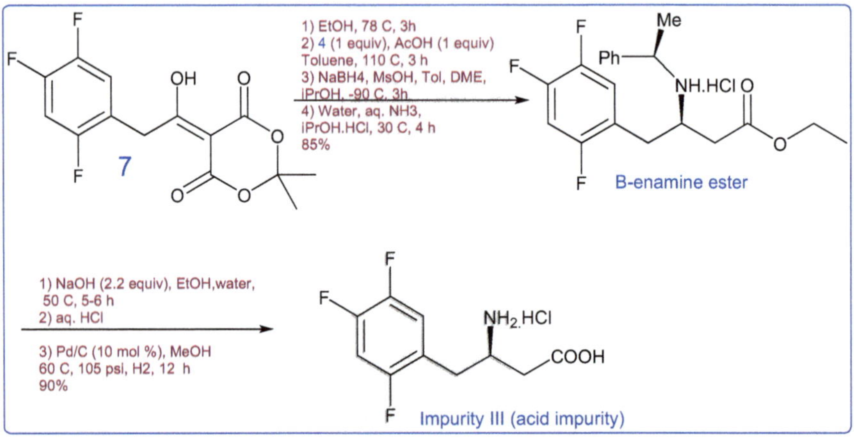

Fig. 5 Synthetic scheme for the Impurity V

followed by hydrolysis in presence of K_2CO_3 and methanol and water. In the next step, addition of 9-BBN, tetrahydrofuran presence of aqueous sodium hydroxide and peroxide hazardous CrO_3-based olefinic oxidation, and ultimately coupling with 1-H-1,2,3-benzotriazol-1-ol in presence of N, N-Diisopropylethylamine and dimethyl formamide to produce the required enantiomer impurity VI [23].

Modified Synthesis

In a synthetic process, 2,4,5-trifluorophenyl acetic acid (**9**) combines with Meldrum's acid (**8**) in the presence of a base to form an intermediate. The essential intermediate, β-ketoamide (**5**), is created by mixing this intermediate with triazole. HCl in the presence of ethyl acetate at 35 °C for six hours as depicted in Fig. 1. To get cyclohexyl impurity, further reactions were carried out with R(-)-Cyclohexyl ethylamine (**4a**), toluene at 110 °C for 3 h. Then, in next step, sodium borohydride as a reducing agent along with dimethoxy ethane (DME) at low temperature for 3 h and at last treated with water and aqueous ammonia (70%) to get desired cyclohexyl **Impurity V**. The steps after intermediate 5 formation are depicted in Fig. 5.

4.3 *Enantiomer Impurity*

Modified Synthesis

The key intermediate β-ketoamide (**5**), from Fig. 6, reacts with toluene, acetic acid, and (S)-phenylethylamine (**4b**). Reacting with $NaBH_4$, DME, isopropyl alcohol at low temperature (-90 °C) for 3 h. Further, water, and aqueous ammonia isopropyl alcohol for 8 h in the next step. Further, reduction using Pd/C (10 mol%) (isopropyl

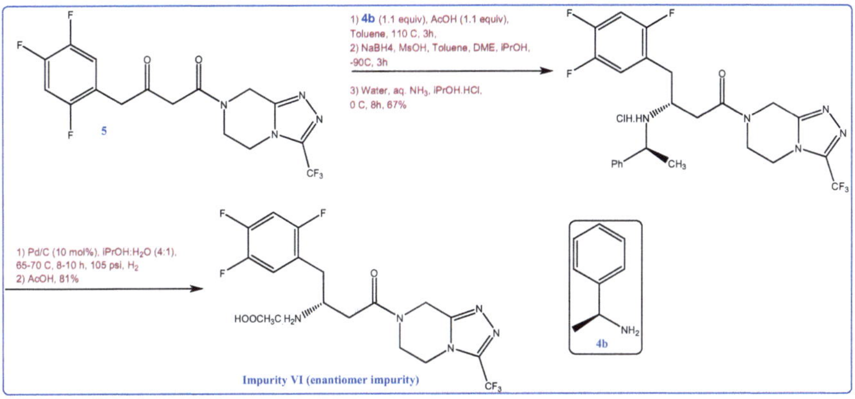

Fig. 6 Synthetic scheme for the Impurity VI

alcohol and water in ratio 4:1) at 65–70 °C,105 psi hydrogen, followed by treatment with acetic acid furnished enantiomeric **Impurity VI**.

5 Detection of Sitagliptin and Its Impurities by HPLC and GC–MS

By co-injecting and comparing retention times with spiked impurities, it is possible to validate all synthetic impurities using HPLC. In most of the references as studied in Table 1, octadecyl silyl silica gel (C18) filled SS (stainless steel) column was employed. Methanol and dihydrogen orthophosphate in water with a pH (2.5 to 4) adjustment make up the mobile phase. The wavelength of detector varied from 210 to 272 nm. Retention times for impurities are often higher than those for active pharmaceutical ingredients. To compute relative retention time, API and impurity values must first be obtained. It is 19.399 min for the following impurities: triazole impurity 4.708 min, acid impurity 6.627 min, alkene impurity 29.582, cyclohexyl impurity 48.398 min, and sitagliptin (API). Enantiomer impurity on a chiral HPLC chromatogram was 15.449, while sitagliptin value was 18.047 [29].

A rapid and simple GC–MS method for identifying sitagliptin in human urine has been reported by Uçaktürk et al. [30]. Sitagliptin was derivatized by N-methyl-trimethylsilyl trifluoroacetamide and transformed into its N-trimethylsilyl amine derivative prior to GC–MS analysis. The LOQ came to be 50 ng/ml with a linear calibration curve between 50 and 600 ng/mL and had a coefficient of determination (R^2) that was above 0.997. It was discovered that the intra- and inter-day precisions and accuracies, respectively, were between 0.83 and 4.53% to 8.76%. The method worked well when used on urine samples collected from people with diabetes [30].

Table 1 Conditions for HPLC isolation of sitagliptin

S. No	Type of stationary phase column	Mobile phase	pH	Wavelength for detector (nm)	Flow rate	References
1	C18 (100 × 4.6 mm, 5 μm)	Methyl alcohol	-	248	1 mL/ Min	[24]
2	C18 (150 × 4.6 mm, 5 μm)	Mixture of 0.01 M of KH_2PO_4: Methanol (50:50% v/v)	2.5	267	0.7 mL/ Min	[25]
3	C18 (250 × 4.6 mm, 5 μm)	Mixture of KH_2PO_4: acetonitrile (35:65% v/ v)	3	210	1 mL/ Min	[26]
4	C18 (250 × 4.6 mm, 5 μm)	Mixture of acetonitrile: Water (60:40% v/v)	-	272	1 mL/ Min	[27]
5	C18 (150 × 4.6 mm, 5 μm)	Mixture of acetonitrile: methanol: buffer (2:3:5% v/v/v)	4	267	1 mL/ Min	[28]

6 Conclusion

Oral incretin-based treatments called gliptins (sitagliptin, vildagliptin, saxagliptin, alogliptin, and linagliptin) are commonly available. According to their approved uses, they can be taken independently or in combination with other drugs like insulin, metformin, sulfonylureas, and thiazolidinediones. DPP-4 inhibitors attach to the DPP-4 molecule's catalytic site to stop the enzymatic breakdown and inactivation of endogenous (Glucagon-like peptide-1) [31]. The first licensed DPP-4 inhibitor, sitagliptin, has demonstrated its safety and well-tolerance throughout the past five years of clinical use. Most of the research has not particularly examined sitagliptin's impact on senior persons. We will eventually require research including this population due to the large numbers of older persons who also have type 2 diabetes mellitus. In the current study, the positive results observed in younger people groups are extended to the senior population, and the treatment's objectives and indications remain the same. The most frequent adverse effects of gliptins include nasopharyngitis, headaches, and upper respiratory tract infections. Even though rhabdomyolysis and renal failure have very seldom been described as major side effects, they are rare enough to warrant individual case reporting [32–34].

Sitagliptin has been shown to successfully control postprandial and fasting blood sugar levels in a range of type diabetes mellitus patients. The indicators of cell function homeostatic model assessment used to quantify insulin, and proinsulin/insulin ratio are also enhanced by sitagliptin medication. The incidence of hypoglycemia has been shown to be significantly decreased by sitagliptin, whether it is used alone or in conjunction with other oral drugs that don't induce it. However, increases to 12% from 1.8% when used alone when combined with a sulfonylurea. Sitagliptin and insulin usage together increases the risk of hypoglycemia (16%) compared to using the medicines separately (8%) and using insulin alone [35].

To conclude, for sitagliptin's four impurities, namely acid, alkene, cyclohexyl, and enantiomer impurity a new, effective, and practical synthetic procedures were developed. A novel cyclohexyl impurity V was discovered, synthesized, and evaluated utilizing the previous and modified approaches. All the above-mentioned contaminants were examined using GC-MS and HPLC. The above-mentioned synthetic approaches are simple, useful, and brief, and they utilize the use of readily available and less costly starting materials to make it easier to generate impurities. We also provided insight into the methods by which these impurities originate, and the method used to manage them during the sitagliptin manufacturing process in this study.

Acknowledgements Authors are thankful to Amity University Uttar Pradesh for giving an opportunity to do this work as a part of NTCC course. Ashima is also thankful to Sun Pharmaceutical for the major project work opportunity.

References

1. Mudaliar S, Edelman SV (2001) Insulin therapy in type 2 diabetes. Endocrinol Metab Clin 30(4):935–982
2. Ganesan K, Rana MBM, Sultan S (2018) Oral hypoglycemic medications
3. Weber AE, Thornberry N (2007) Chapter 7 case history: JANUVIA™ (Sitagliptin), a selective dipeptidyl peptidase iv inhibitor for the treatment of Type 2 diabetes. In: Macor JE (ed) Annual reports in medicinal chemistry. Academic Press, pp 95–109
4. Lipscombe L, Booth G, Butalia S, Dasgupta K, Eurich DT, Goldenberg R et al (2018) Pharmacologic glycemic management of Type 2 diabetes in adults. Can J Diabetes 42:S88–S103. https://doi.org/10.1016/j.jcjd.2017.10.034
5. Herman G, Stein P, Thornberry N, Wagner J (2007) Dipeptidyl peptidase-4 inhibitors for the treatment of type 2 diabetes: focus on sitagliptin. Clin Pharmacol Ther 81(5):761–767
6. Drucker D, Easley C, Kirkpatrick P (2007) Sitagliptin. Nat Rev Drug Discovery 6(2):109–111
7. Kendall DM, Cuddihy RM, RM B (2009) Clinical application of incretin-based therapy: therapeutic potential, patient selection and clinical use. Am J Med 122(6 Suppl):S37–S50. https://doi.org/10.1016/j.amjmed.2009.03.015
8. McIntosh CHS (2008) Dipeptidyl peptidase IV inhibitors and diabetes therapy. FBL 13(5):1753–1773. https://doi.org/10.2741/2797
9. Guideline IHT (ed) Impurities in new drug substances Q3A (R2). In: Proceedings of the international conference on harmonization of technical requirements for registration of pharmaceuticals for human use, Geneva, Switzerland
10. Branch SK (2005) Guidelines from the international conference on harmonisation (ICH). J Pharm Biomed Anal 38(5):798–805
11. Gutierrez O, Metil D, Dwivedi N, Gudimalla N, Chandrashekar ERR, Dahanukar VH et al (2015) Practical, asymmetric route to sitagliptin and derivatives: development and origin of diastereoselectivity. Org Lett 17(7):1742–1745. https://doi.org/10.1021/acs.orglett.5b00520
12. Zeng LL, Ding YJ, Zhang GC, Song HR, Hu WH (2009) A practical synthesis of trifluorophenyl R-amino acid: the key precursor for the new anti-diabetic drug sitagliptin. Chin Chem Lett 20(12):1397–1399

13. Starcevic S, Mrak P, Kopitar G. Inventors; Lek Pharmaceuticals D. D., Ljubljana, assignee. Int. PCT Pub. No. WO/2014/0963752014
14. Hou A, Deng Z, Ma H, Liu T (2016) Substrate screening of amino transaminase for the synthesis of a sitagliptin intermediate. Tetrahedron 72(31):4660–4664
15. Tasnádi G, Forró E, Fülöp F (2010) Improved enzymatic syntheses of valuable β-arylalkyl-β-amino acid enantiomers. Org Biomol Chem 8(4):793–799
16. Hansen KB, Hsiao Y, Xu F, Rivera N, Clausen A, Kubryk M et al (2020) Correction to "highly efficient asymmetric synthesis of sitagliptin". J Am Chem Soc 142(31):13622. https://doi.org/10.1021/jacs.0c06041
17. Hansen KB, Balsells J, Dreher S, Hsiao Y, Kubryk M, Palucki M et al (2005) First generation process for the preparation of the DPP-IV inhibitor sitagliptin. Org Process Res Dev 9(5):634–639
18. Zhou S, Wang J, Chen X, Aceña JL, Soloshonok VA, Liu H (2014) Chemical kinetic resolution of unprotected β-substituted β-amino acids using recyclable chiral ligands. Angew Chem Int Ed 53(30):7883–7886
19. Zhu Y, Xia S, Zhu M, Yi W, Cheng J, Song G et al (2010) Synthesis, biological assay in vitro and molecular docking studies of new imidazopyrazinone derivatives as potential dipeptidyl peptidase IV inhibitors. Eur J Med Chem 45(11):4953–4962
20. Barua NC, Saikia B, Borah P, Baishya G. Inventors; Council of Scientific and Industrial Research, India, assignee. Int. PCT Pub. No. WO/2015/189862. India2015
21. Padi BI PR, Polavarapu S, Padamata S, Nerella K, Ramasamy VA, Vangala RR (2009) Inventor Dr. Reddy's Laboratories Ltd., India, assignee. Int. PCT Pub. No. WO/2009/085990 A3. India 2009
22. Gao Y, He Y, Peng F, Wu Z, Feng B (2015) Inventors; Hefei Huafang Pharmaceutical Sciences and Technology Co. Ltd., China, assignee. CN105130999 A. China 2015
23. Bao H, Bayeh L, Tambar UK (2013) Catalytic enantioselective allylic amination of olefins for the synthesis of ent-sitagliptin. Synlett 24(18):2459–2463. https://doi.org/10.1055/s-0033-1340079
24. Sireesha D, Sravya E, Bakshi V (2017) Development and validation of RP-HPLC method for the estimation of sitagliptin phosphate in tablet dosage form. Int J Appl Pharmaceut Sci Res 2(03):41–45
25. Patil S, Ramesh B, Hareesh A, Patil K, Dhokane A (2010) Development and validation of RP-HPLC method for the estimation of sitagliptin phosphate in tablet dosage form. Asian J Res Chem 3(3):653–655
26. Lavanya R, Yunoos M, Pradesh A (2013) Development and validation of RP-HPLC method for the estimation of sitagliptin phosphate in bulk and its tablet dosage form. J Adv Pharm Educ Res 3(4)
27. Balamurugan K, Mishra K, Suresh R (2018) Sitagliptin: a literature review on analytical and bioanalytical methods. Pharma Innov J 7(8):357–361
28. Kashid AM, Dhange AA, Gawande VT, Miniyar PB, Datar PA, Dhawale SC (2012) RP-HPLC method development and validation for sitagliptin in human plasma. Am J Pharm Res 2:805–811
29. Metil DS, Sampath A, Reddy JR, Chandrashekar ERR, Dahanukar VH, Reddy CVR et al (2018) Efficient and convenient synthetic routes for sitagliptin impurities. ChemistrySelect 3(10):2723–2729. https://doi.org/10.1002/slct.201702552
30. Uçaktürk E (2013) Development of a gas chromatography–mass spectrometry method for the analysis of sitagliptin in human urine. J Pharm Biomed Anal 74:71–76. https://doi.org/10.1016/j.jpba.2012.10.012
31. Bergman A, Ebel D, Liu F, Stone J, Wang A, Zeng W et al (2007) Absolute bioavailability of sitagliptin, an oral dipeptidyl peptidase-4 inhibitor, in healthy volunteers. Biopharm Drug Dispos 28(6):315–322
32. Ahren B (2007) DPP-4 inhibitors. Best Pract Res Clin Endocrinol Metab 21(4):517–533
33. Ahrén B (2007) Dipeptidyl peptidase-4 inhibitors: clinical data and clinical implications. Diabetes Care 30(6):1344–1350

34. Ahrén B (2009) Clinical results of treating type 2 diabetic patients with sitagliptin, vildagliptin or saxagliptin–diabetes control and potential adverse events. Best Pract Res Clin Endocrinol Metab 23(4):487–498
35. Vilsbøll T, Rosenstock J, Yki-Järvinen H, Cefalu W, Chen Y, Luo E et al (2010) Efficacy and safety of sitagliptin when added to insulin therapy in patients with type 2 diabetes. Diabetes Obes Metab 12(2):167–177

Anticancer and Antimicrobial Studies of Dibutyl(hexadecanoyloxy)stannyl Hexadecanoate

Ishika Aggarwal, Rohit Babu Aniyery, Anita Gupta, Harminder Kaur, Megha, Diksha, and Sreeyuktha Sudheer

1 Introduction

The primary goal of organometallic chemistry is to investigate molecules containing at least one metal-to-carbon bond. These specific chemicals, sometimes referred to as stannanes or organotin compounds, are composed of tin atoms with hydrocarbon substituents [1]. Stannanes have several potential uses, including their potential as powerful antituberculosis and anti-tumor agents. Due to their diverse properties, organometallic compounds are widely utilized in both industrial and clinical fields of applications, spanning from economy-friendly catalysts for industrial polymerization reactions to potent therapeutic agents against various health conditions [2]. Organotin compounds are being widely used as antifouling coatings on ships [3]. Organotin compounds are being used as luminescent dyes in silk fibroin scaffolds [4]. They have been examined as hole-injection layers (HILs) in the fabrication of electronic devices as well [5]. Dibutyl(hexadecanoyloxy)stannyl hexadecanoate, a novel chemical, is created in this study and endowed with anticancer, antibacterial, and antituberculosis properties [6].

I. Aggarwal · R. B. Aniyery (✉) · A. Gupta (✉)
Amity Institute of Applied Sciences (AIAS), Amity University, Noida 201301, India
e-mail: rohitaniyery@gmail.com

A. Gupta
e-mail: agupta3@amity.edu

H. Kaur (✉) · Megha · Diksha
Department of Chemistry, Punjab Engineering College (Deemed to Be University), Chandigarh 160012, India
e-mail: hkaur@pec.edu.in

S. Sudheer
Bachelor of Pharmacy, Ramaiah University of Applied Sciences, Bengaluru, Karnataka 560054, India

S. Rattan et al. (eds.), *Emerging Trends in Synthesis and Catalysis in Chemistry*, Springer Proceedings in Materials 51, https://doi.org/10.1007/978-981-97-5169-3_2

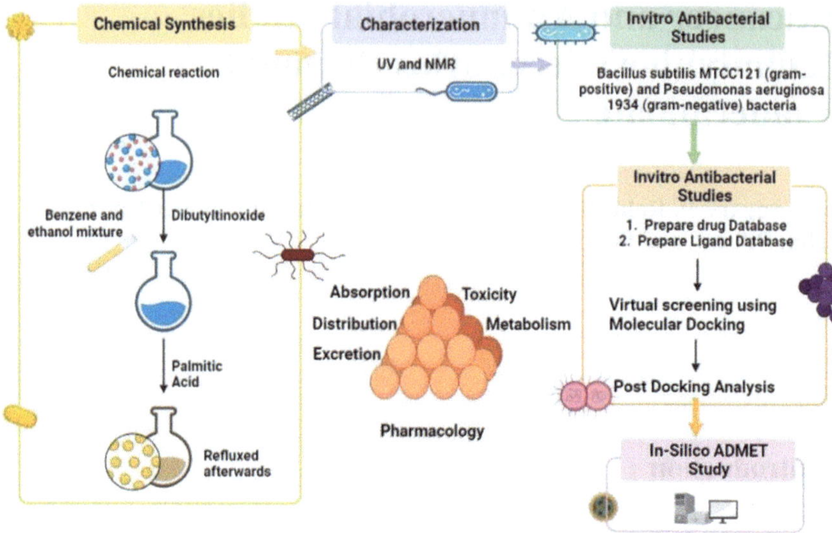

Fig. 1 Schematic presentation of methodology followed by in silico and in vitro investigation of dibutyl(hexadecanoyloxy)stannyl hexadecanoate

Stannanes have a variety of industrial uses, including this one. To manufacture biocides, wood preservatives, marine anti-biofouling agents, and tin dioxide coatings for glass bottles, for instance, they need these substances. In poly vinyl chloride, they are also used as heat stabilizers and hydrochloric acid scavengers. In addition, they function in chemical processes as catalysts, intermediates, miticides, acaricides, bactericides, and fungicides. Stannanes have the potential to be effective anticancer drugs because they can prevent the development of drug resistance [7].

These complexes have also shown promising biological and pharmacological applications in anticancer therapy. However, it should be noted that stannanes, despite their beneficial benefits, may pose risks as potential ecological toxicants, with a particular emphasis on both named pests and a range of cellular structures and enzymatic systems [8]. Current research is focused on developing new organotin compounds with improved properties to create effective and risk-free medicines as shown in Fig. 1. Hence, the present invention aims to synthesize a novel complex dibutyl(hexadecanoyloxy)stannyl hexadecanoate having anticancer, antimicrobial, and antituberculosis activity.

2 Materials and Method

The research methodology and biological studies involved in the dibutyl(hexadecanoyloxy)stannyl hexadecanoate complex is illustrated with the help of a flow chart shown in Fig. 2.

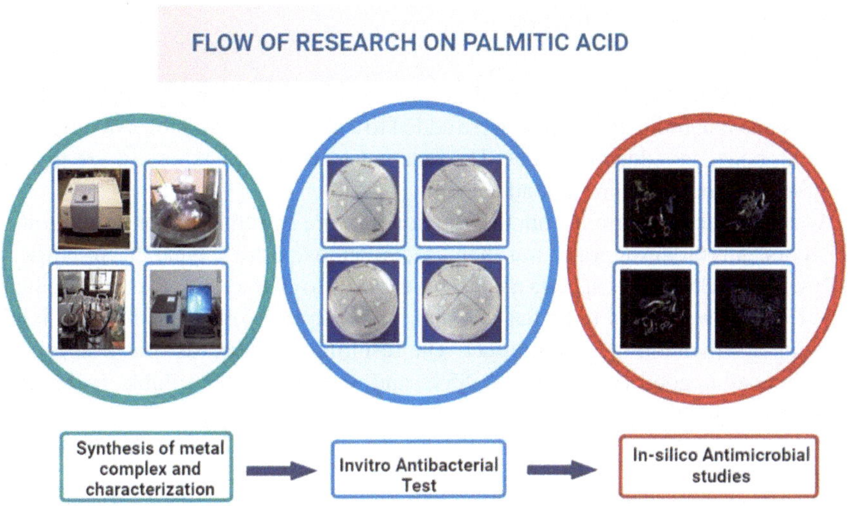

Fig. 2 Overview of the research and the biological studies involved in the Dibutyl(hexadecanoyloxy)stannyl hexadecanoate

2.1 *Chemicals*

Dibutyltinoxide, purchased from Spectrochem, Pvt. Ltd. Mumbai, India, and palmitic acid, obtained from Sigma-Aldrich were used as received without any further purification. The solvents required for the synthesis included AR grade dry benzene (Merck Specialties Pvt. Ltd.) and AR grade chloroform (Fischer Scientific, Qualigen). These solvents were dried before use in the synthesis [9].

2.2 *Instruments*

The Khera Laboratory Autoclave, manufactured by Khera Instrumentation, was used to sterilize samples for the antibacterial investigation. An orbital shaker supplied by PSN Instrumentation Pvt. Ltd. served as the incubator in use. The SHIMADZU UV 1800 equipment was utilized to analyse the UV-visible spectra. The experiments, which covered the wavelength range of 200–600 nm, were carried out in an ethanol medium. The ^{1}H-NMR and ^{119}Sn-NMR spectra were recorded using a 400 MHz FT-NMR Cryo-magnet Spectrometer from Bruker. The computational study was carried out using software Chem3D Ultra Version 7.0.0.

2.3 Experimental Method

Azeotropic distillation, a well-established method, was used for the synthesis [10]. Two moles of palmitic acid ligand were added to a mixture of two moles of dibutyltin oxide in benzene: ethanol solvent. The reaction mixture was exposed to reflux for 15 minutes using a Dean-Stark attachment.

Water was then removed from the reaction mixture by refluxing for an additional six hours. A rotary evaporator was then used to remove extra solvent. ^{119}Sn-NMR, ^{1}H-NMR, and UV-visible spectra analyses were used to characterize the final product. This reaction was found to be feasible and was complete within 5–6 h of refluxing showing a separation of solid product. The resulting complex (as shown in Fig. 3) was obtained in good yield (1.29 gm, 85%) and was white solid, soluble in dimethyl sulphoxide and ethanol.

2.3.1 In Vitro Antibacterial Test

Nutrient agar plates were inoculated with cultures of gram-positive (*Bacillus subtilis* MTCC121) and gram-negative (*Pseudomonas aeruginosa* 1934) bacterial strains. 250 mL bacterial inoculums with an optical density (OD) of 0.168 were created and cultured for 24 h in an incubator. The dibutyl(hexadecanoyloxy)stannyl hexadecanoate introduced put into several wells on the agar plates at varying quantities (0.001 g per 100 mL, 0.0005 g in 100 mL, 0.00025 g in 100 mL, and 0.000125 g in100 mL ethanol). After that, the plates were immediately incubated continuously for 24 h at 37 °C. The antibacterial activity was determined in millimetres by finding the inhibition zones. After 24 h of incubation, the minimum inhibitory concentration (MIC) was established by comparing the experimental findings to a control (chloramphenicol). Here the percentage inhibition of diameter growth (PIDG) was calculated and compared to chloramphenicol to analyse the growth reactions of the bacterial strains to various stannane doses following a 24-h incubation period [11]. The following equation was used to calculate the PIDG:

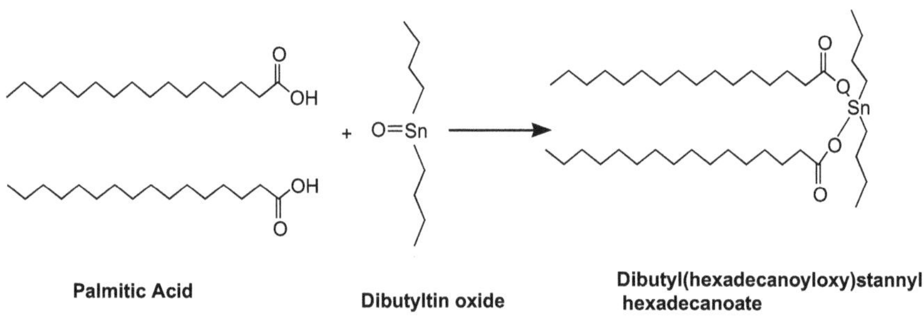

Fig. 3 Synthesis of dibutyl(hexadecanoyloxy)stannyl hexadecanoate

$$\text{PIDG} = \left(\frac{\text{Inhibition zone diameter with stannane}}{\text{Inhibition zone diameter with chloramphenicol}} \right) \times 100$$

2.3.2 IGEMDOCK Studies

The iGemDock v2.1 programme provides profiles of protein interactions by assessing and ranking screening chemicals based on a combination of energy-based scoring algorithms and pharmacological interactions [12]. The following equation is used to calculate the empirical scoring function for iGemDock:

$$\text{Energy} = \text{vdW} + \text{Elec} + \text{Hbond}$$

Van der Waal energy, electrostatic energy, and hydrogen bonding energy are all denoted in this equation by the letters vdW, Elec, and Hbond, respectively [13]. This programme is beneficial in drug development for structure-based virtual screening and post-screening analysis.

Preparation of the Binding Site A literature review was conducted before choosing the binding location. Protein structural files for docking were obtained from the Protein Data Bank (PDB) at http://www.rcsb.org. Bond orders were examined when the co-crystal ligand in the PDB was removed [14].

Ligand Preparation and Docking The structure of the palmitic acid stannane was generated using software and then optimized with advanced Gaussian software. The energy (Ground state) Hartree–Fock technique before docking was utilized. Openbabel software was used to convert the ligand file, which was previously in CDX format, into MOL 2 format [15]. Accurate docking was performed with each PDB file using the slow docking tool. The screening chemicals were then visualized and ranked using a K-means clustering tool in a post-analysis based on a mix of pharmacological interactions and an energy-based scoring algorithm. During the docking procedure, a step-by-step energy optimization was done with an emphasis on hydrogen interactions, side chains, and then the receptor's backbone. Following optimization, several receptor interactions, such as hydrogen bonds, hydrophobic bonds, and van der Waal interactions, were analysed for the complex.

In Silico Pharmacokinetic Properties Study Examining various aspects such as absorption, distribution, metabolism, excretion, and toxicity—collectively known as ADMET—is crucial in identifying and developing novel medications. These characteristics are important in assessing the effectiveness and safety of new drugs. An online-based programme called pkCSM was used to investigate the pharmacokinetic features related to ADMET. Developed by Douglas E. V. Pires and colleagues at the University of Cambridge, pkCSM uses graph-based signatures to create predictive models for key ADMET properties in drug development [16]. The programme uses these models to analyse and forecast the pharmacokinetic properties of potential drug candidates.

Table 1 Physical properties of dibutyl(hexadecanoyloxy) stannyl hexadecanoate

	Physical property	Yield	% Yield
Palmitic acid	Amorphous solid, white colour	–	–
Dibutyl(hexadecanoyloxy)stannyl hexadecanoate	White-coloured solid soluble	1.29 gm	85%

Table 2 Molecular properties of dibutyl(hexadecanoyloxy) stannyl hexadecanoate

Molecular weight	743.787
LogP	14.0778
#Rotatable bonds	36
#Acceptors	4
#Donors	0
Surface area	287.238
Molar mass	743.786 g/mole
Exact mass	744.507863 Da
Chemical formula	C20H32O4Sn
Percentage composition: C (64.59%), H (10.84%), O (8.6%), Sn (15.96%)	

3 Results and Discussion

3.1 Study of Physical Properties

The yield of the resulting complex dibutyl(hexadecanoyloxy)stannyl hexadecanoate, was found to be 1.29 g (85%) and the product is white coloured solid that is soluble in dimethyl sulfoxide and ethanol. The physical properties, molecular properties, and structural properties are enlisted in Tables 1, 2 and 3.

3.2 Characterization

The synthesized complex is characterized using spectral analysis (UV-Vis absorption, ^{1}H NMR, and tin Mossbauer studies).

3.2.1 UV Spectral Studies

The absorption peaks of palmitic acid are observed at λ_{max} 252 nm (absorbance value 1.346762), and are shifted to λ_{max} 286 nm (3.600235) which is observed due to n to pi antibonding transition. This shift towards longer wavelength shown in

Table 3 Structural properties of of dibutyl(hexadecanoyloxy) stannyl hexadecanoate

Atom count	125
Heavy atom count	45
Asymmetric atom count	0
Rotatable bond count	38
Ring count	0
Aromatic ring count	0
Hetero ring count	0
Hydrogen bond donor count	0
Formal charge	0
FSP3	0.95
Topological polar surface area	52.60 Å^2
Molar refractivity	191.78 cm^3/mol
Polarizability	81.20 Å^3

Fig. 4, indicates a 'Bathochromic/Red shift,' attributed to N_b electrons available for interaction with metal ion [17]. A comparative study of UV absorption for ligand (palmitic acid) and complex [dibutyl(hexadecanoyloxy)stannyl hexadecanoate] is depicted in Table 4.

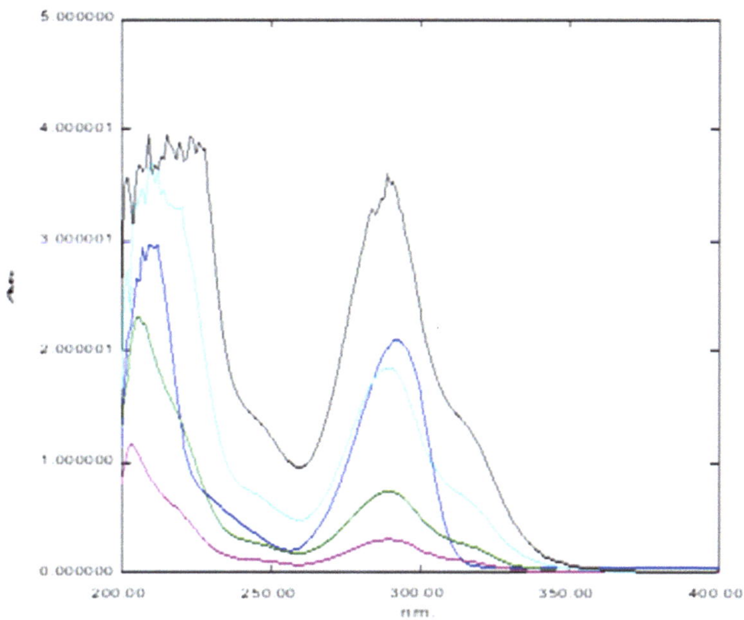

Fig. 4 UV overlay graph of palmitic acid and dibutyl(hexadecanoyloxy)stannyl hexadecanoate

Table 4 UV data for ligand (palmitic acid) and complex [dibutyl(hexadecanoyloxy)stannyl hexadecanoate]

Sample	Wavelength (nm)	Absorbance
Ligand (-----------------)	286.00 252.00	3.600235 1.346762
Tin complex (-------------)	292.00 259.00	2.091965 0.19531
Tin complex dilution 1 (-------------)	290.00 259.00	1.848114 0.171570
Tin complex dilution 2 (-------------)	290.00 259.00	0.729538 0.171570
Tin complex dilution 3 (-------------)	290.00 259.00	0.297668 0.070618

This gives evidence of complex formation of ligands with dibutyl tin oxide.

3.2.2 The ^{1}H NMR Spectral Studies

The ^{1}H Nuclear Magnetic Resonance spectra of the organotin (IV) complex were recorded in DMSO-d_6 using a BRUKER ADVANCE II 400 MHz NMR spectrometer at the Sophisticated Analytical Instrument Facilities (SAIF) at Panjab University (PU), Chandigarh. Here, tetramethyl silane was used as an internal reference. ^{1}H NMR spectra of synthesized complex was recorded and here the n-butyl group protons cluster is observed in the range 0.7469–1.2648 ppm, while the protons adjacent to carbonyl carbon appeared at 7.71–7.7381 ppm.

3.2.3 ^{13}C NMR Spectra

The ^{13}C Nuclear Magnetic Resonance spectra (using DMSO–d6) of the complex of organotin and ligand were recorded in BRUKER ADVANCE II 400 MHz at SAIF centre facility available in Panjab University, Chandigarh. TMS being reference. Significant peaks corresponding to inequivalent carbon atoms are depicted in Fig. 5.

3.2.4 ^{119}Sn NMR Spectral Studies

A singlet is observed at +118.57 (Fig. 6). It indicates the coordination number of 6 around the central metal atom and the geometry would be octahedral.

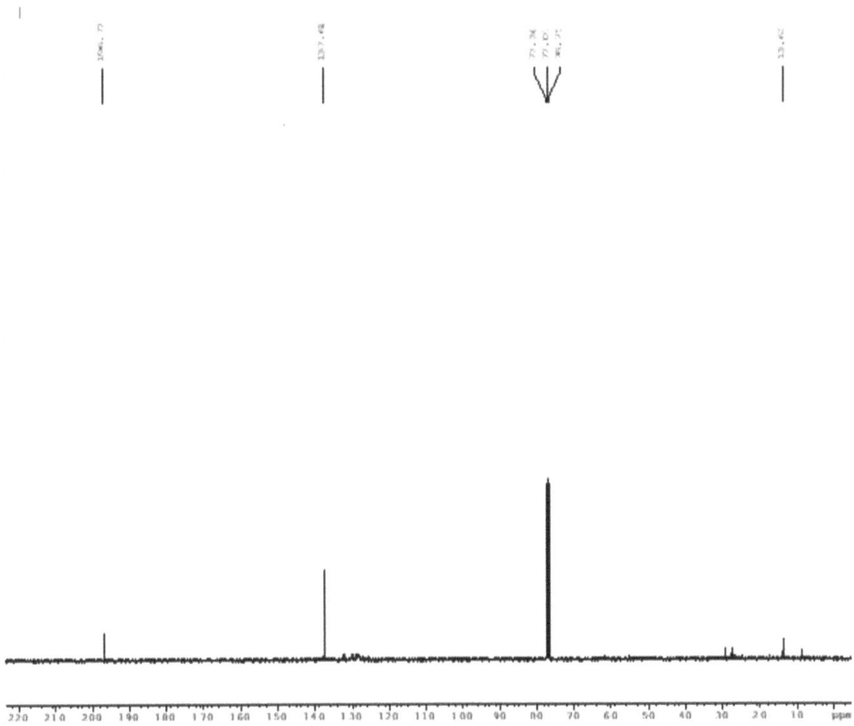

Fig. 5 ^{13}C NMR spectra of synthesized complex

3.2.5 FT-IR Spectra

All the absorption peaks are found by the proposed structure and bonding in the complex as revealed in Fig. 7.

4 Study of Activity andSuggested Applications of Synthesized Complex

4.1 In Vitro Antibacterial Test

Figure 8a shows a developed plate of *Bacillus subtilis* MTCC121 with control-chloramphenicol, palmitic acid (ligand), and dibutyl(hexadecanoyloxy)stannyl hexadecanoate (complex)

Figure 8b shows a developed plate of *Pseudomonas aeruginosa* 1934with control-chloramphenicol, palmitic acid (ligand), and dibutyl(hexadecanoyloxy)stannyl hexadecanoate (complex).

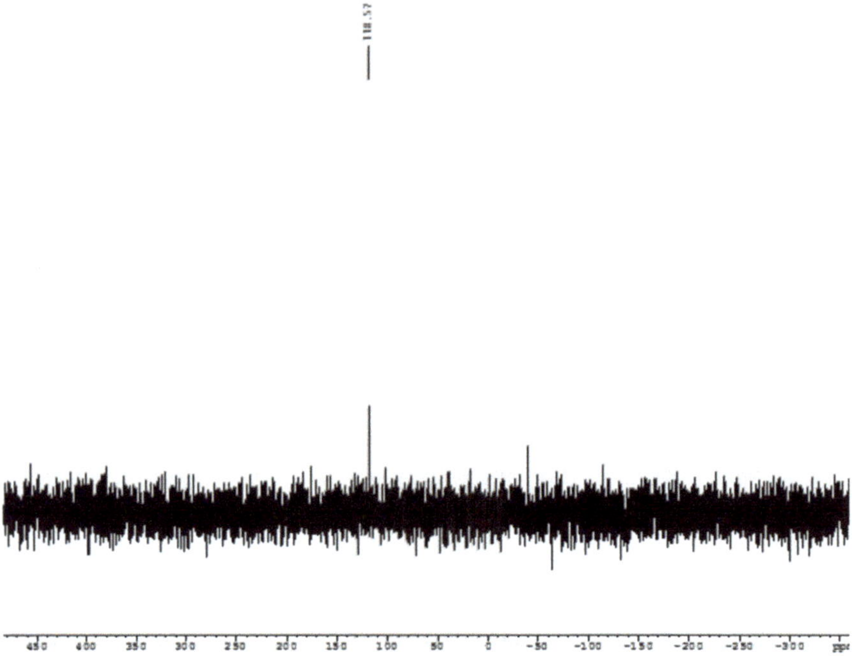

Fig. 6 ^{119}Sn NMR spectra of synthesized complex

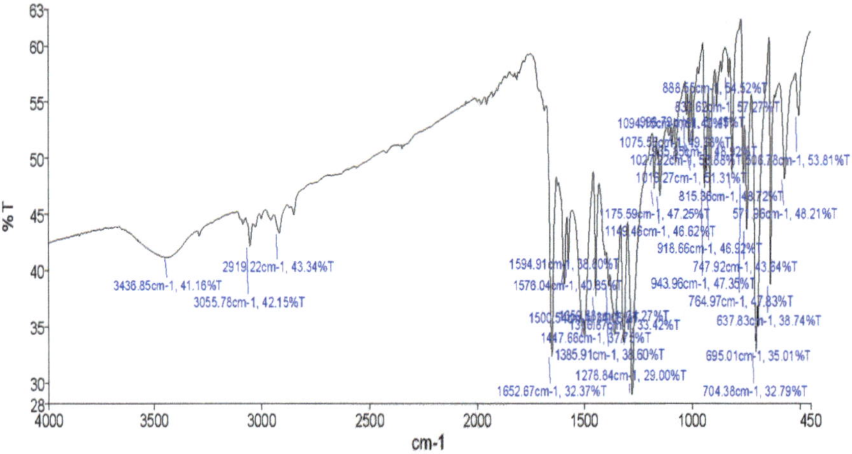

Fig. 7 FT-IR spectra of tin complex of palmitic acid

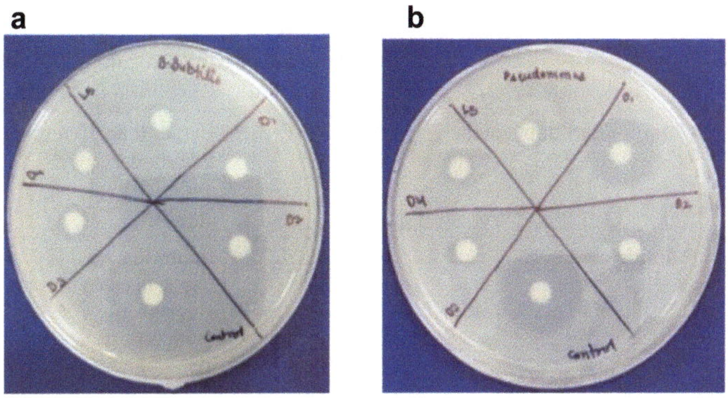

Fig. 8 **a**, **b** Indicate the clear inhibition of bacterial growth

The compound dibutyl(hexadecanoyloxy)stannyl hexadecanoate exhibited higher toxicity towards gram-negative strains compared to gram-positive strains, which can be attributed to the structural disparities in their respective cell walls. Gram-negative cell walls possess a greater degree of complexity in comparison with gram-positive cell walls. The growth inhibition of microorganisms by dibutyl(hexadecanoyloxy)stannyl hexadecanoate suggests a potential impact on enzyme production, thereby impeding the organisms' ability to metabolize nutrients and consequently leading to growth cessation. The results of observed inhibition diameter against ***Bacillus subtilis*** **MTCC12 and** ***Pseudomonas aeruginosa*** **1934** are shown in tabulated form in Tables 5 and 6.

Table 5 Results for Bacterial strain—*Bacillus subtilis* MTCC12

S. No.	Sample	Inhibition diameter (mm)	Interpretation	PIDG (%)
1	Control-chloramphenicol	8 ± 0.8	The diameter inhibition zone is found less than 10 mm; hence, it is a weak antimicrobial agent	–
2	Palmitic acid	10 ± 1.0	Inhibition zone is 10 mm; Hence, it is a weak antimicrobial agent	25
3	Dibutyl(hexadecanoyloxy)stannyl hexadecanoate	11 ± 1.1	Inhibition zone is larger than 10 mm less than 16 mm (>10, < 16); hence, it is a moderate antimicrobial agent	37.5

Table 6 Results for Bacterial strain—*Pseudomonas aeruginosa* 1934

S. No.	Sample	Inhibition zone (mm)	Interpretation	PIDG (%)
1	Control-chloramphenicol	22 ± 2.2	Larger than 16 mm and shows antimicrobial activity	–
2	Palmitic acid	12 ± 1.2	Inhibition zone is larger than 10 mm and less than 16 mm (>10, < 16); hence, it is a moderate antimicrobial agent	45.45
3	Dibutyl(hexadecanoyloxy)stannyl hexadecanoate	18 ± 1.8	Inhibition zone larger than 16 mm and above is an active antimicrobial agent	18.18

Enzymes reliant on free sulfhydryl groups (-SH) for their activity appear to be particularly vulnerable to deactivation by the complex's ions. The antimicrobial activity of the compound is believed to arise from multiple factors: (1) interference with the normal cellular processes of bacteria by the metal, (2) reduced polarity of metal ions upon chelation due to partial charge distribution with donor groups, and (3) enhanced lipophilic properties of the metal complex resulting from π-electron delocalization throughout the entire molecule, thereby compromising the cells' permeability barrier and interfering with their normal functioning. The concept of chelation theory provides a better understanding of the antimicrobial properties of metal complexes, suggesting that a reduction in the metal's polarizability results in enhanced lipophilicity of new complexes [14].

4.2 In Silico Antimicrobial Analysis Utilizing the Molecular Docking Software iGEMDOCK

The PDB files were selected and the docking pose of dibutyl(hexadecanoyloxy)stannyl hexadecanoate; the synthesized complex is depicted here in Fig. 11, which shows the compatibility of the organotin complex.

Biochemical function and process of PDB 3RB9

- Protein file: PDB 3RB9: Crystal structure of the *M. tuberculosis* beta clamp
- Source organism: *Mycobacterium tuberculosis* H37Rv

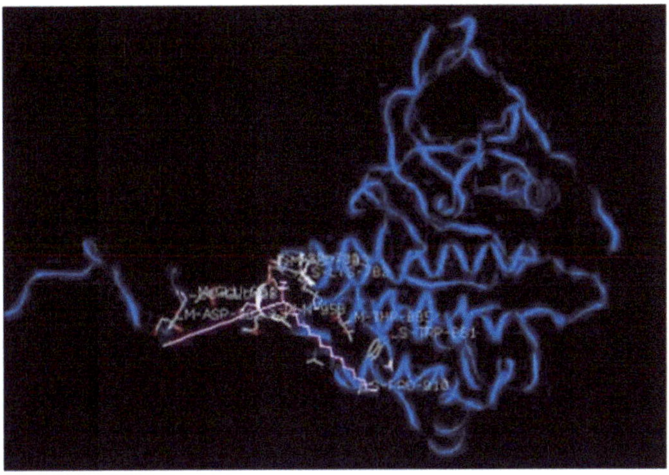

Fig. 9 Docking pose of dibutyl(hexadecanoyloxy)stannyl hexadecanoate with PDB 3RB9

- Biochemical function: Includes transferase activity, identical protein binding, DNA binding nucleotidyl transferase activity, DNA as directed DNA polymerase activity, 3′-5′ exonuclease activity
- Biological process: Involves response to antibiotic, DNA biosynthetic process, DNA strand elongation involved in DNA replication, DNA replication, cellular response to DNA damage stimulus, and nucleic acid phosphodiester bond hydrolysis represented in Fig. 9.

Docking analysis reveals that the complex strongly binds to protein pockets with the most favorable binding energy, indicating its potential to inhibit the growth of *Mycobacterium tuberculosis* H37Rv. The complex shows good docking results with *Mycobacterium tuberculosis* receptors, the observed results supported the antituberculosis activity as presented in Table 7.

The in silico antimicrobial studies for *Pseudomonas aeruginosa* are shown in Fig. 10.

4.3 Biochemical Function and Process of PDB 1IKQ

- Protein file: PDB 1IKQ: *Pseudomonas aeruginosa* Exotoxin A, wild type
- Source: *Pseudomonas aeruginosa* (strain ATCC 15,692/DSM 22,644/CIP 104,116/JCM 14,847/LMG 12,228/1C/PRS 101/PAO1)
- Biochemical function: toxin activity, transferase activity, NAD^{+}-diphthamide ADP- ribosyltransferase activity, transferase activity, transferring glycosyl groups
- Biological process: pathogenesis

Table 7 Interaction of dibutyl(hexadecanoyloxy)stannyl hexadecanoate with the basic amino acid residue of PDB 3RB9

Binding site	Binding energy (kcal/mol)
H-M-GLY-238	− 5.8
H-S-ARG-252	− 3.5
V-S-PRO-133	− 4.3
V-S-GLU-134	− 4.4
V-S-ARG-168	− 5.6
V-M-GLY-235	− 4.3
V-M-LYS-236	− 5.5
V-S-LYS-236	− 5.1
V-M-ASP-237	− 8.1
V-M-GLY-238	− 11
V-S-LEU-239	− 4.6
V-S-ARG-252	− 6.4
V-S-ASP-255	− 4.4

Fig. 10 Docking pose of dibutyl(hexadecanoyloxy)stannyl hexadecanoate with PDB1IKQ

The complex interacted with the PDB 1IKQ protein file, representing the wild-type *Pseudomonas aeruginosa* Exotoxin A, with a total fitness value enlisted below in Table 8.

The docking analysis revealed that the complex exhibited strong binding affinity to specific protein pockets, characterized by highly favourable binding energy. Consequently, the complex possesses the potential to effectively inhibit the growth of

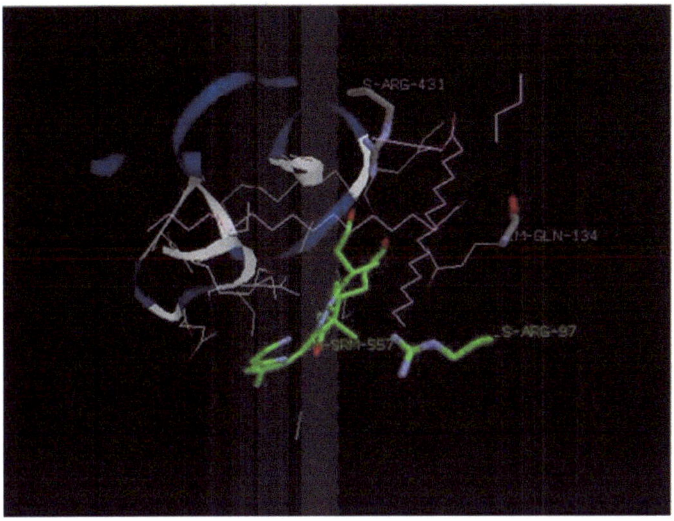

Fig. 11 Docking pose of dibutyl(hexadecanoyloxy)stannyl hexadecanoate with PDB 1ZJ8

Table 8 Interaction of dibutyl(hexadecanoyloxy)stannyl hexadecanoate with the basic amino acid residue of PDB 1IKQ

Binding site	Binding energy (kcal/mol)
H-S-TRP-281	– 6
V-M-SER-192	– 4.7
V-S-HIS-266	– 6.6
V-M-LEU-267	– 5
V-S-ARG-279	– 6.6
V-S-TRP-281	– 17.4

Pseudomonas aeruginosa. By binding to the receptor, the molecules impede its biological function, thus acting as a potent antimicrobial agent against *Pseudomonas aeruginosa.*

Biochemical Function and Process of PDB 1ZJ8

- Protein file: PDB 1ZJ8: Structure of *Mycobacterium tuberculosis* NirA protein
- Source organism: *Mycobacterium tuberculosis* H37Rv.
- Biochemical function: The protein exhibits multiple functions, including haeme binding, metal ion binding, sulphite reductase (ferredoxin) activity, oxidoreductase activity, sulphite reductase activity, and iron-sulphur cluster binding (4 iron, 4 sulphur cluster binding).

Table 9 Interaction of dibutyl(hexadecanoyloxy)stannyl hexadecanoate with the basic amino acid residue of PDB 1ZJ8

Binding site	Binding energy (kcal/mol)
H-M-GLN-134	− 4.1
H-S-ARG-431	− 2.9
V-M-GLN-134	− 7
V-S-GLN-134	− 3.2
V-M-ALA-416	− 2.6
V-S-PHE-422	− 7.7
V-S-ARG-431	− 13.1
V-M-ALA-468	− 5
V-M-ARG-469	− 5.8
V-S-ARG-469	− 7

- Biological process: The protein is involved in various biological processes, including oxidation–reduction processes, sulphate assimilation, and cysteine biosynthetic processes from serine represented in Fig. 11.

The complex interacted with the PDB 1ZJ8 protein structure of *Mycobacterium tuberculosis* NirA represented in Fig. 11.

The interaction of dibutyl(hexadecanoyloxy)stannyl hexadecanoate with the basic amino acid residue of PDB 1ZJ8 is given and represented in Table 9.

4.4 In Silico Anticancer Studies Using Molecular Docking Software (iGEMDOCK)

Biochemical function and process of PDB 1 m14

- PDB 1M14: Tyrosine kinase domain from epidermal growth factor receptor
- Source organism: *Homo sapiens*
- Biochemical function: ATP binding, protein tyrosine kinase activity, protein kinase activity
- Biological process: Protein phosphorylation

The docking pose is illustrated in Fig. 12, and the binding energy values are given in Table 10.

The epidermal growth factor receptor (EGFR) is a member of the ErbB family of receptor tyrosine kinases. EGFR is involved in a variety of cellular processes, including cell growth, proliferation, survival, and differentiation. Dysregulation of EGFR signalling is a common feature in many types of cancer, including non-small cell lung cancer (NSCLC). EGFR is activated through binding to its ligand, epidermal growth factor (EGF). EGF is a small protein that is released by cells in the surrounding

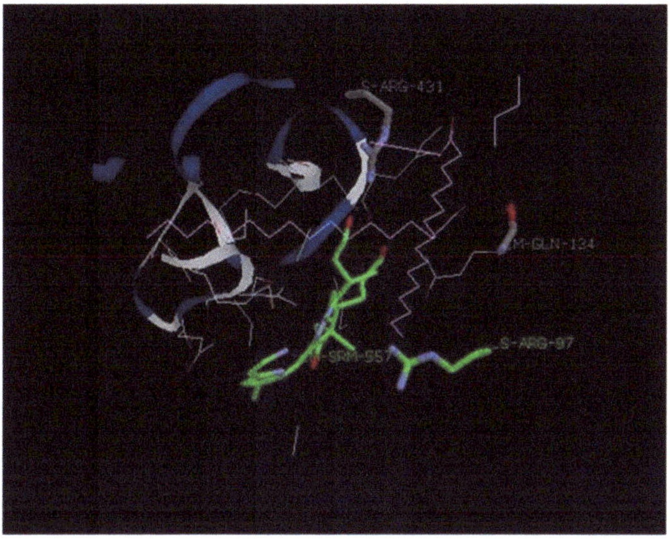

Fig. 12 Docking pose of dibutyl(hexadecanoyloxy)stannyl hexadecanoate with PDB 1M14

Table 10 Interaction of dibutyl(hexadecanoyloxy)stannyl hexadecanoate with the basic amino acid residue of PDB 1M14

Binding site	Binding energy (kcal/mol)
V-S-LYS-782	− 4.3
V-M-ASP-783	− 6.1
V-S-ASP-783	− 6.3
V-S-TRP-881	− 4.6
V-M-THR-885	− 4.8
V-S-PRO-910	− 4.1
V-M-GLU-981	− 4
V-S-GLN-958	− 5.8
V-S-GLU-981	− 5.3
V-M-ASP-982	− 5.3

environment. When EGF binds to EGFR, it triggers a series of events that lead to the activation of EGFR's tyrosine kinase domain. This domain then phosphorylates several downstream proteins, which ultimately leads to the cell's proliferation and survival.

In NSCLC, the most common EGFR mutations occur in the receptor's kinase domain. These mutations result in the receptor becoming constitutively activated, even in the absence of EGF. This leads to uncontrolled cell growth and proliferation, which can eventually lead to cancer. Several targeted therapies are available for the treatment of NSCLC with EGFR mutations. These therapies work by blocking the activation of EGFR, thereby halting the growth and proliferation of cancer cells.

Through molecular docking, it was observed that the tyrosine kinase domain of EGFR effectively binds to dibutyl(hexadecanoyloxy)stannyl hexadecanoate. This complex inhibits the overexpression of EGFR, which is associated with various cancers including breast, lung, ovary, and prostate. This finding suggests that dibutyl(hexadecanoyloxy)stannyl hexadecanoate could potentially prevent protein dimerization, thus demonstrating chemotherapeutic efficacy on the tyrosine kinase domain of the epidermal growth factor receptor. Consequently, this novel complex exhibits promising potential as an anticancer agent.

4.5 *In Silico Pharmacokinetic Properties Study*

The pkCSM software utilizes graph-based signatures to construct predictive models for key ADMET properties, aiding in drug development, as detailed in Table 11. To validate the findings of the pharmacokinetic properties study, a comparative analysis was performed against the well-established reference antibacterial drug, chloramphenicol.

The complex demonstrated a notable $CaCO_2$ permeability, surpassing the threshold of 0.9010–6 cm/s. This model is widely employed to assess the intestinal absorption of drugs and other substances in humans. Since the intestine is a primary site for the absorption of orally administered solutions, the high predicted value suggests that the compound can readily permeate through the human intestine.

Furthermore, the compound exhibited a favourable predicted absorption rate of 86.542% through the human intestine. This implies that the drug can be efficiently absorbed and assimilated within the gastrointestinal tract. Importantly, the synthesized complex did not exhibit any acute toxicity or mutagenic effects based on the results of the AMES test.

5 Conclusion

The novel organotin complex was synthesized and characterized prior to studying its potential applications. The findings derived from both in vitro and in silico experiments present compelling evidence for the synthesized complex's promising potential as a dual-action drug, displaying potent antibacterial activity against both gram-positive and gram-negative bacteria, along with anticancer properties targeting EGFR. These results pave the way for future investigations and highlight the complex's possible applications in the field of pharmaceutical research.

Table 11 Assessment of pharmacokinetic properties of the organotin complex, a comprehensive investigation was conducted employing the software pkCSM, developed by Douglas E. V. Pires and colleagues at the University of Cambridge

Property	Model name	Predicted value of complex	Predicted value chloramphenicol	Unit
Absorption	Water solubility	– 3.889	– 3.213	Numeric (log mol/L)
Absorption	$CaCO_2$ permeability	1.223	– 0.179	Numeric (log Papp in 10^{-6} cm/s)
Absorption	Intestinal absorption (human)	86.542	64.75	Numeric (% absorbed)
Absorption	Skin permeability	– 2.735	– 2.752	Numeric (log K_p)
Absorption	P-glycoprotein substrate	No	Yes	Categorical (Yes/No)
Absorption	P-glycoprotein I inhibitor	No	No	Categorical (Yes/No)
Absorption	P-glycoprotein II inhibitor	Yes	No	Categorical (Yes/No)
Distribution	VDss (human)	– 0.805	– 0.595	Numeric (log L/kg)
Distribution	Fraction unbound (human)	0.189	0.391	Numeric (Fu)
Distribution	BBB permeability	– 1.561	– 1.022	Numeric (log BB)
Distribution	CNS permeability	– 2.814	– 3.207	Numeric (log PS)
Metabolism	CYP2D6 substrate	No	No	Categorical (Yes/No)
Metabolism	CYP3A4 substrate	Yes	No	Categorical (Yes/No)
Metabolism	CYP1A2 inhibitor	No	No	Categorical (Yes/No)
Metabolism	CYP2C19 inhibitor	No	No	Categorical (Yes/No)
Metabolism	CYP2C9 inhibitor	No	No	Categorical (Yes/No)
Metabolism	CYP2D6 inhibitor	No	No	Categorical (Yes/No)
Metabolism	CYP3A4 inhibitor	No	No	Categorical (Yes/No)

(continued)

Table 11 (continued)

Property	Model name	Predicted value of complex	Predicted value chloramphenicol	Unit
Excretion	Total clearance	1.885	0.308	Numeric (log ml/min/kg)
Excretion	Renal OCT2 substrate	No	No	Categorical (Yes/No)
Toxicity	AMES toxicity	No	Yes	Categorical (Yes/No)
Toxicity	Max. tolerated dose (human)	0.091	1.262	Numeric (log mg/kg/day)
Toxicity	hERG I inhibitor	No	No	Categorical (Yes/No)
Toxicity	hERG II inhibitor	Yes	No	Categorical (Yes/No)
Toxicity	Oral rat acute toxicity (LD50)	2.293	2.682	Numeric (mol/kg)
Toxicity	Oral rat chronic toxicity (LOAEL)	1.143	1.871	Numeric (log mg/kg_bw/day)
Toxicity	Hepatotoxicity	No	No	Categorical (Yes/No)
Toxicity	Skin sensitization	No	No	Categorical (Yes/No)
Toxicity	*T.Pyriformis* toxicity	0.285	0.805	Numeric (log ug/L)
Toxicity	Minnow toxicity	− 7.392	1.286	Numeric (log mM)

References

1. Joshi R, Tomar N, Pokharia S, Joshi I (2023) Recent advancements in organotin(IV) complexes of drugs: synthesis, characterization, and application. Results Chem 5:100955. https://doi.org/10.1016/J.RECHEM.2023.100955
2. Rajendran K, Dey R, Ghosh A, Das D (2022) In search of biocatalytic remedy for organotin compounds- the recalcitrant eco-toxicants. Biophys Chem 290:106888. https://doi.org/10.1016/J.BPC.2022.106888
3. Shu SN et al (2023) Characteristics, sources and health risks of organotin compounds in marine organisms from the seas adjacent to the eastern ports of China. Reg Stud Mar Sci 61:102929. https://doi.org/10.1016/J.RSMA.2023.102929
4. Lara-Cerón JA et al (2020) Ultrasound-assisted synthesis of organotin compounds and their application as luminescent dye in silk fibroin scaffolds. Inorganica Chim Acta 505:119490. https://doi.org/10.1016/J.ICA.2020.119490
5. Monzón-González CR et al (2021) Synthesis and characterization of organotin(IV) semiconductors and their applications in optoelectronics. J Phys Chem Solids 150:109840. https://doi.org/10.1016/J.JPCS.2020.109840

6. Shi H et al (2023) Four organotin(IV) complexes derived from 2,6-difluoro-3- (propylsulfonamido)benzoic acid: synthesis, structure, in vitro cytostatic activity and antifungal activity evaluation. Inorganica Chim Acta 551:121485. https://doi.org/10.1016/J.ICA.2023.121485
7. View of in vitro and in silico antimicrobial study of stannane of pyridoxal 5-phosphate. Int J Pharm Pharmaceut Sci. https://journals.innovareacademics.in/index.php/ijpps/article/view/15002/9573. Accessed 20 June 2023
8. Ramzan S et al (2023) Structural characterization, DNA binding study, antioxidant potential and antitumor activity of diorganotin(IV) complexes against human breast cancer cell line MDA- MB-231. J Organomet Chem 990:122671. https://doi.org/10.1016/J.JORGANCHEM.2023.122671
9. (PDF) Synthesis, characterization , biological activities and computational anticancer study of Dibutylbis [(2-isopropyl-5-ethylcyclohexyl) oxy] stannane. https://www.researchgate.net/publication/317691656_Synthesis_characterization_biological_activities_and_computational_anticancer_study_of_Dibutylbis_2-isopropyl-5-ethylcyclohexyl_oxy_stannane. Accessed 20 June 2023
10. (pdf) In silico pharmacological and in vitro biological study of novel organotinsorbate. https://www.researchgate.net/publication/320146672_In_silico_pharmacological_and_in_vitro_biological_study_of_novel_organotinsorbate. Accessed 20 June 2023
11. Sam N et al (2012) Synthesis, spectral characterization and biological activities of Organotin(IV) complexes with ortho-vanillin-2-hydrazinopyridine (VHP). Open J Inorg Chem 2(2):22–27. https://doi.org/10.4236/OJIC.2012.22004
12. Rehman W et al (2014) Organotin (IV) based complexes as promiscuous antibacterials: Synthesis, in vitro, in silico pharmacokinetic and docking studies. J Organomet Chem 767:91–100. https://doi.org/10.1016/J.JORGANCHEM.2014.05.027
13. Mohamed HA, Ammar YA, Elhagali GAM, Eyada HA, Aboul-Magd DS, Ragab A (2023) Discovery a novel of thiazolo[3,2-a]pyridine and pyrazolo[3,4-d]thiazole derivatives as DNA gyrase inhibitors; design, synthesis, antimicrobial activity, and some in-silico ADMET with molecular docking study. J Mol Struct 1287:135671. https://doi.org/10.1016/J.MOLSTRUC.2023.135671
14. Sivanandhan M, Parasuraman A (2023) In-silico molecular docking and ADMET predictions of Pyrido[2,3-d]pyrimidine-2,4(1H,3H)-dione analogues as promising antimicrobial, antioxidant and anticancer agents. Polycycl Aromat Compd. https://doi.org/10.1080/10406638.2023.2191973
15. Day A, Williams A, Batchelor C, Kidd R, Tkachenko V (2012) Utilizing open source software to facilitate communication of chemistry at RSC. Open source software in life science research, pp 63–87. https://doi.org/10.1533/9781908818249.63
16. Khorram M, Tarahhomi A, van der Lee A, Excoffier G (2023) Structural, Hirshfeld surface and molecular docking studies of a new organotin(IV)-phosphoric triamide complex and an amidophosphoric acid ester proposed as possible SARS-CoV-2 and Monkeypox inhibitors. Heliyon e17358. https://doi.org/10.1016/J.HELIYON.2023.E17358.
17. Karpakavalli M, Kumar P, Sanglimuthu A, Kumar EP (2016) Docking, synthesis, characterization and anticancer activity of 4 (4′ hydroxy, 3′-methoxy) phenyl, but-2-one-3-ene, a curcumin analogue precursor. Int J Curr Pharmaceut Rev Res 8(4):1–5. https://doi.org/10.22159/IJCPR.2016V8I4.15266

Quality Control Methods to Check Phthalates Toxicity in Textile Products

Nikita Rawat and Deepshikha Gupta

1 Introduction

Phthalates, a class of organic compounds widely used in various industries, have garnered significant attention in recent years due to their ubiquity in everyday products and potential environmental and health implications. These versatile chemicals have been extensively employed as plasticizers, contributing to the flexibility and durability of plastics, and they also find applications in cosmetics, pharmaceuticals, and numerous consumer goods [1]. Phthalates are plasticizers, often diesters of phthalic acid. They increase the space between the polymer chain, which directly increases the flexibility as well as the mobility of the chain. This makes it suitable for use in flexible plastic tubes and vinyl flooring. They are present in polymeric materials made from polyvinylchloride (PVC) from which they enter the human body [2]. The rise in phthalate levels can be due to containers used for storing food and beverages, which have been found to potentially raise insulin resistance and blood pressure among certain teenage individuals. They are also present in garments such as plastic buttons, fabric coating, and various other products like children's toys, medical devices, packaging of food, cosmetics, etc. [3]. Phthalates have played a crucial role in the formulation of personal care products, ensuring the stable and consistent texture of cosmetics, fragrances, and lotions. Phthalates are used to increase the product's performance, long-lasting wear, and strength. They are often used in automobiles to resist high temperatures.

Although they can be naturally released into the environment during the production of PVC-based materials and waste disposal, they can be easily degraded by bacteria and fungi [4]. Undoubtedly most of the exposure to phthalates is from fabric, toys, and plastic bottles which have drawn significant attention. There are several advantages of using phthalates such as biodegradable, renewable, cheap, and heat

N. Rawat · D. Gupta (✉)
Amity Institute of Applied Sciences, Amity University Uttar Pradesh, Sector 125, Noida, India
e-mail: dgupta2@amity.edu

S. Rattan et al. (eds.), *Emerging Trends in Synthesis and Catalysis in Chemistry*,
Springer Proceedings in Materials 51, https://doi.org/10.1007/978-981-97-5169-3_3

resistant [5]. However, having so many advantages, exposure to phthalates in the long term increases the risk of developing neurological disorders, decreased fertility, and developing breast and skin cancer. Phthalates have become a major route that can disrupt the endocrine system and interfere with hormones of the body. It can also cause allergic reactions when comes in contact with the human body. Studies have linked exposure to specific phthalates with adverse reproductive and developmental effects, leading regulatory agencies to impose restrictions on their use in certain applications. Additionally, concerns have been raised about the endocrine-disrupting properties of some phthalates, potentially affecting hormonal balance in humans and animals. It is important to find suitable alternatives to phthalates which can be an economical challenge. This review paper provides an overview of the toxicity of phthalates and the level of phthalates found in textile products. These products include coated fabrics, labels on the garments, and plastic components such as clothing buttons.

2 Characteristics of Phthalates

Phthalates have low water solubility and low volatility which means they cannot easily evaporate into a gas and has a low boiling point and high vapour pressure [6]. Their consistency is a thick, colourless, and odourless viscous liquid. They show high durability [7]. It is well known that phthalates can be liberated from plastic by heating or separating a compound mixture from an organic solvent. Agitation and prolonged storage are the main reasons for releasing phthalates from the products [8]. It does not cause corrosion to other materials when it encounters other substances, and it is also non-toxic in nature when use limitedly. The costs of phthalates are also the primary reason for choosing it over other options. High molecular weight phthalates are soluble in water to a small extent; whereas, low molecular weight phthalates are reasonably soluble in water [9].

The industry can easily produce phthalate when phthalic anhydride is reacted with ethanol in the presence of concentrated sulphuric acid. The purity lies between 99.70% and 99.97% with some impurities like maleic anhydride. Disposal of phthalates can be done by burning the product at a very high temperature to break it down into gases and ashes or burying them in the ground and layering them with soil or clay to avoid the release of contaminants [10].

There are different pathways by which phthalates undergo degradation, some are natural, and some accelerated pathways designed by research, viz. photocatalytic, biodegradable, and metal-catalysed pathways which are discussed here in detail (Fig. 1).

Photodegradation Pathway with UV/TiO_2

During the degradation of DEP, a mercury lamp with emission at 365 nm is used as a light source and TiO_2 is used as a photocatalyst carried out for 300 min in a photochemical reaction apparatus which results in five by-products under UV/TiO_2.

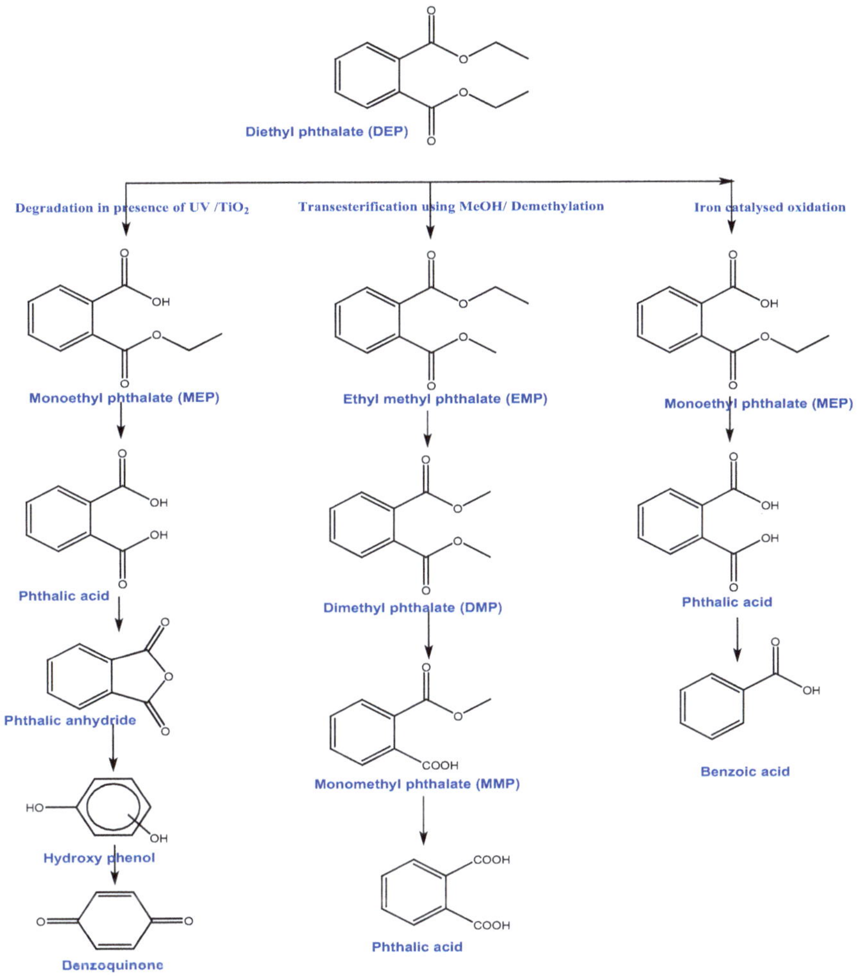

Fig. 1 Degradation of diethyl phthalate (DEP) with different methods

TiO_2 shows strong activity to DEP with no ring-opening by-products. The attack of hydroxyl radical in DEP converts into mono-ethyl phthalate and phthalic acid. Further, it converts it into o-phthalic anhydride which on further oxidation gives benzenediol and p-benzoquinone [11]. Some phthalates undergo photodegradation in the presence of photocatalysts like TiO_2 when exposed to UV light but generally, phthalates are not considered photoactive. The degradation pathway is depicted in Fig. 1.

Biodegradation Pathway with MeOH

Soil containing DEP is dissolved in MeOH for homogeneous solution, it is pretreated with MeOH in phosphate buffer, resulting in the detection of Ethyl methyl phthalate

(EMP), Dimethyl phthalate (DMP), Monomethyl phthalate (MMP) with GC-FID. These target analytes can only be generated when a transesterification reaction takes place. It is a nucleophilic substitution reaction that requires methyl group in the soil to transform it into DEP to EMP and then to DMP. Here MeOH acts as a nucleophile that converts it into EMP. Further water shows greater reactivity than MeOH and converts it into MMP. On further degradation, it transforms it into phthalic acid (PA) [12]. The degradation of phthalates does not directly generate carbon dioxide but there are some indirect ways that contribute to environmental issues. Degraded phthalates can end up in landfills that can leach out to the environment over a period of time. When phthalate products are incinerated, it can release carbon dioxide.

Degradation Pathway with Iron Catalysed Oxidation

DEP removal efficiency increases when iron catalyst dosage increases. This results in the formation of by-products which are detected by LC-TOF-MS. DEP degradation takes place by hydrolysis reaction where cleavage of the ester group converts it into a monoester compound. On further hydrolysis, phthalic acid was formed which on further decarboxylation produces benzoic acid [13]. Phthalates are not considered as catalytic materials. Catalyst is substances that speed up the chemical reaction; whereas, phthalates are organic compound that does not possess inherent catalytic activity.

2.1 Advantages of Phthalates

In recent years, phthalates have been used increasingly because of their high performance and greater stability. They are used to strengthen the products and make them more durable. They are used to increase elongation and flexibility in rigid polymers like PVC which are generally used in wires and cables [14]. However, there are several types of phthalates that show different properties and advantages that are well-suited for specific products. Dimethyl phthalates commonly known as DMP are a methyl phthalate ester used as an insect repellent [15]. DMP is used as a co-solvent for synthetic musk mainly used in perfumes. It is also used in oral and bolus products up to 5% level. The short chains of dialkyl ester show high absorption in rodents as it rapidly hydrolysed into monoester and alcohol and reduces the life cycle. DIBP also known as diisobutyl phthalates is used as a formulation in PVC products as a primary heat stabilizer and it is also a low-cost formulation [16]. Benzyl butyl phthalate (BBP) is used in sheet vinyl flooring because of its ability to show good flexibility at low temperatures. It is used as a solvent for a wide variety of resins because of its excellent tensile resistance, which reduces molten viscosity. It can be used in the manufacturing of packaging materials because it provides films and clear sheets. They are used as plasticizers for vinyl foams. Diisodecyl phthalate (DIDP) plasticizer is produced from a mixture of alcohol and phthalic acid. Because of its ability to show high flexibility, it is used in plastic coating [17]. It is used as an additive and softener in most of the polymer industry for the manufacturing of wires and cables. Di(2-ethylhexyl) phthalate (DEHP) is used in PVC that contains up to 40% DEHP as it is resistant to high temperatures. It shows high weather resilience and can be used in outdoor products. It is also used in medical devices such as blood bags and infusion tubing as it reduces the chances of damaging a tissue [18]. DEHP is also used as a flame-retardant plasticizer. Phthalates are biodegradable as well as photodegradable. Different types of commonly used phthalates with their properties and uses are discussed in Table 1 and their structures are depicted in Fig. 2.

3 Source of Phthalates

Phthalates are synthetic compounds obtained from petroleum that were created in the last century to produce personal care products, solvents, and some daily-use plastic products. It helps to improve the properties of the product [24]. The most preferable method for making phthalate is to combine phthalic anhydride with numerous alcohols ranging in size from simple alcohols to branched alcohols [25]. Toys that are made up of plastics are the source of oral exposure to phthalates in children. These toys include different analytes of phthalates like DMP, DIPP, BBP, DMEP, DEP, and many more.

However, human exposure to clothing that contains chemicals like phthalates has increased due to the size of the contacted surface area and the length of the exposure period [26]. The clothing humans wear covers a large portion of the body that can easily transfer the phthalates present in fabric to the skin. The larger the surface area of the contact, the larger the absorption through the skin. If the exposure period of phthalate is increased, then chemicals can leach out and come in contact with the skin. Due to the sensitive skin of infants, they are most likely to adsorb phthalates easily. The level of DEHP in infant clothing is the main factor contributing to the adverse health effects in newborns. It can be spread with oral exposure or dermal adsorption in human beings which can relatively increase the rate of carcinogenic diseases. However, having so many disadvantages' phthalates are used in the textile industry because they make the textile more flexible, durable and it is also inexpensive.

3.1 Toxicity of Analytes

Over recent years, phthalates in consumer products have become a major issue as it has many health disadvantages that show deleterious endocrine and metabolic side effects. That is why different countries' government has passed rules and regulation on the basis of which a level of phthalate in a consumer product is determined and passed. The exposure of phthalate in human bodies is due to leaching from clothes, children's toys, or waste plastics. The phthalate route can be ingestion, inhalation, and dermal contact. Small kids have the habit of biting their toys which can easily transfer DEHP, DEP, and BBP in their body and can affect the growth and development of

Table 1 Different types of phthalates with properties and uses

Name of phthalate	Molecular weight in (g/mol)	Boiling point (°C)	Characteristics	Uses	Refs
Benzyl butyl phthalate (BBP)	312.365	370	Colourless liquid, non-volatile, mild odour. The water solubility is 2.7 mg/l	It is used in making PVC, vinyl foams, artificial leather, floor tiles, and leather coating	[19]
Diethyl phthalate (DEP)	222.24	298	Colourless oily liquid, bitter taste, insoluble in water, denser than water. The water solubility is 1000 mg/l	It is used in personal care products like shampoos as solvents and fixatives in fragrances	[19]
Diisononyl phthalate (DINP)	418.6	440	Oily viscous liquid, water solubility is < 0.001 mg/l	It is used in the making of toys and plastic items	[20]
Diisodecyl phthalate (DIDP)	446.7	403	A viscous, oily liquid which is soluble in fat, water solubility is < 0.001 mg/l	It is used in construction materials, adhesives, and fuel products, children's care products like toys and food packaging	[21]
Diisooctyl phthalate (DIOP)	390.5	370 °C	It is a colourless viscous liquid with a mild odour, water solubility is 0.09 mg/l	It is used in synthetic rubber and vinyl resins. It is also used in making children toys, food packaging material	[22]
Di-(2-ethylhexyl) phthalate (DEHP)	390.6	385 °C	It is a colourless liquid at room temperature. Soluble in fat but insoluble in water, water solubility is 0.003 mg/l	It is used in medical devices like medical bags, and tubes for dialysis. It is used in rainwear, cable wire, lunch boxes, and backpacks, etc.	[23]

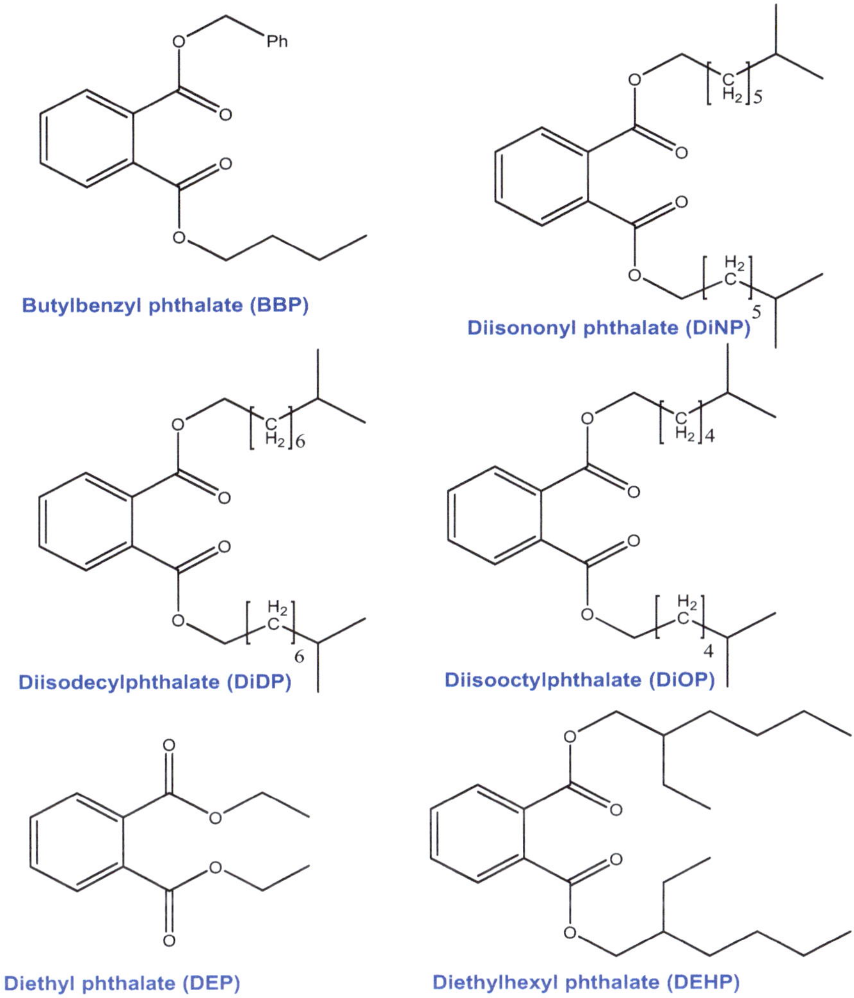

Fig. 2 Structures of some important phthalates used in industries

the child [27]. Infants can also be exposed to DEHP and DINP from drinking breast milk from their mothers.

BBP can be easily found in the environment however their concentration is very low. The intake of higher amounts of BBP is from skin contact or food intake. Phthalates when heated can decrease their interaction and increase flexibility [28]. Phthalates are released from the consumer product due to the weak chemical bond between phthalate and other chemicals due to dipolar interaction. Some studies concluded that DEHP, BBP, and DBP cause breast cancer in women and an increase in blood pressure during prcgnancy. Other phthalates can cause heart disease stroke and the development of cardiometabolic risks, absence of ovulation, thyroid disorders, cell

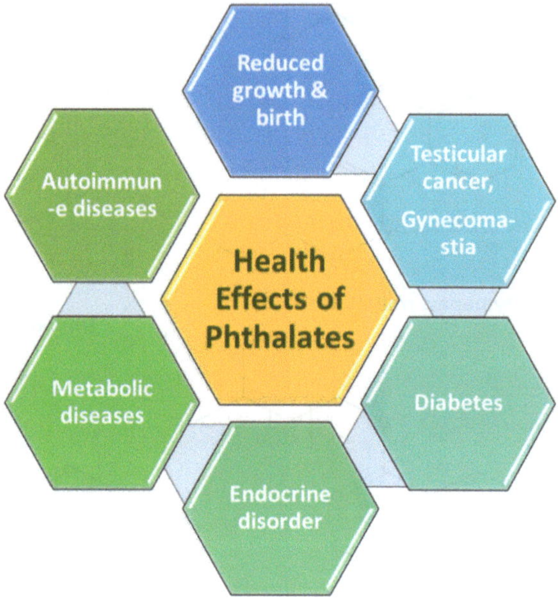

Fig. 3 Health effects of phthalates on humans

differentiation, male infertility, and diabetes, and affect reproductive development. Urine samples were taken from pregnant women and calculated the process density index (PDI). During the evaluation, the scientist gathers the information that the PDI score is decreasing with an increase in phthalate. Exposure to phthalates like DEHP, Monobutyl phthalate (MBP), and Monoethyl phthalate (MEP) during pregnancy can decrease child IQ of children of 3 years old [29]. DEHP is present in the tablecloth, and buttons. DEP and DMP are present in nail polish, perfumes, and body mists; whereas, high molecular weight phthalate like DEHP, DINP, DIDP is present in raincoats, coating in fabric, adhesives, and PVC. Exposure to DEP seems to be linked with damage to nervous tissue in children. Moreover, this type of exposure can also lead to endocrine dysfunction. The major ailments caused by exposure to phthalates in humans are depicted in Fig. 3.

3.2 Impact on Human Health

Over recent years, the decline in human fertility has shown some serious attention worldwide. The decrease in foetal testosterone levels due to modulation in testicular Leydig cells results in reproductive failure. It regulates the function of hormones and binds with progesterone receptors and interferes with the human reproductive system [30]. It can affect the anogenital distance, fertility and can cause testicular cancer. In the female reproductive system, MEHP formation can deplete the fertility of women. It basically mimics the hormone activity by binding with many other human receptors and modulates the hormonal balance.

It also affects the neurological behaviour of children below age 3. DINP affects the psychomotor skills following radiation exposure to the foetus. It gradually decreases the brain development, hyperactivity, and working memory of the child and contributes to low self-esteem, relationships, and trouble in concentrating on tasks. It decreases the immune system of the human by gradually making them weak. Human health plays a very important role, and it should be protected. Humans should avoid using phthalate-containing products, especially pregnant women. However, it is impossible to use plastics in day-to-day life. This is why chemical testing is done to check the limit and concentration of phthalate in consumer products.

3.3 Impact on the Environment

Phthalates have become a rising concern to the environment. It can be easily migrated during the disposal of the consumer product. DEHT is present as a contaminant in various coastal areas. The concentration of DEHP is increasing day by day and the chances of posing health risks. The coastal area of Salakta and Chebba is very much contaminated due to the presence of limited sea currents, anthropogenic sources of phthalates, and increased rate of tourism [31]. Therefore, non-phthalate plasticizers are used as an alternative to control the harmful environmental impact. Phthalates are present in sediments, soil, sludge, seawater, and many others. Thus, an extraction process is needed to extract the concentration of phthalate followed by analysis by GC–MS. During 2018–2019, phthalates are monitored on the Tunisian coast on seawater and sediments with the range of 5–763 μg/g. The concentration level of DEHP and DEHT is the highest among all the phthalates. In sediments, DEHP concentration is 4.59 μg/g, and DEHT concentration is 2.42 μg/g [31]. Another study where the concentration of phthalate in the sediments of the Korean coast is determined. Harbours are the main hotspots that are contaminating seawater and sediments. Household wastewater is also the reason for contaminating the seawater. It can also migrate from the landfill and leach into the fresh water. A sampling of indoor dust is taken from 120 homes 66 compounds which were detected in the indoor environment [32]. In Laizhou bays, phthalate concentration ranges from 813.1 to 11,975.6 μg/g. Investigation has performed where impacts of microplastics and phthalates in the sediments are determined [33]. They are the anthropogenic pollutants which get distributed on sediments containing different analytes of phthalates like DEP, DBP, BBP, DEHP, and many more. Therefore, a person should opt for phthalate free consumer product. In India, urban people wash their clothes in the sea water which can easily leach out phthalate in the environment and disturbs the aquatic life.

3.4 Animal Risks

Animals are not immune to the phthalate, just like people. Exposing them to the things that contain phthalate can have an impact on the animal's liver and kidneys. It can directly affect the reproductive system through DEP exposure. Phthalates induce androgen-independent male reproductive toxicity and growth toxicity during the analysis of 34 experimental studies in animals. Exposure to DEP led to a reduction in postnatal growth. It can easily cause a decline in sperm activity, and testicular cancer [34]. An experiment is conducted where a rat is suffering from testicular dysgenesis due to exposure to phthalate on a daily basis making them weak day by day [35]. It has been discovered that phthalate in aquatic animals is numerous and varied, and their overall concentration in developing nations even approaches 200 g/L. Small molecular weight with an alkyl length of four carbon is acutely toxic to marine life in the range of 0.21–377 mg/l; whereas, large molecular weight is highly toxic [36]. Aquatic animals' endocrine system receptors can be activated by exposure to Phthalates, which can then result in oxidative stress, metabolic diseases, and endocrine disorders. Each system can activate or inhibit the other, which results in genotoxicity and cell death and prevents organism growth and development. Numerous studies using rat models have been done to examine how plasticizers affect the Thyroid hormone system. Rats exposed to DEHP resulted in their TH signalling genes being altered. Rats were dosed with DBP, BPA, and DEHP orally for 27 days which results in the alteration of the thyroid levels. There was a negative correlation between serum levels of T3 and T4 with DBP and BPA [37].

4 Detection Techniques of Phthalate in Textiles

4.1 Sample Preparation and Extraction Process

Phthalates are toxic chemicals that are harmful to the environment, human beings as well as animals. However, to prevent this, situation different types of detection techniques are opted to determine the concentration of phthalate in consumer products. This detection technique consists of three steps: sampling, sample extraction, and analysis by GC-MS. Sampling should be done very carefully for further analysis as it can get easily contaminated by other solvents and plastic consumables in the laboratory. Plastic consumables include pipette tips and syringes that can easily leach out phthalates like DEHP and DiNP. For the determination of phthalate in textile products, the test method opted is CPSC-CH-C1001-09.4. This test method requires the use of hazardous chemicals, so it is very important to handle it with care. Avoid using plastic materials and go for glass wares.

A sample's preparation is essential for a repeatable and predictable analysis. Several factors in the sample preparation for this analysis will have an impact on the conclusion of the analysis. Sample preparation is defined as a process in which a sample is cut into smaller parts and readied for analysis. Here, the accurate weight of the sample is taken according to the method. Then, the sample extraction process is done. It is defined as a process of separating a natural product from the raw material.

In the sample preparation and extraction process, a sample needs to be cut into small pieces. Then take 0.05 g of sample into a sealable test tube and add 5 ml THF. If the sample weight is 1 g, then 10 ml of THF is added and sonicated for 30 min for dissolution. THF is a solvent that is used to dissolve the polymer samples. If the sample is hazy and not easily soluble, then sonicate it for another two hours for complete dissolution. Add 10 ml of hexane drop by drop to 5 ml of THF in the test tube. Hexane is used for the reprecipitation of the sample. Shake the solution to settle down the sample. After that, use a PTFE filter of 0.45 μm to filter the supernatant solution and make vials for analysis on GC–MS. Only dry test tubes are needed in the extraction process. The dry test tube is required in the extraction process otherwise layer formation takes place.

More care must be given in selecting an extraction solvent and procedure if a total phthalate measurement is sought. Cutting the sample in a small and homogeneous way will increase the solubility of the sample in the solvent. It is important to make sure that there is no cross-contamination during the steps, and it can easily be minimized by washing glassware with an alkaline solution for 48 h. This approach holds promise in addressing the issue of phthalate migration from textile products and its potential consequences for health and the environment.

The extraction efficiency will be increased by the particle size, which will increase its surface area. The sample will be dissolved by a solvent mainly hexane is used, which will also aid in the extraction of the phthalates from the entire matrix. The extraction method will also enhance the recovery; ultrasonic-assisted dissolution will help to recover the analytes from the sample. Centrifugation or filtration is suggested as the clean-up procedure for the recovered solution to get rid of any leftover solid matter. A concentration stage may be necessary depending on the amount of sample, solvent, and needed detection limits.

4.2 Principle

The prepared test sample is digested with solvents like tetrahydrofuran and hexane under ultrasonication at room temperature followed by analysis on GC-MS. The extraction efficiency will be increased by the particle size, which will increase its surface area. The sample will be dissolved by a solvent mainly hexane is used, which will also aid in the extraction of the phthalates from the entire matrix. The extraction method will also enhance the recovery; ultrasonic-assisted dissolution will all help to recover the analytes from the sample. Centrifugation or filtration is suggested as the clean-up procedure for the recovered solution to get rid of any leftover solid matter.

Chemical reagents and apparatus used are tetrahydrofuran, hexane, sealable vials, test tubes, and PTFE filters. THF is a useful solvent for environmental monitoring and regulatory compliance since it makes it possible to extract phthalates from a range of matrices efficiently and reliably. THF should, nevertheless, be used in laboratories with the required safety measures in place because it is a volatile and flammable solvent [23]. Determination of phthalates in bottled waters using solid-phase microextraction and gas chromatography-tandem mass spectrometry.

4.3 Detection with GC–MS

The most popular analytical technique for determining phthalates in consumer products is gas chromatography with mass spectrometry. Thus, the selection of the most appropriate analytical methods to distinguish, identify, and quantify a class of target chemicals mostly depends on the sensitivity as well as the physical and chemical properties of the instrument. The target analytes like DEHP, DBP, and DMP have low molecular weight, thermally stable, and low polarity. The mobile phase is He gas with 1 ml/min constant flow and the stationary phase is the column used in the instrument. The injection volume is 1 μl. The oven temperature was programmed from 100 °C to 260°C at 8°C/min and then to 320 °C at 35 °C/min. The analytical column is Phenomex ZB-5MS [38]. Typically, polar fused-silica capillary columns coated with 5% phenyl and 95% dimethylpolysiloxane are used to separate Phthalates, with temperature programmes ranging from 60 to 330 °C [23]. Samples are analysed on both selective ion monitoring and full scan mode. Traditionally, phthalates were detected using a flame ionization detector (FID), but nowadays MS detector shows high sensitivity and specificity, and this is why FID has been replaced [23]. Using an internal calibration curve is essential when using GC–MS.

The International Standard (IS) is used to encourage a more precise quantification of phthalates in a variety of samples. GC–MS shows many more advantages than any other instrument. The cost and the maintenance are low, spilt-less injection is used, and identification of analytes is much easier [23]. The detector used is MS detector which shows high sensitivity, high resolution, and high accuracy. The lowest calibration point served as the limit of quantitation (LOQ), while the limit of detection (LOD) was three times lower than LOQ. Different types of commonly used phthalates with their retention time are discussed in Table 2.

Table 2 List of target analytes and their retention time

Target analyte (phthalates)	Retention time (in min)
Diethyl phthalate (DEP)	7.022
Butylbenzyl phthalate (BBP)	7.750
Diethylhexyl phthalate (DEHP)	10.179
Diisoctyl phthalate (DIOP)	10.360
Diisononyl phthalate (DINP)	11.063
Diisodecyl phthalate (DIDP)	11.718

5 Conclusion

This paper is focused on the migration of phthalates from textile products causing various health problems to humans, animals, and the environment. The growing worry regarding the impact of phthalates on human health is primarily because of their widespread use in various items, with a particular focus on their presence in children's toys. However, the effects of phthalate exposure on health are not yet fully understood. Government intervention and regulation play a crucial role in mitigating the risks associated with phthalates. Banning certain phthalates like DIDP and DINP in various children's products is a very crucial step for safeguarding the population. It is also very necessary that these regulations are updated based on evolving scientific knowledge. A crucial aspect of the research discussed in the paper revolves around the need for precise quantification of these phthalates. To achieve this, researchers must carefully choose an appropriate method for extracting samples and select the right instrumentation for analysis. It is a big task to identify the target compounds due to the sample's complexity and their low concentration in target compounds. The results indicated that the GC–MS is a rapid, efficient, and convenient method for identifying the target compounds that are different types of commercially used phthalates.

References

1. Cacho JI, Campillo N, Viñas P, Hernández-Córdoba M (2015) Direct sample introduction gas chromatography and mass spectrometry for the determination of phthalate esters in cleaning products. J Chromatogr A 1380:156–161. https://doi.org/10.1016/j.chroma.2014.12.067
2. Elwell RJ, Pethrick RA (1990) Positron annihilation studies of poly(methyl methacrylate) plasticized with dicyclohexyl phthalate. Eur Polym J 26(8):853–856. https://doi.org/10.1016/0014-3057(90)90157-Y
3. Trasande L, Sathyanarayana S, Spanier AJ, Trachtman H, Attina TM, Urbina EM (2013) Urinary phthalates are associated with higher blood pressure in childhood. J Pediatr 163(3):747-753.e1. https://doi.org/10.1016/j.jpeds.2013.03.072
4. Gao D-W, Wen Z-D (2016) Phthalate esters in the environment: a critical review of their occurrence, biodegradation, and removal during wastewater treatment processes. Sci Total Environ 541:986–1001. https://doi.org/10.1016/j.scitotenv.2015.09.148
5. Chiellini F, Ferri M, Morelli A, Dipaola L, Latini G (2013) Perspectives on alternatives to phthalate plasticized poly(vinyl chloride) in medical devices applications. Prog Polym Sci 38(7):1067–1088. https://doi.org/10.1016/j.progpolymsci.2013.03.001
6. Cousins I, Mackay D (2000) Correlating the physical–chemical properties of phthalate esters using the 'three solubility' approach. Chemosphere 41(9):1389–1399. https://doi.org/10.1016/S0045-6535(00)00005-9
7. Carothers WH, Arvin JA (1929) Studies on polymerization and ring formation. II. Poly-esters. J Am Chem Soc 51(8):2560–2570. https://doi.org/10.1021/ja01383a042
8. Prokůpková G, Holadová K, Poustka J, Hajšlová J (2002) Development of a solid-phase microextraction method for the determination of phthalic acid esters in water. Anal Chim Acta 457(2):211–223. https://doi.org/10.1016/S0003-2670(02)00020-X

9. Staples CA, Adams WJ, Parkerton TF, Gorsuch JW, Biddinger GR, Reinert KH (1997) Aquatic toxicity of eighteen phthalate esters. Environ Toxicol Chem 16(5):875–891. https://doi.org/10.1002/etc.5620160507
10. Alabi OA, Ologbonjaye KI, Awosolu O, Alalade OE (2019) Public and environmental health effects of plastic wastes disposal: a review. J Toxicol Risk Assessment 5(2). https://doi.org/10.23937/2572-4061.1510021
11. Wang C, Zeng T, Gu C, Zhu S, Zhang Q, Luo X (2019) Photodegradation pathways of typical phthalic acid esters under UV, UV/TiO_2, and UV-Vis/Bi_2WO_6 systems. Front Chem 7. https://doi.org/10.3389/fchem.2019.00852
12. Cartwright CD, Owen SA, Thompson IP, Burns RG (2000) Biodegradation of diethyl phthalate in soil by a novel pathway. FEMS Microbiol Lett 186(1):27–34. https://doi.org/10.1111/j.1574-6968.2000.tb09077.x
13. Yu Y et al (2021) Ferrate (VI)-mediated transformation of diethyl phthalate (DEP) in soil: kinetics, degradation mechanisms and theoretical calculation. Environ Pollut 290:118053. https://doi.org/10.1016/j.envpol.2021.118053
14. Eales J et al (2022) Human health impacts of exposure to phthalate plasticizers: an overview of reviews. Environ Int 158:106903. https://doi.org/10.1016/j.envint.2021.106903
15. Rao SS, Rao KM (1991) Insect repellent *N*,*N*-diethylphenylacetamide: an update. J Med Entomol 28(3):303–306. https://doi.org/10.1093/jmedent/28.3.303
16. Huang S, Qi Z, Ma S, Li G, Long C, Yu Y (2021) A critical review on human internal exposure of phthalate metabolites and the associated health risks. Environ Pollut 279:116941. https://doi.org/10.1016/j.envpol.2021.116941
17. Lensen GJ, Jungbauer FHW, Coenraads PJ, Schuttelaar MLA (2012) Contact allergy to di-isodecyl phthalate. Contact Dermatitis 66(4):230–231. https://doi.org/10.1111/j.1600-0536.2012.02011.x
18. Sampson J, de Korte D (2011) DEHP-plasticised PVC: relevance to blood services*. Transfus Med 21(2):73–83. https://doi.org/10.1111/j.1365-3148.2010.01056.x
19. Giuliani A, Zuccarini M, Cichelli A, Khan H, Reale M (2020) Critical Review on the presence of phthalates in food and evidence of their biological impact. Int J Environ Res Public Health 17(16):5655. https://doi.org/10.3390/ijerph17165655
20. Cadogan D (2002) Health and environmental impact of phthalates. Plastics Additives Compound 4(6):28–29. https://doi.org/10.1016/S1464-391X(02)80091-5
21. Das MT et al (2021) Remediation strategies for mitigation of phthalate pollution: challenges and future perspectives. J Hazard Mater 409:124496. https://doi.org/10.1016/j.jhazmat.2020.124496
22. Baranenko D, Boulkrane MS, Borisova I, Astafyeva B, Lu W, Abd El-Aty AM (2022) Translocation of phthalates from food packaging materials into minced beef. Front Nutr 8(2022). https://doi.org/10.3389/fnut.2021.813553
23. Luís C, Algarra M, Câmara J, Perestrelo R (2021) Comprehensive insight from phthalates occurrence: from health outcomes to emerging analytical approaches. Toxics 9(7):157. https://doi.org/10.3390/toxics9070157
24. Xie H, Han W, Xie Q, Xu T, Zhu M, Chen J (2022) Face mask—A potential source of phthalate exposure for human. J Hazard Mater 422:126848. https://doi.org/10.1016/j.jhazmat.2021.126848
25. Cohen JL, Fong GP (1975) Source of selectivity in the quantitative determination of alcohols by acylation with cyclic anhydrides. Anal Chem 47(2):313–316. https://doi.org/10.1021/ac60352a040
26. Li H-L et al (2019) Phthalates in infant cotton clothing: Occurrence and implications for human exposure. Sci Total Environ 683:109–115. https://doi.org/10.1016/j.scitotenv.2019.05.132
27. Schettler T (2006) Human exposure to phthalates via consumer products. Int J Androl 29(1):134–139. https://doi.org/10.1111/j.1365-2605.2005.00567.x
28. Tumu K, Vorst K, Curtzwiler G (2023) Endocrine modulating chemicals in food packaging: a review of phthalates and bisphenols. Compr Rev Food Sci Food Saf 22(2):1337–1359. https://doi.org/10.1111/1541-4337.13113

29. Whyatt RM et al (2012) Maternal prenatal urinary phthalate metabolite concentrations and child mental, psychomotor, and behavioral development at 3 years of age. Environ Health Perspect 120(2):290–295. https://doi.org/10.1289/ehp.1103705
30. Lee H-R, Jeung E-B, Cho M-H, Kim T-H, Leung PCK, Choi K-C (2013) Molecular mechanism(s) of endocrine-disrupting chemicals and their potent oestrogenicity in diverse cells and tissues that express oestrogen receptors. J Cell Mol Med 17(1):1–11. https://doi.org/10.1111/j.1582-4934.2012.01649.x
31. Jebara A et al (2021) Phthalates and non-phthalate plasticizers in Tunisian marine samples: occurrence, spatial distribution and seasonal variation. Mar Pollut Bull 163:111967. https://doi.org/10.1016/j.marpolbul.2021.111967
32. Rudel RA et al (2001) Identification of selected hormonally active agents and animal mammary carcinogens in commercial and residential air and dust samples. J Air Waste Manage Assoc 51(4):499–513. https://doi.org/10.1080/10473289.2001.10464292
33. Zhang B-T et al (2020) Spatial distribution of phthalate acid esters in sediments of the Laizhou Bay and its relationship with anthropogenic activities and geochemical variables. Sci Total Environ 722:137912. https://doi.org/10.1016/j.scitotenv.2020.137912
34. Weaver JA et al (2020) Hazards of diethyl phthalate (DEP) exposure: a systematic review of animal toxicology studies. Environ Int 145:105848. https://doi.org/10.1016/j.envint.2020.105848
35. Foster PMD, Cattley RC, Mylchreest E (2000) Effects of di-n-butyl phthalate (DBP) on male reproductive development in the rat: implications for human risk assessment. Food Chem Toxicol 38:S97–S99. https://doi.org/10.1016/S0278-6915(99)00128-3
36. Adams WJ, Biddinger GR, Robillard KA, Gorsuch JW (1995) A summary of the acute toxicity of 14 phthalate esters to representative aquatic organisms. Environ Toxicol Chem 14(9):1569–1574. https://doi.org/10.1002/etc.5620140916
37. Tassinari R et al (2021) Metabolic, reproductive and thyroid effects of bis(2-ethylhexyl) phthalate (DEHP) orally administered to male and female juvenile rats at dose levels derived from children biomonitoring study. Toxicology 449:152653. https://doi.org/10.1016/j.tox.2020.152653
38. Li X, Yang Y, Cui X, Li S, Zhu X, Tang S (2015) Determination of Phthalate esters in textiles by solid phase extraction and gas chromatography-mass spectrometry. Anal Lett 48(16):2544–2552. https://doi.org/10.1080/00032719.2015.1043665

Significance of EDTA and Sodium Citrate in Blood Preservation: A Reaction Mechanism with Calcium as Substrate

Nimish Dhakad, Ravi Mishra, Shipra Rohatgi, and Kumar Rakesh Ranjan

1 Introduction

Blood clotting is a crucial phenomenon to prevent excessive blood loss in the event of trauma or injury. At the time of injury, blood stops flowing from the wound. A network of fibrin threads forms the blood clot, known as coagulum [1]. Blood formed elements—which are malformed and dead in form—are caught in this network. The inactive form of thrombin that is seen in plasma is prothrombin. Prothrombin is transformed into active thrombin by thrombokinase, and fibrinogen is then activated to form fibrin. These clotting components all aid in the coagulation of blood. An injury prompts the release of many substances by platelets or thrombocytes that start the blood clotting cascade. Blood coagulation is significantly influenced by calcium ions [2].

Anticoagulation refers to the process of preventing or slowing down the formation of blood clots within the blood vessels. Blood clots, also known as thrombosis, can be dangerous because they can obstruct blood flow and lead to serious medical conditions such as deep vein thrombosis (DVT), pulmonary embolism (PE), stroke, or heart attack. Anticoagulants are medications that are used to inhibit clot formation both in vitro and in vivo. Anticoagulants are routinely added to collection tubes in the field of in vitro diagnostics, either to keep blood fluid for hematological tests or to get appropriate plasma for coagulation and clinical chemistry investigations [3, 4].

N. Dhakad · R. Mishra · S. Rohatgi (✉)
Amity Institute of Forensic Sciences, Amity University, Noida, India
e-mail: srohatgi@amity.edu

K. R. Ranjan
Department of Chemistry, Amity Institute of Applied Sciences, Amity University Uttar Pradesh, Noida, India
e-mail: krranjan@amity.edu

S. Rattan et al. (eds.), *Emerging Trends in Synthesis and Catalysis in Chemistry*, Springer Proceedings in Materials 51, https://doi.org/10.1007/978-981-97-5169-3_4

1.1 EDTA and Other Anticoagulants

1.1.1 Ethylenediaminetetraacetic Acid (EDTA)

Ethylenediaminetetraacetic acid (EDTA) has been a popular anticoagulant and has several benefits over other anticoagulants. By chelating or eliminating calcium from the circulation, it prevents clotting. EDTA is a polyprotic acid that has four carboxylic acid groups and two amine groups with lone-pair electrons that bind calcium and various other metal ions [5, 6]. Calcium is required for a wide range of enzyme processes in the coagulation cascade, and its removal inhibits blood clotting within the collecting tube irreversibly [7]. EDTA has been recommended as the anticoagulant of choice for hematological testing because it allows the best preservation of cellular components and morphology of blood cells [8].

1.1.2 Sodium Citrate

Sodium citrate is another anticoagulant used during blood collection. Plasma products have the greatest final citrate content among blood components. Normally, the liver quickly metabolizes transfused citrate; however, after a large transfusion, the liver's citrate clearance capability may be surpassed. Citrate binds to calcium, resulting in a reduction in ionized calcium and a **hypocalcaemia** condition. Although temporary hypocalcaemia is normally handled successfully, nerve cell membranes can be damaged, resulting in perioral or acral paresthesia. Light headedness, chills, twitching, and tremors are some of the clinical signs. Severe hypocalcaemia can result in persistent muscular contractions and, if left untreated, tetany with spasms in several muscle groups. It is critical to recognize early signs of hypocalcaemia so that therapy can begin [9].

1.1.3 Heparin

Heparin is a popular anticoagulant with an excellent track record for little interference with testing, making it perfect for maintaining the integrity of blood samples. It is an organic blood component that is free of outside pollutants. This acidic mucopolysaccharide inhibits blood coagulation by potentiating antithrombin activity and has a molecular weight of 15,000–18,000 [10]. It is, however, rather more expensive. Heparin promotes antithrombin III action, neutralizing thrombin and prevents fibrin formation from fibrinogen when administered intravenously due to low GI tract absorption. It is essential in anticoagulant therapy because it interacts with thrombin and antithrombin to suppress clot formation for roughly 24 h.

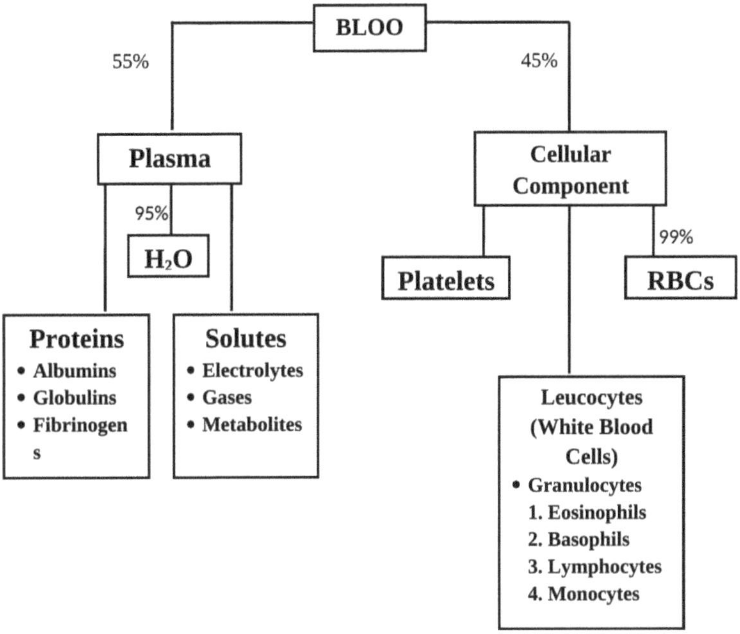

Fig. 1 Depicts the composition of blood

1.2 Composition of Blood

The cellular make-up of blood is extremely diverse. Blood samples are separated into plasma, buffy coat (leukocytes and platelets), and erythrocytes when they are spun in a centrifuge machine. RBC, WBC, platelets, and plasma are therefore components of blood (see Fig. 1).

1.3 Understanding Blood Preservatives: Their Secrets and Pitfalls

Anticoagulants like EDTA, heparin, and sodium citrate are vital in medical applications. EDTA chelates calcium, hindering blood coagulation, while heparin boosts antithrombin to prevent clotting. Sodium citrate binds calcium, impeding its role in coagulation. These ensure accurate results and prevent clotting in sample handling, essential for medical procedures in clinical and laboratory settings [11]. The following table provides a concise comparison of the anticoagulants EDTA, heparin, and sodium citrate, highlighting their mechanisms of action, properties, and drawbacks in laboratory contexts (see Table 1) [7, 10, 12].

Table 1 Comparison of various blood preservatives

Parameters	EDTA	Heparin	Sodium citrate
Properties	• Significant in hematological testing (cell amount and hematocrit) • Highly soluble in blood	• Deep vein thrombosis (DVT) property • Used in – pulmonary embolism – unstable angina	• Used in – coagulation studies – Partial thrombo-plastin time (PTT) and prothrombin time(PT) – Erythrocyte sedimenta-tion rate (ESR) test
Action mechanism	• Calcium work with the help chelating agent • It acts as disodium, dipotassium or tripotassium salt	• Accelerates antithrombin III action, neutralizing thrombin • Prevents fibrin formation from fibrinogen • It forms the thrombin + antithrombin cofactor + heparin complex and prevents fibrin clot formation	• It act as solution • Chelation of calcium inactivates Ca^{++} ions • Prevent the rapid deterioration of factor V and factor VII
Drawbacks	• It inhibits the activities of alkaline phosphatase, creatine kinase, and leucine amminopeptidases • Unsuitable for calcium and Iron estimation	• Costly • Inhibits the acid phosphatase activity • Gives a blue background for Wright's Stain smears • Not good for peripheral blood smear interpretation • Affects the binding of triiodothyronine and thyroxin to their carrier protein and produces a higher free concentration of these hormones	• Used in liquid form (liquid anticoagulant) • Not a good anticoagulant for a complete blood examination • Not good for the estimation of calcium

1.4 Mechanism of Coagulation

1.4.1 Understanding Blood Clotting: How Your Body Stops Bleeding

When you get a cut or injury, your body has an amazing way to stop the bleeding. This process is called blood coagulation, or clotting. Imagine it like a team of tiny superheroes inside your body, working together to seal off the wound and prevent excessive bleeding [13].

Step 1: Injury Alert
The process starts when you get a cut or injury. Your body recognizes that something is wrong and sends out a signal to start the clotting process.

Step 2: Platelet Rush
Platelets are small, sticky cells in your blood. They rush to the scene of the injury and stick together to form a plug. Think of them as the first responders who quickly arrive to seal off the breach [14].

Step 3: Fibrin Net
To make the clot stronger and more stable, your body deploys a net-like substance called fibrin. It is like building a scaffold around the platelet plug. Fibrin strands crisscross and creates a sturdy mesh that traps more platelets and forms a solid plug.

Step 4: Reinforcements-Clotting Factors
To ensure the clotting process works smoothly, your body relies on special proteins called clotting factors. They are like the commanders giving orders to make sure everything happens in the right order and at the right time.

Step 5: Control Valve-Anticoagulants
While it is essential to stop bleeding, it is equally important to prevent excessive clotting. To avoid this, your body has natural "control valves" known as anticoagulants. These keep the clotting in check, so it doesn't get out of hand.

Step 6: Clot Dissolution-Fibrinolysis
Once your injury has healed, your body needs a way to remove the clot. Enter fibrinolysis, which is like the cleanup crew. It breaks down the fibrin net and dissolves the clot, restoring normal blood flow.

1.4.2 Unraveling the Clotting Cascade

(a) **The Intrinsic Route**

Foreign substances such as collagen, sub-endothelium, or phospholipids trigger the process, which involves contact factors and factor XI. The damaged endothelium, factor XII, and other proteins create a complex. The complicated form of VIII + XI + X activated Xa is XII XIIa XI to XIa. The shared road then begins.

(b) **The Extrinsic Route**

It is the result of a complicated interaction between tissue factor (factor III or thromboplastin) and factor VII. Factor VIIa is activated, which stimulates factor X. Factor VIIa can also activate factors IX and X.

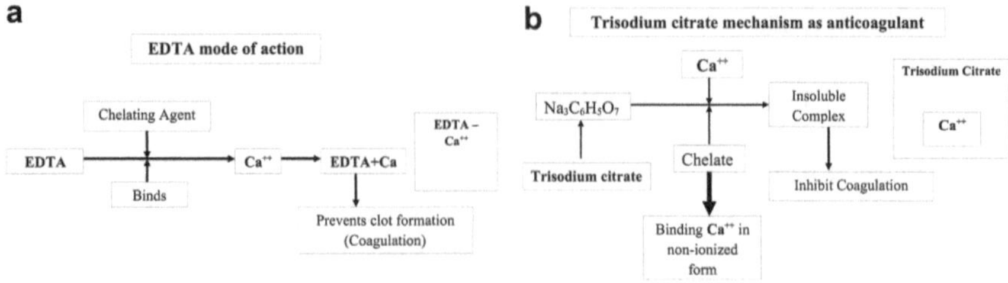

Fig. 2 **a** EDTA mode of action. **b** Sodium citrate mode of action

1.5 Action Mechanism of EDTA and Sodium Citrate

Comparison of action of EDTA and sodium citrate on blood.

EDTA acts as a powerful anticoagulant by chelating or binding to divalent cations, particularly calcium (Ca^{2+}). In the presence of EDTA, calcium ions are sequestered, rendering them unavailable for the coagulation cascade. This prevents the activation of coagulation factors that rely on calcium ions, effectively inhibiting blood clot formation [5, 15]. EDTA is particularly useful for preserving whole blood and cellular components because it prevents coagulation within the blood sample, allowing it to remain in a liquid state [see Fig. 2a]. This property also makes EDTA suitable for molecular biology applications, such as DNA and RNA extraction, as it maintains the integrity of nucleic acids [8].

Sodium citrate functions as an anticoagulant primarily by binding to calcium ions. It forms calcium citrate complexes, reducing the concentration of free calcium in the blood. This is crucial in coagulation studies, where accurate measurements of clotting times are necessary [16]. Sodium citrate is especially employed in tests like prothrombin time (PT) and activated partial thromboplastin time (APTT) to preserve the stability of coagulation factors and inhibit clotting [see Fig. 2b]. While it serves as an excellent anticoagulant for coagulation studies, it may not be suitable for preserving cellular components or nucleic acids as it affects calcium-dependent cellular processes [17].

2 Discussion

The various scientific reports show that the hematological and serological tests of blood have immense importance in the field of forensics. The use of blood preservatives such as EDTA, sodium citrate, and heparin is wide. The criteria for the best among them are to be the availability, cost effective and most importantly their effect on the blood sample. The effectiveness of sodium citrate and EDTA as anticoagulants depends on the specific use case. Sodium citrate is commonly employed in coagulation studies due to its ability to preserve coagulation factors without interfering with tests. However, it may not be suitable for preserving whole blood or cellular components as it chelates calcium ions. In contrast, EDTA is preferred for preserving whole blood, cells, and molecular components like DNA and RNA [18]. It excels in hematology tests but can interfere with coagulation studies. Therefore, the choice between these anticoagulants hinges on the intended purpose of the blood sample, with sodium citrate catering to coagulation studies and EDTA serving broader applications. Sodium citrate is found to be the best of all of them. It prevents the shrinkage of RBCs and gives better results in the serological tests. There are a few more anticoagulants and blood preservatives which are in use by professionals who are practicing in the field of medicine. Heparins, both inpatient and outpatient settings frequently used low molecular weight heparins (LMWH), such as enoxaparin and unfractionated heparin (UFH). Warfarin, a long-used oral anticoagulant, is a vitamin K antagonist used to treat a variety of diseases. Due to its limited therapeutic range, its dose calls for cautious supervision. These two are the future of anticoagulants [19].

Conflict of Interest No conflicts of interest exist that could potentially bias the findings or interpretation of this study.

References

1. Proescher F, Nolan J. Progress in blood preservation. PubMed Central
2. Hassock SR, Zhu MX, Trost C, Flockerzi V, Authi KS (2002) Expression and role of TRPC proteins in human platelets: evidence that TRPC6 forms the store-independent calcium entry channel. Blood J Am Soc Hematol 100(8):2801–2811
3. York MJ. Clinical pathology. In: A comprehensive guide to toxicology in preclinical drug. Science Direct
4. Hardwick J (2008) Blood storage and transportation. ISBT Sci Ser 3(2):177–196
5. Banfi G, Salvagno GL, Lippi G (2007) The role of ethylenediamine tetraacetic acid (EDTA) as in vitro anticoagulant for diagnostic purposes. Clin Chem Lab Med 45(5):565–576. https://doi.org/10.1515/CCLM.2007.110. PMID: 17484616
6. Bergmann T, Leberecht C, Labudde D (2021) Analysis of the influence of EDTA-treated reference samples on forensic bloodstain age estimation. Forensic Sci Int 325:110876
7. Putra AAMS, Hernaningsih Y (2022) Comparison of complete blood count parameters using EDTA, sodium citrate, and heparin anticoagulants. Res J Pharm Technol 15(10):4687–4691. https://doi.org/10.52711/0974-360X.2022.00786
8. Banfi G, Salvagno G, Lippi G (2007) The role of ethylenediamine tetraacetic acid (EDTA) as in vitro anticoagulant for diagnostic purposes. 45(5):565–576
9. Mann KG, Whelihan MF, Butenas S, Orfeo T (2007) Citrate anticoagulation and the dynamics of thrombin generation. J Thromb Haemost 5(10):2055–2061
10. Hirsh J, Anand SS, Halperin JL, Fuster V (2001) Mechanism of action and pharmacology of unfractionated heparin. Arterioscler Thromb Vasc Biol 21(7):1094–1096
11. Lima-Oliveira G, Brennan-Bourdon LM, Varela B, Arredondo ME, Aranda E, Flores S, Ochoa P (2021) Clot activators and anticoagulant additives for blood collection. A critical review on behalf of COLABIOCLI WG-PRE-LATAM. Crit Rev Clin Labor Sci 58(3):207–224
12. Mohri M, Rezapoor H. Effects of heparin, citrate, and EDTA on plasma biochemistry of sheep: comparison with serum. Science Direct
13. Chatterjee MS, Denney WS, Jing H, Diamond SL (2010) Systems biology of coagulation initiation: kinetics of thrombin generation in resting and activated human blood. PLoS Comput Biol 6(9):e1000950

14. Mohri M, Allahyari L, Sardari K (2007) Effects of common anticoagulants on routine plasma biochemistry of horse and comparison with serum. J Equine Vet 27(7):313–316
15. Chae H, Kim M, Lim J, Oh EJ, Kim Y, Han K (2012) Novel method to dissociate platelet clumps in EDTA-dependent pseudothrombocytopenia based on the pathophysiological mechanism. Clin Chem Labor Med 50(8):1387-
16. Mollison PL (2000) The introduction of citrate as an anticoagulant for transfusion and of glucose as a red cell preservative. Br J Haematol 108(1):13–18
17. Li K, Xu Y (2015) Citrate metabolism in blood transfusions and its relationship due to metabolic alkalosis and respiratory acidosis. Int J Clin Exp Med 8(4):6578
18. Wright A. Characterizing the presence of EDTA in blood samples. Master of Science in Forensic Science Directed Research Projects
19. Cerón JJ, Martínez-Subiela S, Hennemann C, Tecles F (2004) The effects of different anticoagulants on routine canine plasma biochemistry. Vet J 167(3):294–301

Modern Approaches in the Synthesis of Optically Stimulated Luminescent (OSL) Materials for Dosimetry Applications

Seeza, Rohit Verma, and Tejendra K. Gupta

1 Introduction

Radiation dosimetry is a wide topic that includes all the domains of science and is an essential part of radiation protection and radiation therapy. It is the measurement of the absorbed dose in a medium as a result of interaction with ionizing radiation in the ionization chamber. Dosimetry aims to accurately determine the amount of radiation delivered to a specific target or absorbed by an object, such as a patient, during a medical procedure or radiation exposure. While having the dosimetry, it is very important to have accurate knowledge about the importance of accurate dose measurement to precisely have the determination of the absorbed dose with minimal uncertainty. Accurate dosimetry involves using calibrated instruments and reliable measurement techniques to obtain accurate results. It is crucial in radiation therapy to ensure that the prescribed dose is delivered to the tumor while minimizing the dose to surrounding healthy tissues [1, 2].

In the 1950s and 1960s, optically stimulated luminescence (OSL) was first suggested as a dosimetry tool [3–6]. But today, it has become a successful commercial tool in personal radiation dosimetry, medical dosimetry, diagnostic imaging, clinical dosimetry, optical data storage and security, environmental monitoring, and many more. Optically stimulated luminescence (OSL) is a process in which a material that has already been exposed to ionizing radiation, when subjected to the proper optical stimulation, emits a light signal proportional to the absorbed dose, and the wavelength of the emitted light is characteristic of the OSL material. Optically stimulated luminescence dosimeters (OSLDs) are most commonly made up of carbon-doped aluminum oxide (Al_2O_3:C) and were first used in space dosimetry, and it is an alternative to the thermoluminescent (TLD) dosimeter. Along with the (Al_2O_3:C), europium-doped potassium bromide (KBr: Eu), BeO also continues to influence the

Seeza · R. Verma (✉) · T. K. Gupta
Amity Institute of Applied Sciences, Amity University Uttar Pradesh, Noida 201313, India
e-mail: rverma85@amity.edu

S. Rattan et al. (eds.), *Emerging Trends in Synthesis and Catalysis in Chemistry*, Springer Proceedings in Materials 51, https://doi.org/10.1007/978-981-97-5169-3_5

dosimetry applications which are used in medical dosimeters. Some of the OSL materials which are currently being used in the industry are as follows:

- Aluminum oxide (Al_2O_3:C)—This is one of the most widely used OSL materials. It has a high sensitivity to ionizing radiation and is commonly used in medical dosimetry and geological dating.
- Beryllium oxide (BeO)—BeO is another material that can be used in OSL dosimeters.
- Silicon dioxide (SiO_2)—This material, in various forms, can also exhibit OSL properties and has been used in certain applications.
- Feldspar minerals—Feldspar minerals like potassium feldspar (K-feldspar) and sodium feldspar (Na-feldspar) are commonly used in geological dating.
- Quartz (SiO_2)—Quartz can also be used in OSL dating.
- Lithium Fluoride (LiF:Mg,Ti)—While more commonly associated with thermoluminescence dosimetry (TLD), LiF can also be used in OSL dosimetry.

1.1 Principle of OSL

Optically stimulated luminescence (OSL) is a phenomenon in which certain materials emit light when stimulated by optical radiation after being previously exposed to ionizing radiation [7]. The physical mechanisms behind OSL involve a combination of trapping and recombination processes within the material. Figure 1 shows three stages involved in the OSL process, and the physical mechanisms involved in OSL are given below.

Exposure to Ionizing Radiation Initially, the material is exposed to ionizing radiation, such as X-rays or gamma rays. This radiation ionizes atoms or molecules within the material, creating electron–hole pairs (free electrons and positively charged holes).

Trapping of Electrons and Holes Some of the free electrons and holes generated during the radiation exposure become trapped within defects or impurities in the crystal lattice of the material. These defects act as trapping centers, capturing and immobilizing the charges.

Decay of Trapped Charges Over time, the trapped charges decay through various processes, such as thermal emission or recombination. Recombination occurs when an electron and a hole occupying neighboring trapping centers combine, neutralizing each other.

Optical Stimulation When the material is subsequently exposed to optical radiation, typically in the form of visible or infrared light, the trapped electrons can be stimulated and promoted to higher energy levels.

Releasing of Trapped Charges The absorbed optical energy is sufficient to overcome the trapping energy levels. This allows the trapped electrons to escape from their respective trapping centers and return to the conduction band.

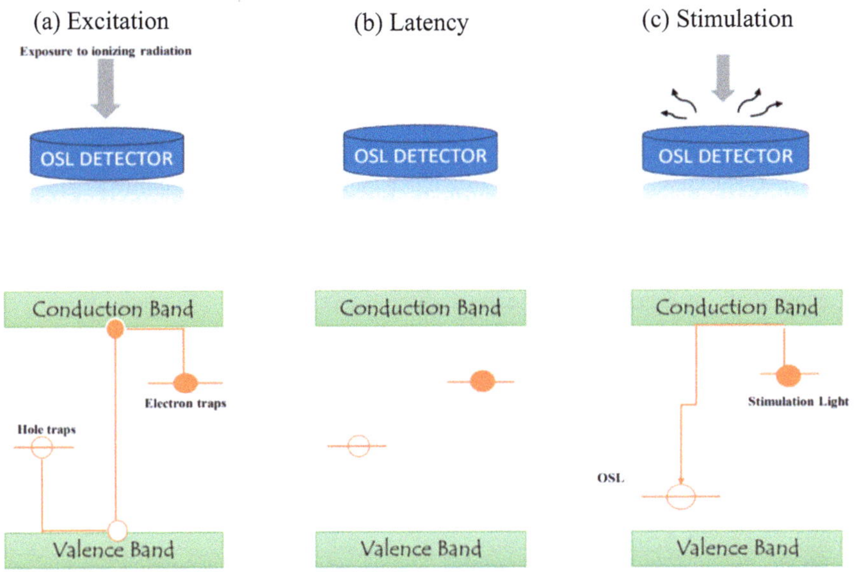

Fig. 1 Three different stages involved in the OSL processes: **a** excitation of the OSL detector by ionizing radiation creating free electrons (black circle) and holes (white circle); **b** the state of latency with metastable electrons and holes captured by traps in the host; and **c** stimulation of the detector with light, leading to the release of charge carriers along with light emission (OSL)

Luminescence Emission As the trapped electrons return to the conduction band, they recombine with holes, resulting in the emission of light. This light emission is referred to as luminescence and can be detected and measured. The intensity of the luminescence emitted during the optical stimulation is directly proportional to the amount of ionizing radiation previously absorbed by the material. Therefore, by measuring the intensity of the emitted light, it is possible to estimate the radiation dose received by the material or determine the age of the sample (in the case of dating applications).

1.2 Properties of Materials Used for OSL Dosimetry

OSL Dosimetry

OSL dosimetry is a technique used to measure the radiation doses. It is also a common technique to be used in various fields, such as medical radiation therapy, industrial radiation monitoring, and environmental radiation assessment. OSL dosimeters typically use crystalline materials, such as aluminum oxide (Al_2O_3) or lithium fluoride (LiF), as the active material. These materials have excellent luminescence properties and can efficiently trap and store radiation energy. Therefore, the key properties of the materials used in OSL dosimetry are given below [8–10].

Luminescent Efficiency The material should exhibit high luminescent efficiency, meaning it should emit a strong and measurable luminescent signal when stimulated by light after radiation exposure. When the dosimeter is exposed to ionizing radiation, the radiation energy is stored in the crystal lattice of the dosimeter material. After that, when the dosimeter is illuminated with a specific wavelength of light, the stored energy is released as luminescence, which can be measured and quantified.

Sensitivity The material should be highly sensitive to the ionizing radiation and able to capture and retain the radiation energy. Higher sensitivity allows for accurate dose measurements, as compared to the low radiation levels [8].

Linear Response The luminescent response of the material should be linearly proportional to the absorbed dose over a wide range of radiation doses. Linearity ensures accurate dose measurements across different radiation levels. It covers the radiation doses ranges from a few milligrays (mGy) to several kilograys (kGy). The dose range can be adjusted by choosing the appropriately sensitive dosimeter materials.

Stability The material should exhibit long-term stability, meaning it should retain its luminescent properties over time and not degrade or lose sensitivity. This stability ensures that the dosimeters can be stored and used for extended periods without significant changes in their performance. The materials used in OSL are also reusable, which allow the multiple measurements and dose assessments.

Tissue-Equivalence The material should have similar radiation absorption and scattering properties to human tissue. This property is important for accurately measuring the dose received by human tissue during radiation exposure.

Based on the above properties of the materials used, OSL dosimetry can provide the accurate and reliable measurements of radiation doses in a wide range of applications, which makes it a valuable technique in radiation monitoring, and protection.

2 Techniques Used for OSL Detection and Analysis

Sample Preparation

The first step is to collect the sample of interest, which could be a sediment sample, a mineral grain, or any material exposed to ionizing radiation. The sample is then processed in the laboratory to extract the desired material, such as quartz or feldspar grains, which are commonly used in OSL dating.

Irradiation

The extracted material is then irradiated with a known dose of ionizing radiation, such as gamma rays or beta particles. This step aims to create trapped charge centers within the crystal lattice of the material.

Bleaching

After irradiation, the sample is exposed to intense light, usually from a light source rich in short wavelengths (e.g., ultraviolet light or blue light). The purpose of bleaching is to remove any residual charge that might have been stored in the material due to environmental or background radiation.

Stimulation

Once the sample has been breached, it is placed in a measurement chamber equipped with a stimulation source, which emits light of a specific wavelength. The stimulation light interacts with the trapped charge centers, causing them to release the stored energy as luminescent light.

Detection and Analysis

The emitted luminescent light is detected using a photomultiplier tube (PMT) or other light-sensitive detectors. The detected light signal is converted into an electrical signal and amplified. Various parameters of the OSL signal, such as intensity, decay time, and spectral characteristics, are measured and recorded [9].

3 Overview of Different Types of OSL Materials

Optically stimulated luminescence (OSL) materials are used in various applications. These materials possess the ability to store energy when exposed to ionizing radiation and release it as visible light when stimulated by light or heat. Some of the different types of OSL materials are as follows:

Alkali Halides Alkali halide crystals, such as potassium bromide (KBr) and sodium iodide (NaI), are commonly used OSL materials. They exhibit excellent sensitivity and have been widely employed in radiation dosimetry and medical imaging applications [12, 13].

Lithium Fluoride (LiF) LiF is a widely studied OSL material due to its high sensitivity and stability. It can be found in two forms: LiF: Mg, Ti (MTS) and LiF: Mg, Cu, P (MCP). LiF-based detectors are used for personal dosimetry, environmental monitoring, and retrospective dosimetry.

Aluminum Oxide (Al_2O_3) Aluminum oxide, also known as alumina, is a versatile OSL material. It exists in different crystal structures, such as corundum and sapphire. Al_2O_3:C is a commonly used type of aluminum oxide, which is doped with carbon. It is employed in medical dosimetry, radiation therapy, and retrospective dosimetry.

Quartz Quartz is a natural mineral with OSL properties. It is composed of silicon dioxide (SiO2) and is often used in geological dating and archeology. Quartz-based OSL dating techniques have been utilized to determine the age of sediments and geological formations.

TLD-OSL Hybrid Materials Some materials combine the properties of thermoluminescence (TL) and OSL to provide enhanced dosimetric capabilities. For example, TLD-OSL hybrid materials like LiF:Mg,Ti and BeO are used to measure both TL and OSL signals, allowing for a wider dose range and more accurate dose measurements.

Table 1 provides the different types of compounds used in OSL dosimetry and their families.

Table 1 Different types of OSL materials compound families

Compound family	Host materials	Dopant and activators	Z_{eff}	Emission wavelength, λ_E (nm)	Stimulation wavelength, λ_S (nm)	References
Halides	CaF_2	Mn, AlCe, Tb, Gd, Dy, Eu, Nd, Ln^{3+}, Al^{3+}/Li^+	16.9	430	850	[17]
	NaCl	Eu^{2+}	15.3	353	500	[13]
		Cu^+			470	
		Eu^{2+}	31.5	420		[17]
	KBr	Sm^{3+}		200–400	470	[10]
		Eu^{2+}				
	LiF	Mg, Cu, P	8.3	360	470	[10]
Aluminates	$CaAl_2O_4$	Eu^{2+}, Nd^{3+}, Dy^{3+}, Tm^{3+}	14.8	440	980	[10]
	$MgAl_2O_4$	Mn^{2+}	11.2	520	800	[17]
	$ZnAl_2O_4$	Tb	22.4			[10]
	$SrAl_2O_4$	Eu^{2+}, Dy^{3+}	29.3	510	630–850	[8]
	$LaAlO_3$	Cr^{3+}, Sm^{3+}		734	625, 808, 977	[8]
Sulfides	CaS	Eu^{2+}, Sm^{3+}	18.5	650	800–1600	[8]
	SrS	Yb^{2+}, Eu^{2+}, Sm^{3+}	34.6	550–700	980	[8]
				590	1020	[11]
	MgS	Ce^{3+}, Eu^{2+}, Sm^{3+}	14.6	510–590	1050	[10]
Gallates	$MgGa_2O_4$	Mn	26.7			[13]
Oxides	Al_2O_3	C	11.3	420	450–550	[10]
	BeO	Na^+			470	[8]
	ZrO		7.2	480	808/980	[17]

4 Synthesis Methods of OSL Dosimetry

OSL involves the stimulation of luminescent materials by light, which causes the stored energy from ionizing radiation to be released as visible light. This light emission can be detected and quantified to determine the radiation dose received by the material. The synthesis of materials used in OSL dosimetry involves the preparation of luminescent substances capable of trapping and releasing energy upon stimulation. Here are a few important synthesis methods which are commonly employed for OSL materials.

4.1 Sol–gel Method

The sol–gel technique involves the synthesis of materials from a precursor solution or "sol" that undergoes gelation to form a solid network. Luminescent species are typically incorporated into the sol by dissolving suitable precursor compounds. The sol is then subjected to controlled drying and heat treatment, leading to the formation of a porous glass-like material containing the desired luminescent centers [14]. Figure 2 displays the reaction pathway for the production of OSL materials by the sol-gel method. The steps involved in the synthesis are stepwise described as follows.

Step 1: Preparation of Sol A precursor solution is prepared by dissolving suitable precursors in a solvent, such as alcohol or water. The precursors may consist of metal alkoxides or metal salts, depending on the desired composition of the OSL material.

Step 2: Gelation The precursor solution is then subjected to gelation, which can be achieved through various methods such as chemical reactions, acid–base reactions, or hydrolysis. This step leads to the formation of a three-dimensional network of interconnected nanoparticles.

Step 3: Aging and Drying The gel is allowed to age for a specific duration to promote further particle growth and stabilization. After aging, the gel is dried to remove the solvent and obtain a solid material.

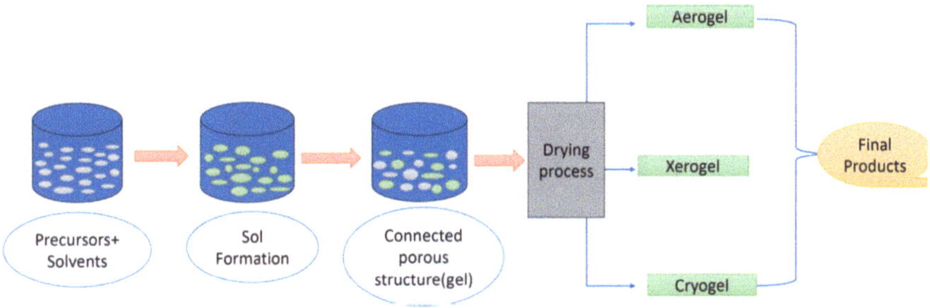

Fig. 2 Reaction pathway for the production of OSL materials by the sol-gel method

Step 4: Annealing The dried gel is subjected to an annealing process at a high temperature, typically between 400 and 1000 °C. Annealing helps in the removal of any residual organic compounds and enhances the structural and luminescent properties of the OSL material.

Step 5: Grinding and Pellet Formation The annealed material is ground into a fine powder and then compacted into pellets or discs for OSL measurements.

4.2 Co-Precipitation Method

This method involves the simultaneous precipitation of luminescent materials with suitable host compounds. Precursor salts or solutions are mixed under controlled conditions, leading to the formation of a solid material containing the desired luminescent centers. The composition and properties of the resulting material can be tailored by adjusting the reaction parameters. Figure 3 shows the reaction pathway for the production of OSL materials by the co-precipitation method. The steps involved in the synthesis are stepwise described as follows [14]:

Step 1: Selection of Precursors Choose appropriate precursor materials based on the desired luminescent properties. Common precursors for OSL materials include metal oxides, rare earth elements, and dopants.

Step 2: Solution Preparation Dissolve the selected precursors in a suitable solvent to create a homogeneous solution. The choice of solvent depends on the nature of the precursors and their solubility characteristics.

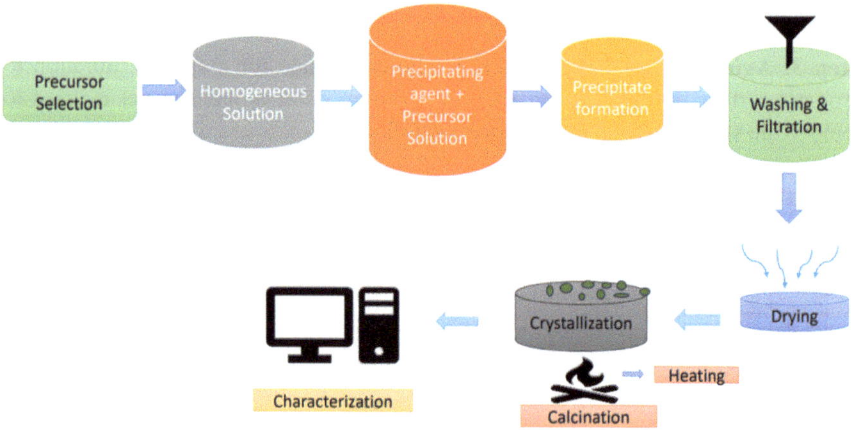

Fig. 3 Reaction pathway for the production of OSL materials by the co-precipitation method

Step 3: Mixing and Precipitation Combine the precursor solutions while continuously stirring to ensure proper mixing. This step promotes the formation of a homogeneous mixture and prevents the precipitation of individual components. Adjust the solution pH, temperature, and other parameters as required for the precipitation process.

Step 4: Co-Precipitation Initiate the co-precipitation process by introducing a precipitating agent into the precursor solution. The choice of precipitating agent depends on the desired properties of the final OSL material. Common precipitating agents include ammonium hydroxide, sodium hydroxide, or other alkali metal hydroxides.

Step 5: Precipitate Formation As the precipitating agent reacts with the precursors, the desired OSL material begins to form a solid precipitate. This stage involves nucleation and growth processes, where the formation of stable particles occurs. The reaction conditions, such as temperature, pH, and reaction time, influence the particle size and morphology.

Step 6: Washing and Filtration After the co-precipitation is complete, the resulting precipitate needs to be separated from the solution. Use filtration to collect the precipitate, and then wash it several times with a suitable solvent to remove impurities, unreacted precursors, and byproducts.

Step 7: Drying Once the precipitate is adequately washed, remove the excess solvent by drying. Common drying methods include air drying, vacuum drying, or oven drying. Ensure the temperature and time used for drying is appropriate to avoid any phase transformations or material degradation.

Step 8: Calcination In some cases, a calcination step is necessary to convert the dried precipitate into the desired OSL material. Calcination involves heating the precipitate at a controlled temperature for a specific duration. This process promotes crystallization, phase formation, and elimination of residual impurities.

4.3 Chemical Vapor Deposition (CVD)

CVD is a vapor-phase synthesis method widely used for the deposition of thin films. In the context of OSL materials, CVD can be employed to grow luminescent layers on suitable substrates. The process involves the reaction of precursor gases in a heated chamber, resulting in the deposition of a thin film on the substrate [15]. This method offers precise control over film thickness and composition. The steps involved in the synthesis are stepwise described as follows.

Step 1: Selection of Precursor Choose a suitable precursor material that contains the desired elements for OSL. This precursor should have the capability to decompose and deposit the desired elements onto a substrate during the CVD process.

Step 2: Preparation of Substrate Prepare the substrate on which the OSL material will be deposited. The substrate is typically a solid material such as a silicon wafer or glass slide. The substrate should be cleaned and prepared to ensure good adhesion of the deposited material.

Step 3: Reactor Setup Set up a CVD reactor system. The reactor typically consists of a reaction chamber where the deposition will take place. The chamber is equipped with temperature control, gas flow control, and a vacuum system.

Step 4: Precursor Delivery Introduce the precursor material into the CVD reactor. The precursor can be in the form of a liquid, solid, or gas, depending on its properties. It is usually delivered into the reactor as a vapor or a gas by using a carrier gas or through sublimation.

Step 5: Activation Activate the precursor by providing suitable energy. This can be achieved by heating the precursor to a specific temperature, which causes it to decompose and release the desired elements onto the substrate surface.

Step 6: Deposition Allow the activated precursor to deposit onto the substrate. The released elements react or condense on the substrate surface, forming a thin film or coating. The deposition process can be controlled by adjusting the temperature, gas flow rates, and other parameters.

Step 7: Annealing and Post-Treatment After deposition, the deposited material may require annealing or other post-treatment steps to enhance its crystallinity, remove defects, or improve its luminescent properties. This step may involve subjecting the substrate to specific temperature and atmosphere conditions.

4.4 Microwave-Assisted Synthesis

It is another approach that has gained attention in recent years. This technique involves the use of microwaves to promote the rapid and efficient synthesis of OSL materials. Microwave-assisted synthesis has been shown to produce high-quality materials with excellent luminescent properties [16]. The steps involved in the synthesis are stepwise described as follows:

Step 1: Material Selection Identify the specific material or compound that is suitable for OSL studies. This could be a phosphor material or a composite material containing luminescent components.

Step 2: Precursor Selection Choose appropriate precursor compounds that can be used to synthesize the desired material. The precursors should have suitable reactivity and compatibility with the microwave-assisted synthesis method.

Step 3: Reactant Preparation Weigh and mix the precursor compounds in the desired stoichiometric ratio. This mixture will serve as the reactant for the synthesis process.

Step 4: Reaction Vessel Setup Transfer the reactant mixture into a suitable reaction vessel that is compatible with microwave irradiation. The vessel should be microwave-safe and allow efficient heating and mixing of the reactants.

Step 5: Microwave Irradiation Place the reaction vessel in a microwave reactor and apply microwave irradiation. The microwave energy will induce rapid heating of the reactants, promoting faster reaction kinetics compared to conventional heating methods.

Step 6: Reaction Monitoring Monitor the reaction progress during microwave irradiation. This can be done using in-situ techniques such as temperature monitoring, pressure monitoring, or spectroscopic analysis (if applicable). Adjust the microwave power and irradiation time if necessary to optimize the reaction.

Step 7: Cooling and Quenching Once the desired reaction is complete, remove the reaction vessel from the microwave reactor, and allow it to cool down. Some reactions may require quenching to halt the reaction process. Quenching can be achieved by adding a suitable quenching agent or by cooling the reaction vessel rapidly.

Step 8: Product Isolation After cooling, the synthesized product needs to be isolated from the reaction mixture. This typically involves filtration, washing, and drying steps to remove impurities and obtain the desired material in a purified form.

Step 9: OSL Measurement Finally, perform optically stimulated luminescence measurements on the synthesized material to evaluate its luminescent properties, such as luminescence intensity, decay characteristics, and response to optical stimulation.

4.5 Hydrothermal Synthesis

This technique involves the use of high-pressure, high-temperature water to promote the formation of crystalline materials. Hydrothermal synthesis has been shown to produce high-quality OSL materials with high luminescence efficiency and excellent dosimetric properties [17, 18].

Here are the steps involved.

Step 1: Selection of Precursors Choose suitable precursor compounds that will provide the desired composition and properties for the OSL material. These precursors typically consist of metal salts or complexes containing the necessary elements.

Step 2: Preparation of Reaction Mixture Dissolve the precursors in a suitable solvent to create a reaction mixture. The choice of solvent depends on the precursors and reaction conditions. Water is commonly used as the solvent for hydrothermal synthesis.

Step 3: Sealing the Reaction Vessel Transfer the reaction mixture into a sealed reaction vessel, such as a Teflon-lined autoclave. The vessel is sealed to maintain high-pressure and temperature conditions during the hydrothermal reaction.

Step 4: Heating and Reaction Place the sealed reaction vessel in a high-temperature furnace or hydrothermal reactor. The reaction vessel is heated to the desired temperature and maintained for a specific duration. Typical reaction temperatures range from 100 to 300 degrees Celsius but can vary depending on the specific materials and conditions.

Step 5: Cooling and Isolation After the desired reaction time, the reaction vessel is slowly cooled to room temperature. This gradual cooling helps to prevent rapid pressure changes and potential damage to the synthesized material. Once cooled, the product can be isolated by filtration or other suitable methods.

Step 6: Washing and Drying Wash the synthesized material to remove any impurities or residual reactants. This is typically done by rinsing with distilled water or suitable solvents. After washing, the material is dried under controlled conditions, such as in an oven or desiccator.

5 Characterization of OSL Materials

Characterizing the optically stimulated materials involves studying their structural and optical properties. X-ray diffraction (XRD) is a widely used technique to analyze the crystal structure and phase composition of materials. It involves irradiating a sample with X-rays and measuring the resulting diffraction pattern. XRD can provide information about the crystal structure, lattice parameters, and the presence of impurities or structural defects in OSL materials. Electron microscopy (EM) surface morphology, topographic image, and elements present in the synthesized materials can be studied via scanning electron microscopy (SEM), transmission electron microscopy (TEM), and energy-dispersive X-ray spectroscopy (EDS), respectively. SEM can be used to examine the microstructure and surface features of OSL materials. TEM can provide information about the crystalline structure, grain boundaries, defects, and composition of the material at the atomic scale. EDS can identify the elements present in the OSL material and provide quantitative information about their relative concentrations. EDS is useful for determining the elemental composition and mapping the distribution of different elements within the material [19].

6 Applications of OSL Dosimetry

OSL dosimetry is a technique used for measuring and assessing radiation doses. It is commonly employed in various applications where accurate and precise radiation dose measurements are required. Here are some of the applications of optically stimulated luminescence dosimetry:

6.1 Radiation Therapy

OSL dosimeters are used to measure the radiation doses delivered to patients during radiation therapy treatments. These dosimeters provide accurate and reliable measurements of the dose received by the patient, ensuring that the prescribed dose is properly delivered while minimizing the risk to healthy tissues.

6.2 Personal Radiation Monitoring

OSL dosimeters are widely used in occupational radiation monitoring programs. Workers who are occupationally exposed to radiation, such as nuclear power plant employees, radiographers, and medical personnel, wear OSL dosimeters to measure and record their radiation exposure levels. This helps ensure that radiation doses are within acceptable limits and that proper safety measures are implemented.

6.3 Environmental Monitoring

OSL dosimeters are used to measure radiation levels in the environment, such as around nuclear power plants, uranium mines, or other areas where radiation sources may be present. These dosimeters can be deployed in various locations to monitor and assess potential radiation hazards, enabling timely intervention and appropriate safety measures if required.

6.4 Radiological Emergencies

In the event of a radiological accident or emergency, OSL dosimetry plays a crucial role in assessing the radiation doses received by individuals, including emergency workers and the general public. OSL dosimeters can provide rapid and accurate measurements of radiation exposure, assisting in the effective management and response to such incidents.

6.5 Medical Imaging

OSL dosimeters can be used to measure the radiation doses received by patients undergoing diagnostic imaging procedures, such as X-rays and CT scans. By quantifying the radiation exposure, healthcare professionals can optimize imaging protocols to minimize patient radiation dose while maintaining image quality.

6.6 Research and Development

OSL dosimetry is extensively used in research and development related to radiation, dosimetry, and radiation protection. It helps scientists and researchers study the behavior of different types of radiation, assess the effectiveness of shielding materials, and develop new dosimetric techniques.

7 Current Challenges and Limitations of OSL Dosimetry

Some of the limitations of OSL dosimetry are as follows:

Sensitivity to Environmental Factors OSL dosimeters can be sensitive to environmental conditions such as temperature, humidity, and light. These factors can affect the stability and reproducibility of the OSL signal, leading to potential errors in dose measurements [20–25].

Fading of OSL Signal OSL signals can decay over time, even in the absence of radiation exposure. This fading effect can lead to an underestimation of the measured dose if not properly accounted for. Various correction techniques, such as pre-and post-irradiation annealing, are employed to mitigate this issue, but they add complexity to the dosimetry process.

Energy Dependence OSL dosimeters can exhibit energy-dependent responses, meaning that their sensitivity to different types of radiation may vary. Calibration factors specific to the radiation type and energy need to be established for accurate dose measurements. Additionally, some OSL materials may have limited sensitivity to low-energy radiation, which can be a limitation in certain applications.

Dose Rate Dependence OSL dosimeters may exhibit dose rate dependence, which means that their response can vary with the rate at which the radiation is delivered.

This can introduce uncertainties in dose measurements, particularly when dealing with high dose rates or dynamic radiation fields.

Dose Range Limitations The dose range over which OSL dosimeters can provide accurate measurements may be limited. For extremely high or low doses, the response of the dosimeter may deviate from linearity, necessitating the use of alternative dosimetry techniques or specialized dosimeters.

8 Conclusions

Advancement in optically stimulated luminescent (OSL) materials has increased since last a decade. Lot of progress have been made in terms of research publications to understand the mechanism of OSL dosimetry, selection of new materials and developing new techniques, and upgradation of instruments. In the present work, we have discussed the few important methods such as sol–gel method, hydrothermal synthesis, microwave-assisted synthesis, co-precipitation method, and chemical vapor deposition (CVD) method to synthesize OSL dosimetry materials. OSL has wide range of applications including environmental monitoring, radiation protection and radio therapy, computed radiography, space dosimetry, and medical physics. Selection of the materials for OSL dosimetry is very important. Therefore, aluminum oxide (Al_2O_3) or lithium fluoride (LiF) are generally used materials because these materials are active materials, high luminescent efficiency, sensitive to the ionizing radiation, linear luminescent response proportional to the absorbed dose over a wide range of radiation doses and long-term stability. The current developments indicate that reliable OSL dosimeter systems could be commercialized soon in the near future for radiotherapy, radiation protection, and point dose measurements.

References

1. McKeever SWS (2022) A course in luminescence measurements and analyses for radiation dosimetry. Wiley
2. McKeever SWS, Chen CY, Halliburton LE (1985) Point-defects and the predose effect in natural quartz. Nucl Tracks Radiat Meas 10:489–495
3. Daniels F, Boyd CA, Saunders DF (1953) Thermoluminescence as a research tool. Science 117:343–349
4. Antonov-Romanovskii VV, Keirum-Markus IF, Poroshina MS, Trapeznikova ZA (1955)
5. Yukihara EG, Mittani J, McKeever SW, Simon SL (2007) Optically Stimulated Luminescence (OSL) of dental enamel for retrospective assessment of radiation exposure. Radiat Meas 42(6):1256–1260. https://doi.org/10.1016/j.radmeas.2007.05.038
6. Dieterich S, Ford E, Pavord D, Zeng J (2016) Chapter 3—In-Vivo dosimetry. Dieterich S, Ford E, Pavord D, Zeng J (eds) Practical radiation oncology physics. Elsevier, pp 30–39. ISBN 9780323262095. https://doi.org/10.1016/B978-0-323-26209-5.00003-1

7. Kalnins AGC, Heidepriem EH, Spooner AN, Monro MT (2012) Radiation dosimetry using optically stimulated luminescence in fluoride phosphate optical fibres. Opt Mater Express 2(1):1648–1656
8. Nick C (2019) What is OSL dosimetry—Optically stimulated luminescence dosimeter.
9. Sahil RK, Yadav MK, Kumar P (2022) OSL and TA-OSL properties of Li2B4O7:Al for radiation dosimetry. J Alloys Compounds 908:164628, ISSN 0925-8388. https://doi.org/10.1016/j.jallcom.2022.164628
10. Oliveira L, Yukihara E, Baffa O (2016) MgO:Li, Ce, Sm as a high-sensitivity material for Optically Stimulated Luminescence dosimetry. Sci Rep 6:24348. https://doi.org/10.1038/srep24348
11. Duller GAT (2008) Luminescence dating. Quaternary Sci (2008)
12. Nanto H, Okada G (2023) Jpn J Appl Phys 62:010505
13. Teng TP, Hung YH, Teng TC, Chen JH (2011) Nanoscale research letters
14. Yuan L, Jin Y, Su Y, Wu H, Hu Y, Yang S (2020) Optically stimulated luminescence phosphors: principles, applications, and prospect. Laser Photonics Rev 2000123
15. Kanies B, Hayes R, Yang GE (2022) Thermoluninescence and optically stimulated luminescence response of Al_2O_3 coatings deposited by mist-chemical vapor deposition. Science Direct
16. Kunzel R, Umisedo NK, Okuno E, Yoshimura EM, Marques APDA (2020) Effects of microwave-assisted hydrothermal treatment and beta particles irradiation on the thermoluminescence and optically stimulated luminescence of $SrMoO_4$ powders. Science direct, 15018–15026 (2020).
17. Aslani F, Bagheri R, Shirazi, A (2017) Synthesis and characterization of hydrothermal derived OSL Mater Dosimetry Appl (2017)
18. Hosseini SH, Faghihi R, Ghafoori-Fard H (2015) Hydrothermal synthesis of OSL materials for dosimetry applications (2015)
19. Vokhmintsev AS, Weinstein IA, Karabanalov MS, Smorodinskii YaG (2015) OSL diagnostics of luminescent materials in a scanning electron microscope. Nondestructive testing
20. ChenR (2001) Advantages and disadvantages in the utilisation of thermoluminescence and optically stimulated luminescence for radiation dosimetry
21. Eduardo GY, Adrie JJB, Pawel B, Stephen WS (2022) The quest for new thermoluminescence and optically stimulated luminescence materials: needs, strategies and pitfalls
22. Ayaka H, Kosuke M, Yusuke M, Masayasu K, Thunyarat C, Akihiro T (2021) Use of optically stimulated luminescence dosimeter and radiophotoliminescent glass dosimeter for dose measurement in dual-source dual-energy computed tomography
23. Muhammad Y, Nazeeh A, Abdulraheem (2017) Organ dose measurement using Optically Stimulated Luminescence Detector (OSLD) during CT examination
24. Paul AJ (2010) Changes in optically stimulated luminescent dosimeter (OSLD) dosimetric characteristics with accumulated dose
25. Bossin L, Plokhikh I, Christensen JB, Gawryluk DJ, Kitagawa Y, Leblans P, Tanabe S, Vandenbroucke D, Yukihara EG (2023) Addressing current challenges in OSL dosimetry using MgB_4O_7:Ce,Li: state of the art limitations and avenues of research. Materials 16:3051. https://doi.org/10.3390/ma16083051

Recent Advances in the Synthesis of Thermoluminescent Materials for Dosimetry Applications

Shalini Kumari, Rohit Verma, and Tejendra K. Gupta

1 Introduction

The world of material science continues to push the boundaries of innovation, bringing us closer to a future filled with cutting-edge technologies. One fascinating area of research is the synthesis of thermoluminescent materials, which have seen remarkable advancements in recent years. The unique property of these materials is the ability to emit light when heated. This remarkable characteristic has captivated scientists and engineers alike, as it holds great potential for a wide range of applications such as spanning from radiation dosimetry and medical imaging to environmental monitoring and advanced lighting systems [1].

Thermoluminescence, the phenomenon behind this breakthrough, occurs when certain materials absorb energy from ionizing radiation and store it in the form of trapped charge carriers within their lattice structure. When subjected to heat, these trapped charges are released, resulting in the emission of light. Figure 1 is a schematic diagram displaying all these steps. The ability to harness and control this phenomenon is crucial in the development of efficient and reliable thermoluminescent materials.

Considering TL material for personal dosimetry purpose. The development of dosimetry systems has received a lot the attention, and over the past 35 years, a lot has changed in the globe. Thermoluminiscence dosimeters (TLDs) built using solid-state technology have proven to be one of the finest options for delivering the necessary precision, dependability, toughness, and automation potential for managing a large number of dosimeters.

S. Kumari · R. Verma (✉) · T. K. Gupta
Amity Institute of Applied Sciences, Amity University Uttar Pradesh, Noida 201313, India
e-mail: rverma85@amity.edu

S. Rattan et al. (eds.), *Emerging Trends in Synthesis and Catalysis in Chemistry*,
Springer Proceedings in Materials 51, https://doi.org/10.1007/978-981-97-5169-3_6

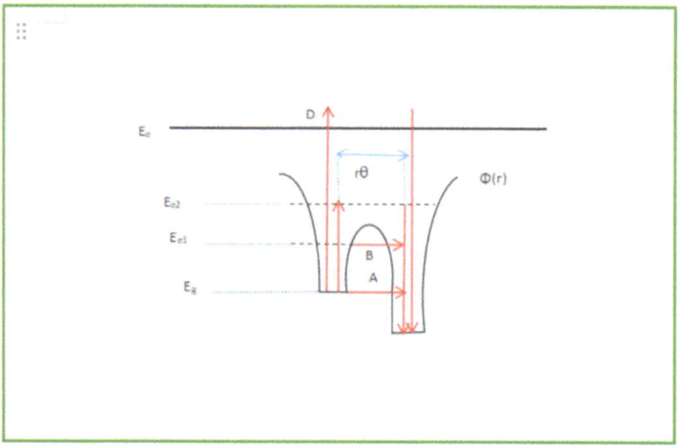

Fig. 1 Schematic diagram of TLD working principle

2 Synthesis Techniques of Thermoluminescent Materials

In recent years, significant strides have been made in the synthesis of thermoluminescent materials, driven by advancements in material design, synthesis techniques, and characterization methods. Researchers have explored a variety of materials, including inorganic compounds, organic compounds, and hybrid materials, in order to enhance the thermoluminescent properties and expand the potential applications of these materials.

The development of advanced synthesis techniques has played a pivotal role in achieving improved control over the structure, composition, and morphology of thermoluminescent materials [2, 3]. Nanoscale engineering, for instance, has allowed scientists to tailor the properties of these materials at the atomic and molecular levels, leading to enhanced sensitivity, efficiency, and stability. Additionally, the use of novel dopants and modifiers has opened up new avenues for tuning the luminescent properties of these materials, enabling a wider range of color emission and response to specific radiation types. Some common thermoluminescent materials and their synthesis methods are described as follows and are shown in Table 1.

Table 1 Some common thermoluminescent materials and their synthesis methods

Material	Synthesis method	References
$CaSO_4$:Dy	Solution growth	[9]
MgB_4O_7:Dy	Solid-state reaction	[4]
LiF:Mg,Ti	Co-precipitation	[5]
Al_2O_3:C	Sol–gel	[7]
$Li_2B_4O_7$:Cu,P	Sol–gel	[8]

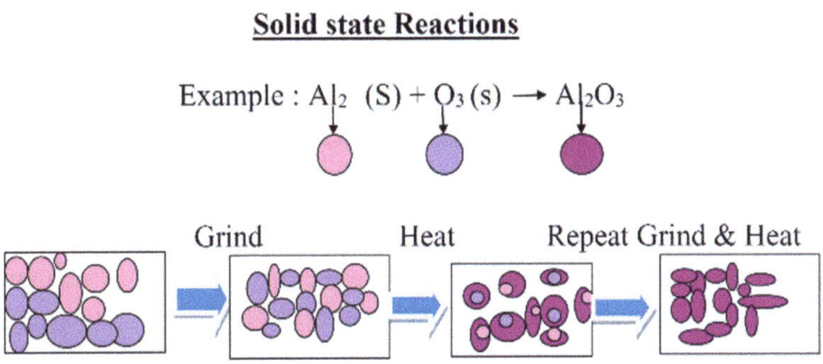

Fig. 2 Pathways of synthesis of thermoluminescent materials through solid-state reaction method

I. **Solid-State Reaction Method**: This is a common method for synthesizing thermoluminescent materials. It involves mixing and grinding the precursor materials followed by heating at high temperatures to induce a chemical reaction and form the desired compound [4]. As shown in Fig. 2, the reaction leads to the formation of a crystalline thermoluminescent material. Solid-state reactions are widely used for the synthesis of inorganic thermoluminescent materials such as metal oxides and sulfides.

II. **Co-precipitation Method**: In this method, two or more precursor salts (or desired metal ions) are dissolved in a suitable solvent, and then a precipitating agent is added to form a solid. The resulting precipitate is then separated, washed, and heated to convert it into the desired thermoluminescent material [5]. Co-precipitation can be used to synthesize various types of thermoluminescent materials, including phosphors and scintillators. The Dy Doped $Caso_4$ and Tm doped $Caso_4$ TLD materials are synthesized by co-precipitation method $Caso_4$ through the following reaction [6]:

$$(CH_3COO)_2Ca_{(S)} + (NH_4)_2SO_4 + Tm_2(SO_4)_{3(s)}(0.01\%) \xrightarrow{\text{E to H + H}_2\text{O}} CaSO_4{:}Tm_{(s)} + 2CH_3\ COONH_{4(aq)} \tag{1}$$

$$(CH3COO)_2Ca_{(s)} + (NH4)_2SO_4(S) \xrightarrow{\text{E to H+H}_2\text{O}} CaSO_4(s) + 2CH_3\ COONH_{4(aq)} \tag{2}$$

III. **Sol–gel Method**: The sol–gel process involves the synthesis of materials from a colloidal suspension (sol) that gradually transforms into a gel-like network. This method offers good control over the composition and structure of the material and can be used to prepare various thermoluminescent oxides [7, 8]. This is a versatile approach for synthesizing various kinds of materials, including thermoluminescent materials. In this method, precursor compounds form a sol on hydrolysis and condensation, which is then transformed into a gel by solvent evaporation method. The gel is subsequently heat-treated to obtain the desired thermoluminescent material. The sol–gel method allows for fine control over the composition, purity, and morphology of the resulting material.

IV. **Hydrothermal Synthesis**: Hydrothermal methods involve the synthesis of materials under high-pressure, high-temperature aqueous conditions. This approach allows for the controlled growth of crystalline thermoluminescent materials with specific morphologies and properties. Precursor compounds are dissolved in a solvent, and the solution is sealed in a pressure vessel and heated to the desired conditions. The hydrothermal environment facilitates the controlled formation of thermoluminescent crystals with specific morphologies and properties.

V. **Microwave-assisted Synthesis**: Microwave irradiation can be used to facilitate the synthesis of thermoluminescent materials. This method offers advantages such as rapid heating, enhanced reaction rates, and improved product properties.

VI. **Chemical Vapor Deposition (CVD) Method**: CVD is commonly used to deposit thin films of various materials, including thermoluminescent materials (see Fig. 3). It involves the reaction of volatile precursor compounds in a gas phase, which then deposits onto a substrate to form a thin film. CVD allows for the deposition of high-quality, uniform films with precise control over composition and thickness.

VII. **Nanomaterial Synthesis**: The synthesis of thermoluminescent nanomaterials has gained significant attention. Various techniques, such as solvothermal, hydrothermal, and template-assisted methods, have been employed to produce nanoscale thermoluminescent materials with unique properties.

These are just a few examples of the synthesis methods used for thermoluminescent materials. The choice of method depends on factors such as the desired material composition, structure, purity, and application requirements. Researchers

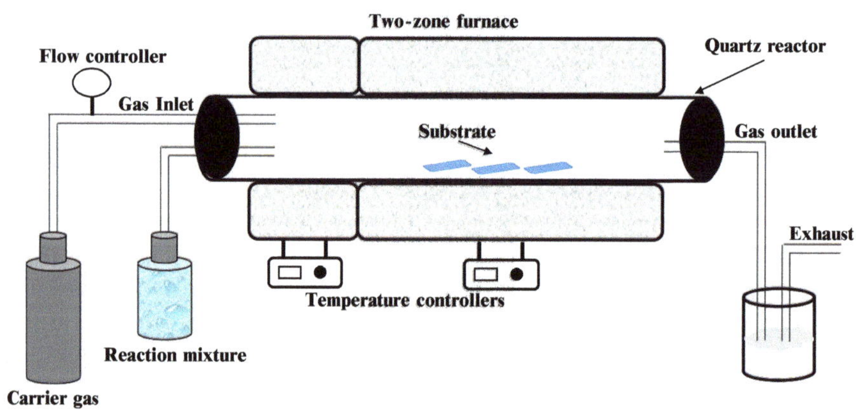

Fig. 3 Schematic diagram of CVD method for the synthesis of thermoluminescent materials

continue to develop and refine synthesis techniques to improve the performance of thermoluminescent materials in various fields.

Furthermore, advancements in characterization techniques have provided researchers with deeper insights into the fundamental processes occurring within thermoluminescent materials. Sophisticated imaging techniques, such as atomic force microscopy (AFM), scanning tunneling microscopy, spectroscopic analysis, and computational modeling, have shed light on the intricate mechanisms of charge trapping, release, and recombination, aiding in the design and optimization of these materials for specific applications.

Thermoluminescent materials (TLMs) are used in radiation dosimetry, and their synthesis involves several steps. Synthesis of TLMs typically involves the following steps:

i. **Selection of materials**: The first step in the synthesis of TLMs is the selection of materials. The materials should have a high sensitivity to ionizing radiation, and they should be able to emit light when heated. Common materials used in TLMs include LiF, CaF_2, MgF_2, and TLD-100 (a proprietary material consisting of LiF, Mg, and Ti). TLMs can be made from a variety of materials, including phosphors, ceramics, and glasses. The material selected should have the desired properties, such as a high sensitivity to radiation and a well-defined thermoluminescent glow curve.

ii. **Preparation of the material**: The material is typically prepared by mixing the appropriate chemicals or compounds and heating them to a high temperature to form a homogeneous mixture. The mixture is then cooled and ground into a fine powder.

iii. **Formation of the TLM**: The TLM is typically formed by pressing the powder into a pellet or other desired shape using a hydraulic press. The pellet is then heated to a high temperature to sinter the particles together.

iv. **Characterization of the TLM**: The TLM is characterized using various techniques, such as X-ray diffraction, scanning electron microscopy, and thermoluminescence measurements.

v. **Preparation of materials**: The selected materials are then prepared. This involves grinding the materials into a fine powder and then mixing them with a small amount of dopant material, such as Mg or Cu. The dopant material enhances the sensitivity of the TLMs to radiation.

vi. **Pressing of materials**: The prepared materials are then pressed into pellets or discs under high pressure. This ensures that the TLMs have a uniform density, which is important for accurate dosimetry.

vii. **Annealing of materials**: The pressed TLMs are then annealed at a high temperature (typically 400–500 °C) for several hours. This process helps to remove any defects in the crystal structure of the TLMs, which can affect their sensitivity.

viii. **Cutting and polishing of materials:** After annealing, the TLMs are cut and polished to a smooth surface. This ensures that the TLMs have a uniform thickness, which is important for accurate dosimetry.

ix. **Packaging of materials:** The final step in the synthesis of TLMs is the packaging of the materials. The TLMs are typically packaged in a light-tight container to protect them from ambient light. The container may also include a small amount of material that emits a known amount of radiation, which can be used to calibrate the TLMs.

3 Applications of Thermoluminescent Materials

The potential applications of thermoluminescent materials are vast. In the field of radiation dosimetry, these materials can accurately measure and quantify ionizing radiation exposure, ensuring the safety of medical professionals, nuclear workers, and individuals undergoing radiation therapy. They also hold promise in environmental monitoring, enabling the detection and quantification of various radiation sources in air, water, and soil. Table 2 describes the different applications of these materials.

Moreover, thermoluminescent materials have the potential to revolutionize the field of medical imaging. By harnessing their unique luminescent properties, researchers can develop highly sensitive and low-dose imaging techniques for early disease detection and precision medicine. Additionally, the integration of thermoluminescent materials into advanced lighting systems could yield energy-efficient and environmentally friendly illumination solutions.

Table 2 Applications of thermoluminescent materials

Applications	Description	References
Radiation dosimetry	Used to measure and monitor radiation exposure	[10]
Archeological dating	Determines the age of ancient artifacts and materials	[11]
Environmental radiation monitoring	Monitoring radiation levels in the environment	[12]
Medical imaging	Used in medical imaging techniques such as CT scanners	[13]
Nuclear power plant safety	Monitoring radiation levels and doses in nuclear facilities	[14]
Personal radiation dosimeters	Compact devices for personal radiation monitoring	[15]

4 TLD Badge Application as Personal Dosimetry

In India $CaSO_4$:Dy thermoluminescent phosphor is used in TLD badges as a radiation detector [6, 16–34]. Its highly sensitive thermoluminescent (TL) phosphor (approximately 40 times that of the most widely used LiF TLD-100) is made possible by the Dy dopant concentration of 0.05 mol% in $CaSO_4$. (Only 3.8 mGy equivalent Co-60 gamma ray reaction per 10^{10} thermal neutron cm^{-2}). The neutron response is still insignificant. With extremely minor satellite glow peaks on the upper and lower temperature sides of the main peak, the phosphor exhibits a primary TL glow peak at about 220 °C (the TL peak of $CaSO_4$: Dy Teflon Disc occurs at around 240 °C). In the dose range of importance for radiation protection, the glow curve shape and TL response are both linear for both low and high LET radiation. The reaction and chemical form are extremely resistant to climate changes. Bright sunshine causes the TL signal to fade more quickly, whereas exposure to everyday room light has no effect on the response. The response from TLDs packaged in paper wrappers, however, and sunlight exposure has no effect on polythene bags. $CaSO_4$:Dy phosphor was placed in teflon (polytetrafluoroethylene PTFE) to create TLD discs, which allowed for the production of a large number of integrated dosimeters with the same TL sensitivity. Teflon is a material that can resist the temperatures required for TL reading and the annealing process for repeated usage, making it ideal for use in field dosimeters [16].

The TLD badge consists of a plastic cassette or container and a TLD card. Three teflon TLD discs (13.3 mm in diameter and 0.8 mm thick) are mechanically attached to a circular hole (12.0 mm in diameter) cut in an aluminum plate with dimensions of $52.5 \times 30 \times 1$ mm^3 to form a TLD card. The card has an uneven "V" cut to guarantee its alignment. Before being loaded into the plastic cassette, the card, which is wrapped in a thin paper wrapper with printed information about the user, the service duration, the type of radiation, etc., is placed in a thin plastic bag with a thickness of 3–4 mg/cm^2. The plastic cassette/holder has three clearly defined sections that correspond to the three TLD discs on the TLD card. The card's asymmetric "V" cut allows for only one orientation of loading in the plastic cassette and ensures that the three discs are properly positioned as follows: (i) Disc 1 is sandwiched between two 1.0 mm thick Al and 1.0 mm thick Cu filters (the copper filter is closer to the disc); (ii) Disc 2 is sandwiched between two 1.6 mm thick (180 mg/cm^2) plastic filters; and (iii) Disc 3 is positioned beneath a circular open window. Permanent placement of the user's photo on the central transparent area of the badge is also an option for identification purposes [19]. Three different TLD badge/cassette types are in use: the head badge, the wrist badge, and the chest badge. All three varieties of TLD badges have the same dosimeter and design, but depending on their intended function, they have a different attachment (clip or strap) for wearing [30–34].

5 Conclusions

In conclusion, we have discussed the recent methods for the synthesis of thermoluminescent materials. The advancement in these methods has paved the way for transformative applications across various fields. The ability to harness the thermoluminescent properties of materials opens up exciting possibilities for radiation dosimetry, medical imaging, environmental monitoring, and beyond. As researchers continue to delve deeper into the synthesis and characterization of these materials, we can expect further breakthroughs that will illuminate our future in ways we have yet to imagine.

Acknowledgements Authors are grateful to the Amity University Uttar Pradesh, Noida, India, for motivation and encouragements.

References

1. Horowitz YS, Gerstenberg HM, Akselrod MS, Bizarri GA (2015) Thermoluminescent materials. In: Handbook of radiation effects. CRC Press, pp 151–187
2. Chen R, Liu F, Cheng Y, Zhang Y (2019) Synthesis and thermoluminescence properties of $CaSO_4$:Dy, Tm, Gd phosphors. J Lumin 207:156–162
3. Khaidukov NM, Kulkina AV, Mironov AF, Novikov LP (2017) Synthesis and characterization of new thermoluminescent materials based on aluminum oxide doped with rare earths. J Lumin 192:425–429
4. Tsirlin TV et al (2016) Effect of synthesis conditions on thermoluminescent properties of MgB_4O_7:Dy ceramics. J Lumin 178:97–103
5. Kumar G et al (2011) Characterization of LiF:Mg, Ti (TLD-100) thermoluminescent dosimeter material prepared by different co-precipitation techniques. Radiat Meas 46:1607–1610
6. Karatikasari D et al (2018) Synthesis of thermoluminescence dosimeter (TLD) using calcium sulfate ($CaSO_4$) with variations of dysprosium (Dy) and thulium (Tm) dopants. AIP Conf Proc 2023:020084
7. Kim CW et al (2015) Effects of synthesis temperature and thermal treatment on thermoluminescence properties of Al_2O_3:C. J Lumin 168:111–114
8. Aguilar-Morales JA et al (2014) Thermoluminescence properties of $Li_2B_4O_7$:Cu, P synthesized by sol-gel method. Radiat Meas 67:78–83
9. Rizzutto MD et al (2001) Thermoluminescent properties of $CaSO_4$:Dy crystals grown by the slow evaporation method. Nucl Instrum Methods Phys Res Sect B 181:237–241
10. Furetta C (2011) Thermoluminescence dosimetry materials: properties and uses. CRC Press
11. Aitken MJ (1990) Science-based dating in archaeology. Routledge
12. Amrani D et al (2008) Environmental thermoluminescence dosimetry: basic concepts and applications. Radiat Prot Dosimetry 130(1):52–58
13. Kwon J, Shin S (2017) A review of the thermoluminescent dosimeters used in radiation therapy. J Radiat Prot Res 42(1):7–14
14. Barjaktarovic Z et al (2016) Exposure to ionising radiation induces persistent alterations in human and rat cells. Int J Radiat Biol 92(5):229–239
15. Piesch E et al (2013) A novel organic thermoluminescent dosimeter for personal radiation monitoring. Radiat Prot Dosim 157(2):183–188

16. Juan Azorin et al (2014) Preparation methods of thermoluminescent materials for dosimetric applications. J Appl Radiat Isotopes, 83:187–191. Esparza A et al (2002) Luminescent properties of Al_2O_3:Tb thin films deposited by spray pyrolysis. Superficies y vacio (2000)
17. ANSI (1983) American National Standard for Dosimetry "personnel dosimetry radiation performance—criteria for testing, American National Standards Institute (ANSI) Inc.", New York, HPSN 1311-1983
18. ANSI (2001) American National Standard for Dosimetry "personnel dosimetry radiation performance—criteria for testing, American National Standards Institute (ANSI) Inc.", New York, HPSN 1311-2001, 11/2000 Version, 2001
19. Bhatt BC, Srivastava JK, Sanaye SS, Shinde SS, Patil AS, Supe SJ (1994) Personal dosimeter intercomparisons as indices of performance—a quality assurance procedure. Radiat Prot Dosim 54:343–346
20. Pradhan AS, Bakshi AK (1997) Role of post-irradiation annealing for instant evaluation of personal monitoring badges based on $CaSO_4$:Dy Teflon TLD Discs. In: Proceedings of the national conference symposium on thermoluminescence and its application, Raipur, pp 196–199
21. Nagpal JS (1993) TL monitoring system, dose evaluation procedure and quality assurance workshop on individual monitoring, July 19–23, BARC, Bombay
22. Kramer HM, Bohn J, Iles WJ, Thompson IMG (1994) On the current status of an ISO working document on the calibration and type testing of radiation protection dosimeters for photons. Radiat Prot Dosim 54:267–272
23. Yukihara EG et al (2022) The quest for new thermoluminescence and optically stimulated luminescence materials: needs, strategies and pitfalls. Radiat Meas 158
24. Pradhan AS, Adtani MM, Varadhamjan G, Bakshi AK, Srivastava K, Bihari RR (2002) Hand book on the use of TLD badge based on CaSO:Py Teflon TLD Discs for individual monitoring. Bhabha Atomic Research Centre Mumbai, India
25. Azorin J (2014) Preparation methods of thermoluminescent materials for dosimetric applications: an overview. Appl Radiat Isotopes 83:187–191
26. Zaragoza E (2013) Effect of thermal treatment on TL response of $CaSO_4$:Dy obtained using a new preparation method. Appl Radiat Isotopes
27. Omanwar SK et al (2013) Recent advances and opportunities in TLD materials: a review. Mater Sci Eng, Defect Diffus Forum 347:75–110
28. Murthy KVR, Virk HS (2014) In: Defect and diffusion forum, Virk HS (ed), vol 347. Trans Tech Publications, Switzerland, pp 1–34
29. Yamashita T, Nada N, Onishi H, Kitamura S (1971) Health Phys 21:295
30. Salah N, Sahare PD, Lochab SP, Kumar P (2006) Radiat Meas 41:40
31. Salah N (2015) Radiat Phys Chem 106:40
32. Chagas MAP, Nunes MG, Campos LL, Souza DN (2010) Radiat Meas 45:550
33. Kamarudin N, Abdullah WSW, Hamid MAA (2017) Mater Sci Forum 888:333
34. Shinsho K, Harada K, Yamamoto Y, Urushiyama A (2008) Radiat Meas 43:236

Simulation Study of Some Characteristics of Single and Binary Stars Using MAdSTAR, MESA, and NAAP Astronomical Tools

Tanmay, Lakshmi, Vijay Kumar Soni, and Adarsh Kumar

1 Introduction

Stars and their evolution have been important research from decades to understand the universe, its existence, nature and to study the nature and formation of planets [1–3, 6, 10]. There are several stages of any star's evolution [7]. Stellar evaluation or the evolutionary track of a star tells us about the formation of a star with its initial changes and processes comprising chemical reactions involved, and how the star undergoes transformation at its near end life and after its death [5]. Till today's date about 88% of stars are a part of binary systems and some of them being a part of a high order system [4]. Mass of a star is a significant component which provides information how a star evolves during its whole lifetime and what the fate holds for the star in the formation of any stars which was initiated in a Nebula [8]. As high mass stars have lower lifetime as compared to stars with lower mass due to amount of fusion reaction occurring and the amount of energy it creates in its main sequence phase [5]. It is known from recent studies that stellar gases are massive clouds composed of materials like dust, gas in the form of hydrogen and helium and plasma in the interstellar medium [9]. In addition to the gases, dust particles present in the stellar cloud have a wavelength of about 2540 Å, and even in this dimension, the dust particles are effective in scattering the light of distant stars and thereby rendering our visibility to get a clear idea [10]. Therefore, nebula formation occurs when a considerable amount of interstellar medium is acted upon by gravity and undergoes gravitational collapse due to which the all the matter gets accumulated to

Tanmay · Lakshmi · A. Kumar (✉)
Department of Physics, Amity Institute of Applied Sciences, Amity University Uttar Pradesh, Noida 201313, India
e-mail: adarsh_phy@yahoo.co.in; akumar25@amity.edu

Lakshmi · V. K. Soni
Environment Monitoring and Research Center (EMRC), India Meteorological Department, Ministry of Earth Sciences, Mausam Bhawan, Lodi Road, New Delhi 110003, India

S. Rattan et al. (eds.), *Emerging Trends in Synthesis and Catalysis in Chemistry*, Springer Proceedings in Materials 51, https://doi.org/10.1007/978-981-97-5169-3_7

form clumps forming regions around interstellar medium with higher densities [11]. However, the density of these stellar gas clouds is very less due to its abundance and the enormous size of galaxy [12]. Depending on the size and amount of stellar gas cloud, it can consist of up to 1/10 of the total mass of a galaxy, and this much amount of stellar gas clouds can support the formation of stars [8]. In the initial phase of the star formation, 'dark globules' are formed due to the influence of gravity in the gas cloud. Further, due to formation of denser regions, a protostar is formed, and due to influence of gravity, it undergoes contraction. In this process, about one half of the protostars' stored potential energy gets converted in the form of heat that increases the protostar's central temperature, and the remaining portions energy is just radioed in the interstellar medium [13]. As the protostar is undergoing contraction, it would not forever be under the process of contraction, but as the temperature rises and when this temperature reaches a certain temperature limit also known as the critical temperature, due to this energy generation starts in the interior of the protostar, this newly generated energy compensates for the energy being radiated away from the surface and now we can observe a state of equilibrium, beyond which there is no further contraction and a gigantic sphere is obtained with the star's core temperature about 15 Million K now according to the mass of the protostar and the main sequence star [11]. It was further suggested that as the star enters its main sequence the stars remain here for a long time, and this stage of the total lifespan of star covers 80–90% of its life [12, 14]. In different stages of star's life, various kinds of chemical cycles occur inside the star such as the PP cycle, CNO cycle, and He fusion [15].

Mass, luminosity, and radius of any star are reared as the fundamentals parameters of star affecting its evolution [13, 14]. All these parameters were observed to be time-dependent and change during time. Besides these three parameters, various parameters like the angular momentum, magnetic field, the mass loss rate of star and pulsation period become important regarding stellar evolution [15]. Moreover, the metallicity of a star is also a significant criterion for stellar evolution [16]. Metallicity is basically the riches of elements existing in the universe that have more atomic weight than the elements Hydrogen and helium which are the most abundant visible and detectable matter in the universe [10]. Metallicity measures the abundance of the elements heavier than hydrogen and helium such as carbon, oxygen, iron, and others [17]. In stellar astrophysics, it usually signifies the abundance of iron and is defined as the iron-to-hydrogen ratio [8]. It was found that stars, which have higher metallicity than the Sun are more probable to support planets as compared to the stars having lower metallicity [17]. More recent studies support that the metallicity of any star has direct influence on its luminosity, mass, and velocities, etc. [18]. Galaxies can have different amounts of mass of interstellar gas clouds [11]. The lower mass galaxies can be visualized as having a comparably high interstellar gas ratio star mass which in short indicated the richness of metallic quantity that happened to be low [12]. It was found that if the size of the galaxy changes by a factor of two, the metallicity of the galaxy can change by 0.1 decimals [6, 17]. In this paper, we shall discuss the evolution stages of 10 and 40 solar mass stars along with other parameters and the characteristic properties of eclipsing binary stars.

2 Methodology

The characteristics of star evolution were retrieved from online astronomical tools MAdSTAR, MESA, and NAAP [4]. MAdSTAR is the quasi-experimental website of Rich Townsend. It sheds light into massive-star astrophysics, in the form of papers, codes, data, etc. The Web site is also a framework for playing around with computer technology. The Modules for Experiments in Stellar Astrophysics (MESA) Web tool provides a modern software infrastructure for sustained innovations in the stellar evolution. MESA solves the complex compositional equations governing stellar evolution. MESA is a standard computational tool used to understand stars. It generates stellar evolution code, which is a piece of software that can construct a model for the interior of a star, and then evolve it over time [6]. Its state-of-the-art modules provide various astronomical data involving timestep controls, equation of state, opacity, nuclear reaction rates, element diffusion, and changes in the mass of the star [9]. They provide online tools for performing experiments on stars and discovering significant factors that contribute towards the formation of red giants. They also shed light on different stages of stellar evolution. The Nebraska Astronomy Applet Project, NAAP, provides online simulation tools targeting astronomical star like objects [17].

3 Results

We have divided the results into two sections involving single stars with variable solar masses and eclipsing binary stars.

3.1 Single Stars with Variable Solar Masses

Figure 1a–c shows the thermodynamic profiles for a single star with three different parameters, namely pressure (top), density (middle), and temperature (lower) as a function against the mass coordinate within the 10 solar mass stars, whereas Fig. 2a–c shows the same thermodynamic plot profiles for 40 solar mass stars. The logarithm normal plot against mass coordinate was plotted for temperature, density, and pressure of single star. The thermodynamic variables (temperature, density, and pressure) were found to decrease steely at the beginning after its evolution till single star mass reaches nearly two-unit single star mass. After then, the thermodynamic variables became almost constant to 12-unit single star mass.

Fig. 1 Figure shows the variation of thermodynamic variables of single star with solar mass 10

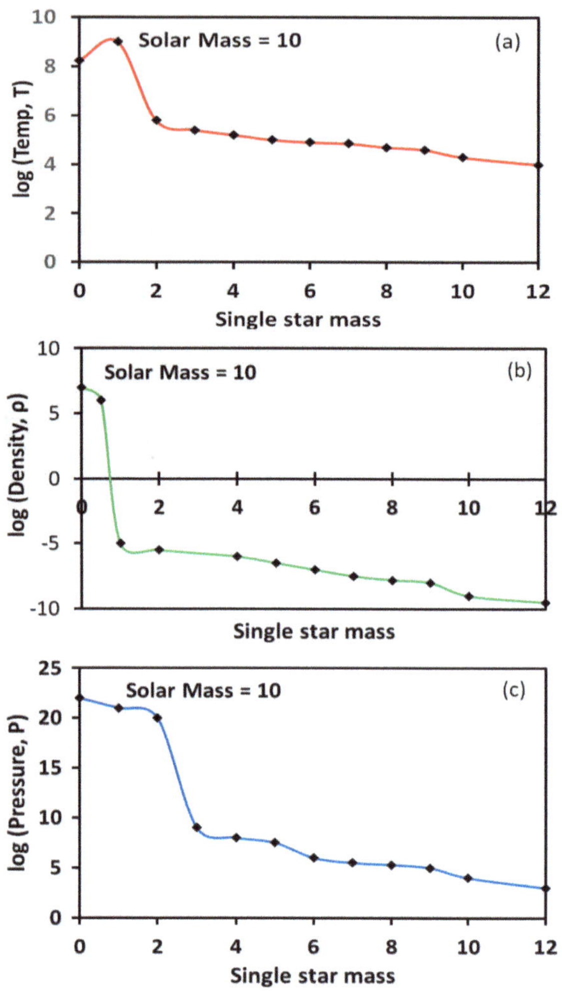

3.2 *Eclipsing Binary Stars*

When one looks at the night sky, many points are twinkling light which in fact might be that the single twinkling point are two stars orbiting one another. Some of them being a part of a high order system. An example of binary system is Sirius.

Binary stars can be of different kinds depending upon their visuality. They could be visual, spectroscopic, eclipsing, and astrometric binaries. An individual can spot the visual paired stars using telescope. One must note that only about 5–10% of visible stars are visual binaries and it is not possible to view binary stars with the naked eyes. From the spectroscopique perview of binary system, the stars appear close even when telescope is used to view or study them. Eclipsing binaries are the

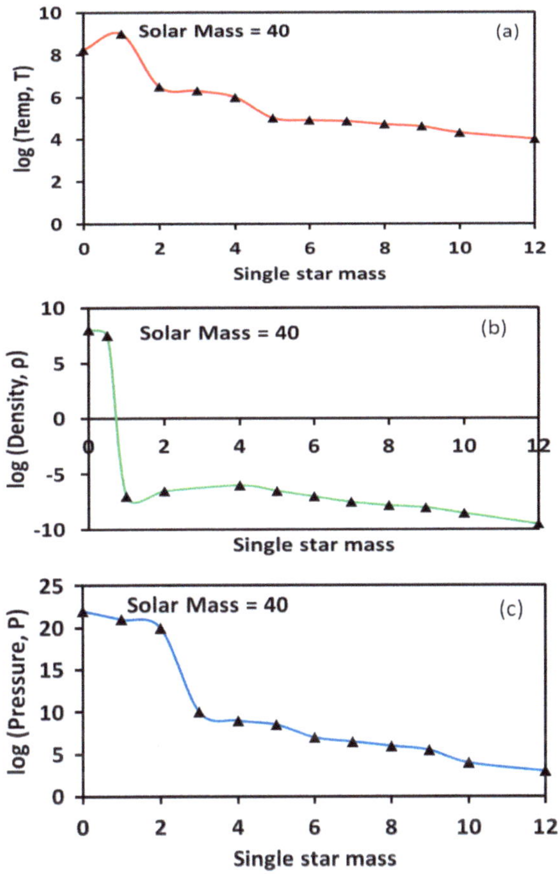

Fig. 2 **a–c** Figure is same as that of Fig. 1 except for solar mass 40

two stars of system have an orbit at angle, whereas astrometric binaries are the system the stars, which appear as if they are dancing in an empty space [6, 9].

It is from this system that empirical mass–luminosity relationship (MLR) can be obtained which is useful in finding the masses of single stars being minority in universe.

To understand evolution in binary stars, first we took two stars in a binary system and fixed their mass in the simulator to get the HR diagram of the binary system. A generalized diagram of binary system with low and high masses is shown in Fig. 3, and the luminosity dip can be seen in Fig. 4.

It is clear from Fig. 4 that the STAR 1 had a mass of 10 solar mass along with other parameters such as radius which was bout 1.5Ro and a surface temperature of about 8700 K. As far as the STAR 2 is concerned in the binary system, it had a mass value of 5 solar mass, radius was set to be around 1.5 Ro (same radius as STAR 1), and it had a surface temperature of about 5000 K. The distance between the two stars was around 10 Ro, and the eccentricity of the binary system was taken approximately

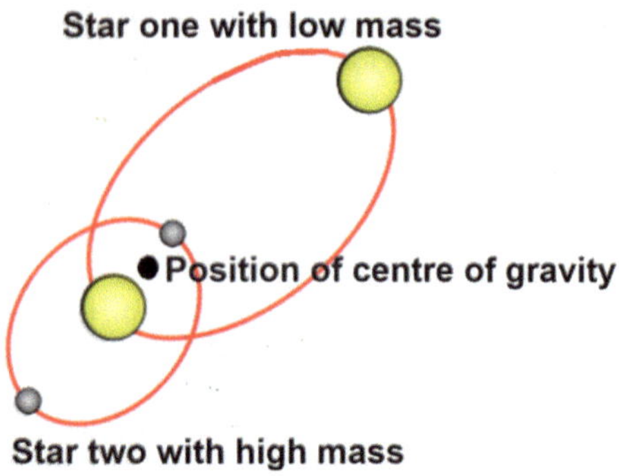

Fig. 3 Generalized diagram of binary star [11]

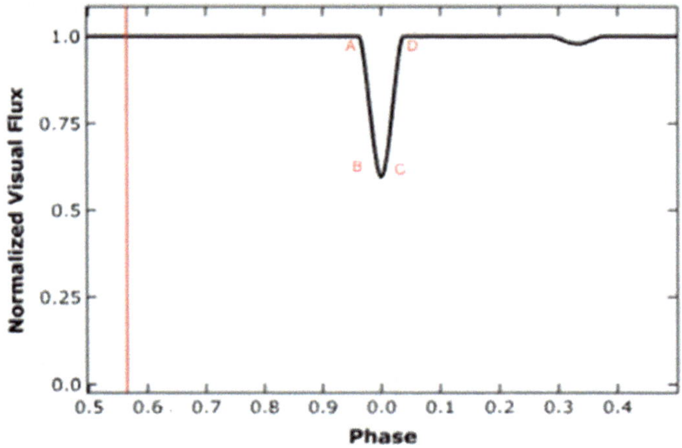

Fig. 4 This figure shows the magnifying image of luminosity dip in response to phase angle

0.30. Looking at the simulation, we can see dips in the binary system. These dips are due to the action of the stars eclipsing each other. If the larger star (here star 1) eclipses star 2, then a larger dip is observed as compared to the dip which will be observed in the eclipse when star 2 eclipses star 1. A lot of information can be obtained from the spectrum of the eclipsing binary stars. Let us see what all can be obtained from the eclipsing binaries.

In Fig. 4, A, B, C, and D represent time at start and end intervals of dips. Therefore, by finding the width of the dips, we can find the duration of eclipse for the binary system. After then, the radius and masses of binary stars can be computed [10].

4 Discussion of the Results

The results of the simulation study show that the mass of a star has a significant impact on its evolution. More massive stars have shorter main sequence lifetimes and are more likely to end their lives in a supernova explosion. The mass of a star also affects its chances of merging with another star. More massive stars are more likely to merge, which can lead to the formation of even more massive stars. The results of the simulation study have implications for our understanding of the evolution of stars and galaxies. More massive stars are more common in the early universe, and their supernova explosions can play a role in the formation of new stars and galaxies. The merger of massive stars can also lead to the formation of black holes, which can have a significant impact on the evolution of galaxies. From our study, the following important investigations were obtained.

a. Single stars with 10 solar masses have a main sequence lifetime of about 100 million years. They will then evolve into red giants and eventually supernovae.
b. Single stars with 40 solar masses have a main sequence lifetime of only a few million years. They will then evolve into blue giants and eventually supernovae.
c. Binary stars with 10 solar masses can have a variety of outcomes. If the stars are close enough together, they can merge and form a more massive star. If the stars are not close enough together, they can continue to evolve independently.
d. Binary stars with 40 solar masses are more likely to merge than single stars with the same mass. This is because the strong gravitational pull of the two stars can pull them closer together.

It is to be noted that the simulation study was conducted using three different stellar evolution codes, namely MAdStar, NAAP, and MESA. All three codes were able to reproduce the observed characteristics of single stars and binary stars with different masses. However, there were some differences in the results from the three codes. For example, MAdStar and NAAP predicted that binary stars with 10 solar masses were more likely to merge than MESA did. These differences may be due to the different assumptions and numerical methods used by the three codes. Overall, the results of the simulation study show that stellar evolution codes are a valuable tool for understanding the evolution of stars and galaxies. However, it is important to be aware of the limitations of these codes and to compare the results from different codes to get a more accurate picture of stellar evolution.

5 Conclusions

The simulation study showed that the mass of a star has a significant impact on its evolution. More massive stars have shorter main sequence lifetimes and are more likely to end their lives in a supernova explosion. The mass of a star also affects its chances of merging with another star. More massive stars are more likely to merge, which can lead to the formation of even more massive stars. Stellar evolution codes are a valuable tool for understanding the evolution of stars and galaxies. It is important to be aware of the limitations of stellar evolution codes and to compare the results from different codes to get a more accurate picture of stellar evolution.

Acknowledgements The corresponding author is thankful to the reviewers for their valuable comments on the manuscript. Authors are also thankful to different scientists of National Aeronautics and Space Administration (NASA) for utilizing open-source online astronomical simulation tools. The authors are also thankful to different scientists involved in the creation of various online tools like MAdSTAR, MESA, and NAAP.

References

1. Asplund M, Grevesse N, Sauval AJ, Scott P (2009) The chemical composition of the sun. Ann Rev Astron Astrophys 47(1):481–522
2. Bertone, De Lucia G, Thomas PA (2007) The recycling of gas and metals in galaxy formation: predictions of a dynamical feedback model. Mon Not R Astron Soc 379(3):1143–1154
3. Collison FM, Poe K (2013) “Astronomical tourism”: the astronomy and dark sky program at Bryce Canyon National Park. Tour Manage Perspect 7:1–15
4. Dave R, Finlator K, Oppenheimer BD (2011) Galaxy evolution in cosmological simulations with outflows—II. Metallicities and gas fractions. Mon Not R Astron Soc 416(2):1354–1376
5. Ellison SL, Patton DR, Simard L, McConnachie AW (2007) Clues to the origin of the mass-metallicity relation: dependence on star formation rate and galaxy size. Astrophysical J 672(2):L107–L110
6. Huang L, Adams FC, Grohs E (2019) Sensitivity of carbon and oxygen yields to the triple-alpha resonance in massive stars. Astropart Phys 105:13–24
7. Hawken AJ, Bridle SL (2009) Gravitational flexion by elliptical dark matter haloes, pp 1132–1138
8. Kroupa P (2001) On the variation of the initial mass function. Mon Not R Astron Soc 322(2):231–246
9. Lagos C del P, L, Lacey CG, Baugh CM, Bower RG, Benson AJ (2011) On the impact of empirical and theoretical star formation laws on galaxy formation. Mon Not R Astron Soc 416(2):1566–1584
10. Laplace ET (2022) A tool for understanding the lives, interiors, and physics of stars. Astron Comput 38:100516
11. Mannucci F, Cresci G, Maiolino R, Marconi A, Gnerucci A (2010) A fundamental relation between mass, star formation rate and metallicity in local and high-redshift galaxies. Mon Not R Astron Soc 408(4):2115–2127
12. Nagao T, Maiolino R, Marconi A (2006) Gas metallicity diagnostics in star-forming galaxies. Astron Astrophys 459:85–101
13. Pinzke A, Pfrommer C (2010) Simulating the γ-ray emission from galaxy clusters: a universal cosmic ray spectrum and spatial distribution. Astrophys J 409(1):449–480
14. Salaris M, Cassisi S (2005) Evolution of stars and stellar populations. John Wiley, and Sons, pp 119–121
15. Swinbank AM, Webb TM, Richard J, Bower RG, Ellis RS, Illingworth G, Jones T, Kriek M, Smail I, Stark DP, Dokkum PV (2009) A spatially resolved map of the kinematics, star formation and stellar mass assembly in a star-forming galaxy at z= 4.9:1121–1131
16. Wittkowski M (2005) Fundamental stellar parameters. Bull de la Soc R des Sci de Liège 74(1–3):165–181

17. Yates RM, Kauffmann G, Guo Q (2012) The relation between metallicity, stellar mass and star formation in galaxies: an analysis of observational and model data. Mon Not R Astron Soc 422(1):215–231
18. Yin SY, Liang YC, Hammer F, Brinchmann J, Zhang B, Deng LC, Flores H (2007) Empirical strong-line oxygen abundance calibrations from galaxies with electron-temperature measurements. Astron Astrophys 462(2):535–546

Variation of In-Situ Partial Pressures of Different Gases Over Southern Polar Region of Moon Utilizing Chandrayaan-2 Data

Lakshmi, Yashvi Gupta, Vijay Kumar Soni, and Adarsh Kumar

1 Introduction

The Moon has always been the canter of human attention, more than any other celestial body in the night sky of outer space [1, 4]. It has been a subject of study to scientists and researchers all over the world [2, 8, 11, 21]. Understanding the Moon would give us a way to reveal the first appearance of the solar system and that of the planet Earth [3, 7, 9, 10]. The traditional Chandrayaan-2 mission contained an orbiter, a lander (Vikram), and a rover (Pragyan) [5, 6]. With the payloads of fully home-grown technology, it helped us to better understand the origin and evolution of the Moon by conducting comprehensive geographical and wide-ranging mineralogical studies [13]. The Chandrayaan-2 mission not only helped us to better understand the origin and evolution of the Moon by conducting topographical studies, comprehensive mineralogical and morphological analyses, and many other studies in the lunar field but also enhancing our capabilities by some of the questions raised by the discovery of Chandrayaan-1 and other lunar missions [14]. The main purpose of these missions was to demonstrate the ability to soften the lander and use the rover in the South polar region of the Moon [15, 19]. The Moon is considered a tenuous atmosphere, a non-collision atmosphere all envelopes weighing less than 100 tons [3, 22]. It has its origins based on external and internal processes, namely solar protons, plasma particles, and micro-meteors [12, 23, 24, 27]. When solar wind plasma of MeV order energy enters the lunar surface, it is only absorbed to remain neutral, warm, and released to form oxygen molecules from the surface [16, 25].

Lakshmi · Y. Gupta · A. Kumar (✉)
Department of Physics, Amity Institute of Applied Sciences, Amity University Uttar Pradesh, Noida 201313, India
e-mail: adarsh_phy@yahoo.co.in; akumar25@amity.edu

Lakshmi · V. K. Soni
Environment Monitoring and Research Center (EMRC), India Meteorological Department, Ministry of Earth Sciences, Mausam Bhawan, Lodi Road, New Delhi 110003, India

S. Rattan et al. (eds.), *Emerging Trends in Synthesis and Catalysis in Chemistry*, Springer Proceedings in Materials 51, https://doi.org/10.1007/978-981-97-5169-3_8

This may be attributed to the presence of hydrogen (H), helium (He), neon (Ne), and argon (Ar) atoms in the lunar exosphere [17]. Recent spectroscopic measurements on the lunar exosphere minerals showed the presence of sodium (Na), potassium (K), and H_2O [18]. On the other hand, X-ray fluorescence spectroscopy is one of the most important methods of examining the structure of the extra-terrestrial planets, which can be used to obtain quantitative information about the multiplicity of major substances such as Mg, Al, Si, K, Ca, Ti, and Fe [20, 26]. In the present work, we have analysed the partial pressure of various gases and compounds like Mg, H_2O, Si, H_2, Ar, H_2SO_4, Na, and Al over the southern polar region of Moon using Chandrayan-2 mission pay load data for four seasons (JF, JJAS, MAM, ON) of 2020 through the linear trend statistical technique. Contour plotting maps over the lunar surface have also been attempted.

2 Data Used and Methodology

In the present work, Chandrayan-2 Atmospheric Compositional Explorer 2 (CHACE-2), which provided the new insights into the lunar exosphere, was used to extract the data in the form of partial pressure exerted by different gases on the lunar surface [26]. The partial pressure (torr) was first recorded by hour and then converted into morning, noon, and evening which was named as day data. On the other hand, night data was also recorded in tabular form. Day and night data was converted into weeks and further converted to months and then annual form. The data recorded was then modified into different seasons, namely JF, MAM, JJAS, and ON, respectively. Along with the extraction of the data, 2-D and 3-D graphs were plotted to analyse the lunar surface and the factors affecting the partial pressure of different gases. The variation of partial pressure for different season, i.e. JF, MAM, JJAS, and ON for different gases, were plotted. A flow chart showing the methodology is given in Fig. 1. The highest and the lowest partial pressure was identified after which the linear trend analysis was done and nature of trend of each gas was determined. With help of the data recorded in tabular form, the average values of different months were calculated, and daily variation in partial pressure was plotted in the form of seasons in Origin pro-2022. For detailed analysis, contour plotting was done which helped in signifying the change and variation of partial pressure with respect to time and seasons. The contour plotting depicted the dependence of partial pressure with respect to temperature.

3 Results and Discussion

Partial pressure data of different gases and compounds like Mg, H_2O, Si, H_2, Ar, H_2SO_4, Na, and Al over southern region of Moon was divided into different months of four seasons (JF, JJAS, JJAS, ON) of 2020. Table 1 shows the observed seasonal averaged partial pressure of different trace gases, whereas the average day time

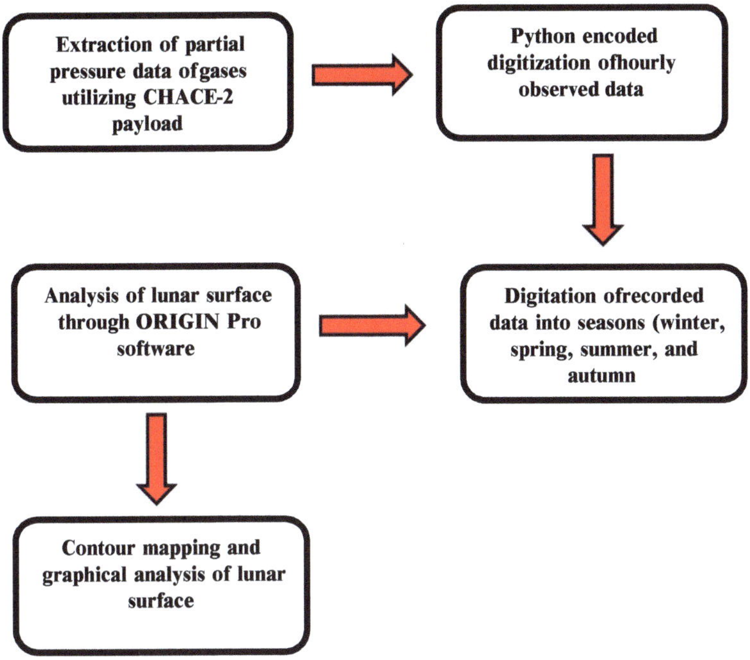

Fig. 1 Flow chart showing the methodology for the analysis of partial pressure data of different gases

concentration of some chemical elements at lunar surface is shown in Table 2. It is to be noted that the observed molecules at the Earth's surface are nearly 10^{19} per cc in the same volume [20]. The seasonal and month-wise data has been presented in the following subsequent sections.

3.1 *Variation of Partial Pressure for JF Season*

During the JF season, highest partial pressure of magnesium (Figs. 2a, 3, 4, 5 and 6a) was observed on 30 January, i.e. 9.7×10^{-9} torr, whereas the lowest partial pressure was -0.3×10^{-9} torr on 1 January. The highest partial pressure, i.e. 5.01×10^{-9} torr was observed on 25 February, whereas the lowest partial pressure was observed at 0.08×10^{-9} torr on 11 February. The linear trend line was drawn, and the slope was calculated as 1.77 ± 0.46 and 1.1 ± 0.21 for January and February, respectively. The graph was seen increasing till 26 January and gained a sudden peak of 8.69×10^{-9} torr. Partial pressure was observed to be declined in the last month of February. The variation of partial pressure of H_2O (Figs. 2b and 6b) was plotted. Highest partial pressure of H_2O was observed on 30 January, i.e. 21.5×10^{-9} torr, whereas the lowest partial pressure was -1.5×10^{-9} torr on 1 January. The highest partial pressure, i.e.

Table 1 Statistical analysis of partial pressure of different gaseous elements for different seasons over the southern pole of Moon

Season	JF			MAM			JJAS			ON		
Trace gases	Max	Min	*t*-test	Max	Min	*t*-test	Max	Min	*t*-test	Max	Min	*t*-test
Mg	55	2.5	1.25E−06	247	5.25	1.07E−10	1	0.01	1.96E−07	224	2	4.80E−06
H_2O	215	0.04	0.64	269	0.03	9.22E−10	2.22	0.011	0.032	218	0.07	0.0016
Si	191	0.72	0.091	421	0.03	1.51E−09	258	0.01	2.50E−11	681	0.01	0.0099
H_2	19.5	0.0034	0.14	47	4.75	1.63E−11	0.28	0.0022	0.00019	43	0.02	0.00031
Ar	296	0.62	0.42	549	26	1.76E−09	2.11	0.11	6.70E−03	605	0.14	0.0035
H_2SO_4	447	0.62	0.37	557	0.08	1.38E−11	4.12	0.03	4.72E−10	980	0.86	0.006
Na	300	0.065	0.87	352	0.01	1.60E−10	1.98	0.04	2.00E−04	372	0.1	0.0087
Al	374	0.23	0.97	314	0.02	1.90E−12	1.72	0.05	1.00E−03	419	0.03	0.0082

Table 2 Average day time concentration of some gaseous elements at the lunar surface [20]

S No.	Gaseous element	Average day time concentration in per cc (approximately)
1	Argon	20,000–1,00,000
2	Helium	5000–30,000
3	Neon	20,000
4	Sodium	70
5	Potassium	17
6	Hydrogen	14
7	Molecular concentration at lunar surface	10^6

6.93×10^{-9} torr was observed on 19 February, whereas the lowest partial pressure was observed at 1.6×10^{-10} torr on 1 February. The linear trend line was drawn, and the slope was calculated as 3.13 ± 0.77 and 2.5 ± 0.25 for January and February, respectively. The graph was seen increasing till 25 January and gained a sudden peak in last week of January.

In the JF season, highest partial pressure of silicon (Si) (Figs. 2c and 6c) was observed on 31 January, i.e. 19.05×10^{-9} torr, whereas the lowest partial pressure was -1.12×10^{-9} torr on 1 January. The highest partial pressure, i.e. 9.85×10^{-9} torr was observed on 26 February, whereas the lowest partial pressure was observed at 0.13×10^{-9} torr on 1 February. The linear trend line was drawn, and the slope was calculated as 2.8 ± 0.68 and 3.24 ± 0.37 for January and February, respectively. The highest peak in partial pressure between January and February was seen in the month of January. The variation of partial pressure for JF season of H_2 (Figs. 2d and 6d) gas was plotted. Highest partial pressure of hydrogen (H_2) was observed on 31 January, i.e. 1.95×10^{-9} torr, whereas the lowest partial pressure was -0.25×10^{-9} torr on 1 January. The highest partial pressure, i.e. 0.92×10^{-9} torr was observed on 27 February. The linear trend line was drawn, and the slope was calculated as 0.39 ± 0.084 and 0.12 ± 0.06 for January and February, respectively. The highest peak in partial pressure between January and February was seen in the month of January. During the JF season, highest partial pressure of argon (Figs. 2e and 6e) was observed on 31 January, i.e. 29.63×10^{-9} torr, whereas the lowest partial pressure was -2.52×10^{-9} torr on 1 January. The highest partial pressure, i.e. 10.8×10^{-9} torr was observed on 25 February, whereas the lowest partial pressure was observed at 0.12×10^{-9} torr on 2 February. The linear trend line was drawn, and the slope was calculated as 4.21 ± 1.04 and 3.90 ± 0.37 for January and February, respectively. The highest peak in partial pressure between January and February was seen in the month of January.

The variation of partial pressure for JF of sulphuric acid (Figs. 2f and 6f) was plotted. During the JF season, highest partial pressure of (H_2SO_4) was observed on 31 January, i.e. 44.72×10^{-9} torr, whereas the lowest partial pressure was -4.2×10^{-9} torr on 1 January. The highest partial pressure, i.e. 26.84×10^{-9} torr was

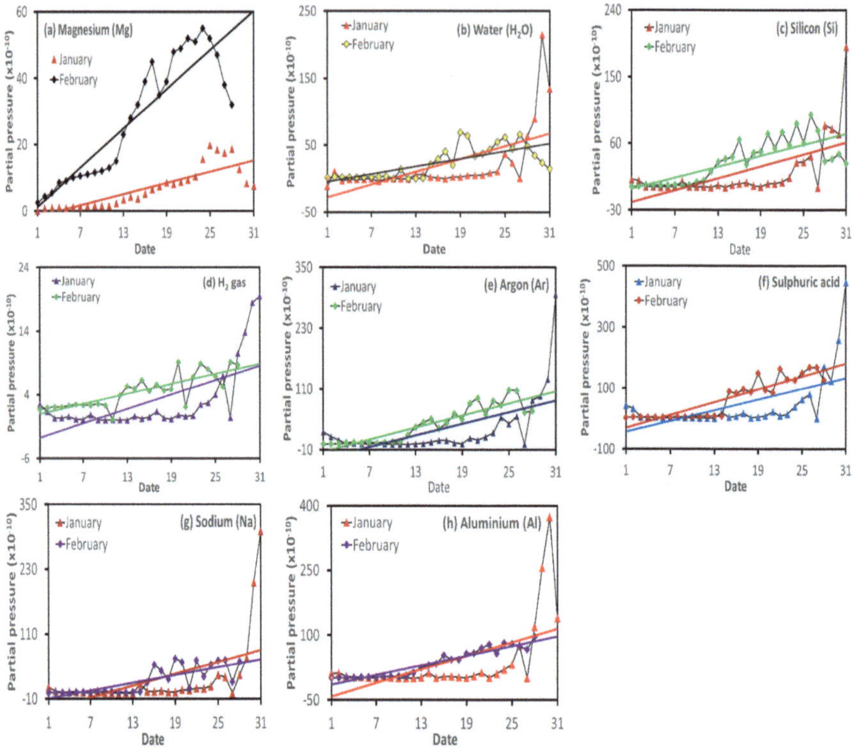

Fig. 2 This figure shows the daily variation in partial pressure ($\times 10^{-10}$) in torr of different gases on southern polar region of Moon during JF season

observed on 26 February, whereas the lowest partial pressure was observed at 0.3×10^{-9} torr on 1 February. The linear trend line was drawn, and the slope was calculated as 6.6 ± 1.69 and 6.96 ± 0.54 for January and February, respectively. The highest peak in partial pressure between January and February was seen in the month of January. During the JF season, highest partial pressure of sodium (Na) (Figs. 2g and 6g) was observed on 31 January, i.e. 30.04×10^{-9} torr, whereas the lowest partial pressure was -1.13×10^{-9} torr on 1 January. The highest partial pressure, i.e. 6.47×10^{-9} torr was observed on 19 February, whereas the lowest partial pressure was observed at 0.1×10^{-9} torr on 1 February. The linear trend line was drawn, and the slope was calculated as 3.98 ± 1.21 and 2.33 ± 0.37 for January and February, respectively. The highest peak in partial pressure between January and February was seen in the month of January. The variation of partial pressure for JF season of aluminium (Figs. 2h and 6h) was plotted. During the JF season, highest partial pressure of aluminium was observed on 30 January, i.e. 37.3×10^{-9} torr, whereas the lowest partial pressure was -1.07×10^{-9} torr on 1 January. The highest partial pressure, i.e. 9.66×10^{-9} torr was observed on 28 February. The linear trend line was made, and the slope was calculated as 4.61 ± 1.38 and 2.08 ± 0.57 for January

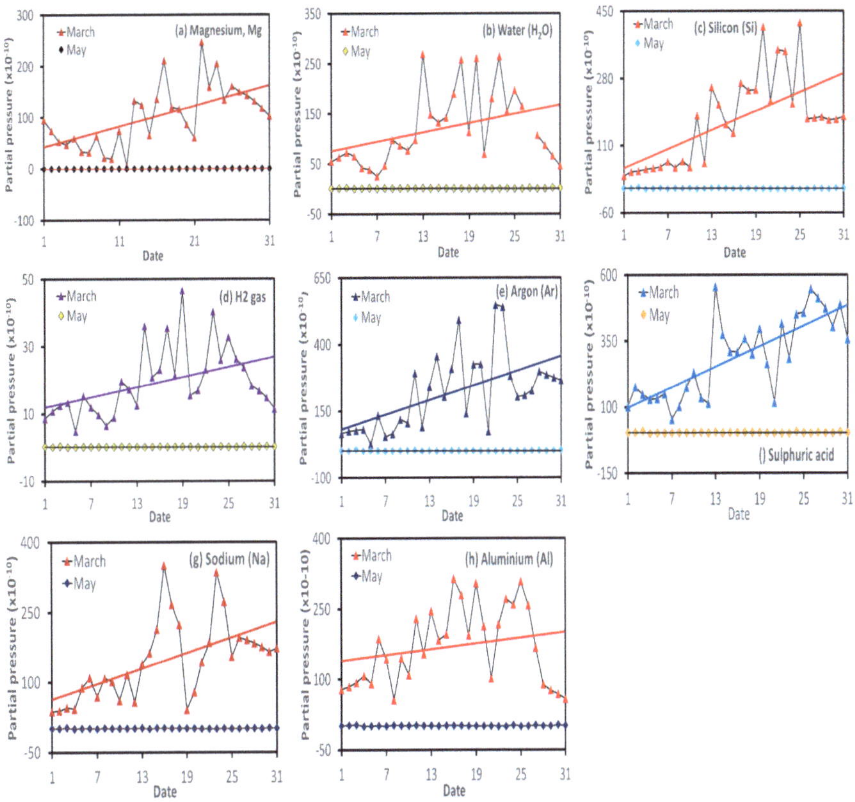

Fig. 3 This figure shows the daily variation in partial pressure ($\times 10^{-10}$) in torr of different gases on southern polar region of Moon during MAM season

and February, respectively. The highest peak in partial pressure between January and February was seen in the month of January.

3.2 *Variation of Partial Pressure for MAM Season*

In the MAM season, a peak for partial pressure of magnesium (Figs. 3a and 7a) was found to be 24.702×10^{-9} torr on 22 March, and the lowest value of partial pressure was observed as 0.52×10^{-9} torr on 12 March. The highest partial pressure in the month of May is found to be 0.14×10^{-9} torr, whereas the lowest partial pressure was found at 0.0012×10^{-9} torr. The linear trend line for the month of March and May was drawn, and the slope was calculated which came to be 4.17 ± 0.93 and -0.011 ± 0.0082, respectively. A lot of variations in partial pressure were observed in the month of March while partial pressure in May was observed to be almost

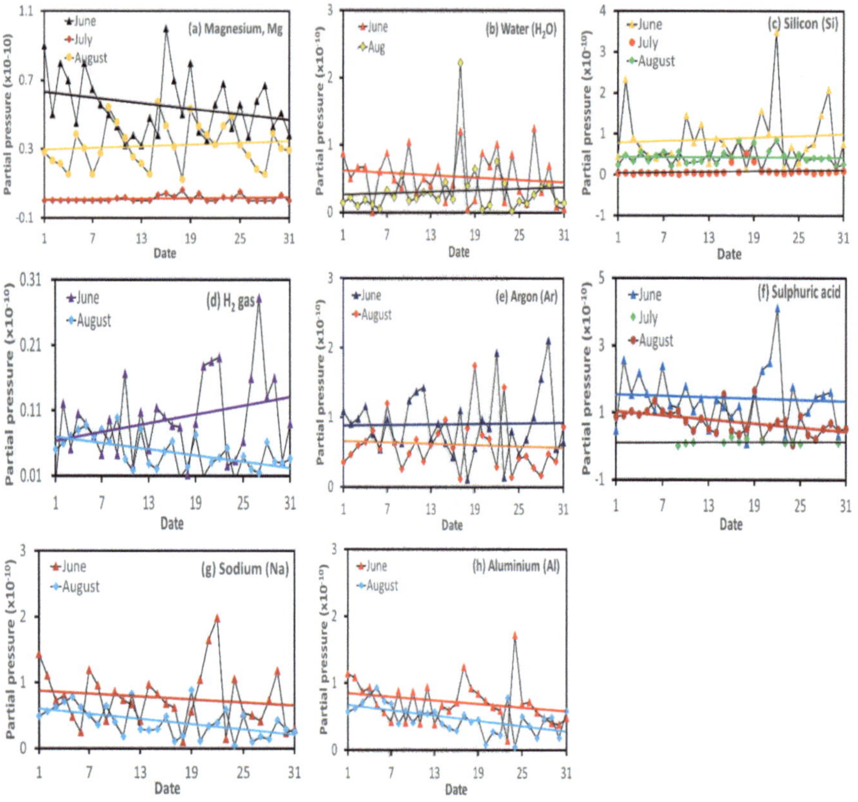

Fig. 4 This figure shows the daily variation in partial pressure ($\times 10^{-10}$) in torr of different gases on southern polar region of Moon during JJAS season

similar. During MAM season, a peak for partial pressure of water (H_2O) (Figs. 3b and 7b) was found to be 26.9 × 10^{-9} torr on 13 March, and the lowest value of partial pressure was observed as 2.5 × 10^{-9} torr on 7 March. The highest partial pressure in the month of May is found to be 0.25 × 10^{-9} torr on 3 May, whereas the lowest partial pressure was found at 0.0026 × 10^{-9} torr on 23 May. The linear trend line for the month of March and May was drawn, and the slope was calculated which came out to be 6.27 ± 1.84 and − 0.0088 ± 0.013, respectively. A lot of variations in partial pressure were observed in the mid-week of March while partial pressure in May was observed to be almost similar. Graph for MAM season was plotted, a peak for partial pressure of silicon (Si) (Figs. 3c and 7c) was found to be 42.04 × 10^{-9} torr on 25 March, and the lowest value of partial pressure was observed as 3.34 × 10^{-9} torr on 1 March. The highest partial pressure in the month of May is found to be 0.48 × 10^{-9} torr on 3 May, whereas the lowest partial pressure was found at 0.003 × 10^{-9} torr on 23 May. The linear trend line for the month of March and May was drawn, and the slope was calculated which came out to be 12.09 ± 2.35 and −

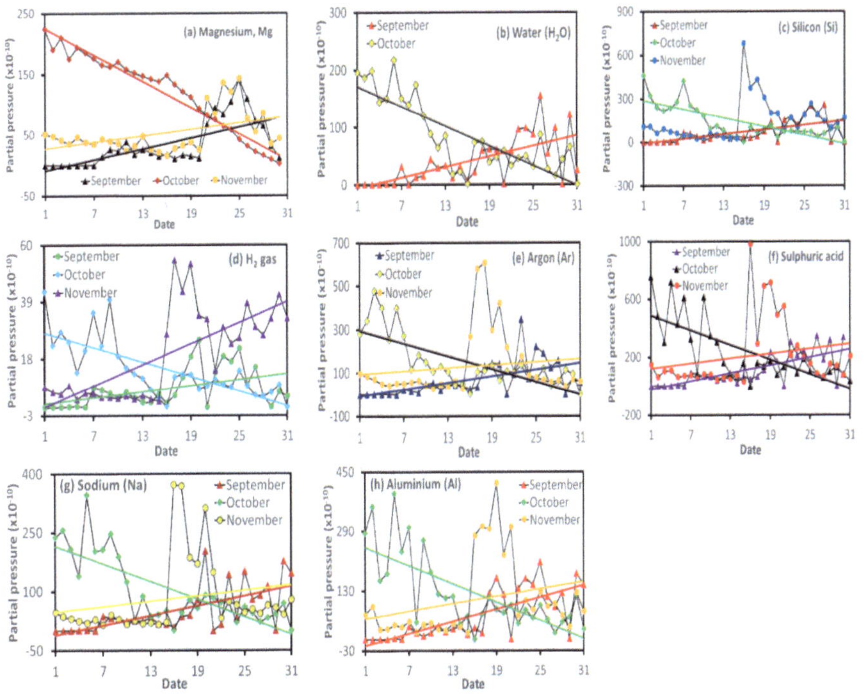

Fig. 5 This figure shows the daily variation in partial pressure ($\times 10^{-10}$) torr of different gases on southern polar region of Moon during ON season

0.032 ± 0.021, respectively. The partial pressure in May was observed to be almost similar, and the slope was negative, while the slope in March was positive. In the MAM season, a peak for partial pressure of H_2 gas (Figs. 3d and 7d) was found to be 4.66 × 10^{-9} torr on 19 March. The highest partial pressure in the month of May is found to be 0.028 × 10^{-9} torr on 1 May, whereas the lowest partial pressure was found at 0.00035 × 10^{-9} torr on 23 May. The linear trend line for the month of March and May was drawn, and the slope was calculated which came out to be 0.56 ± 0.31 and − 0.003 ± 0.002, respectively. The partial pressure in May was observed to be almost similar, and the slope was negative, while the slope in March was positive.

During MAM season, a peak for partial pressure of argon (Ar) (Figs. 3e and 7e) was found to be 54.8 × 10^{-9} torr on 22 March, and the lowest value of partial pressure was observed ~ 0.07 × 10^{-9} torr on 29th March. The highest partial pressure in the month of May is found ~ 0.58 × 10^{-9} torr on 3 May, whereas the lowest partial pressure was found at 0.004 × 10^{-9} torr on 4 May. The linear trend line for the month of March and May was drawn, and the slope was calculated which came out to be 3.11 ± 3.89and − 0.027 ± 0.025, respectively. The partial pressure in May was observed to be almost similar, and the slope was negative, while the slope in March was positive. In the MAM season, a peak for partial pressure of sulphuric acid (H_2SO_4) (Figs. 3f and 7f) was found to be 55.6 × 10^{-9} torr on 13 March. The

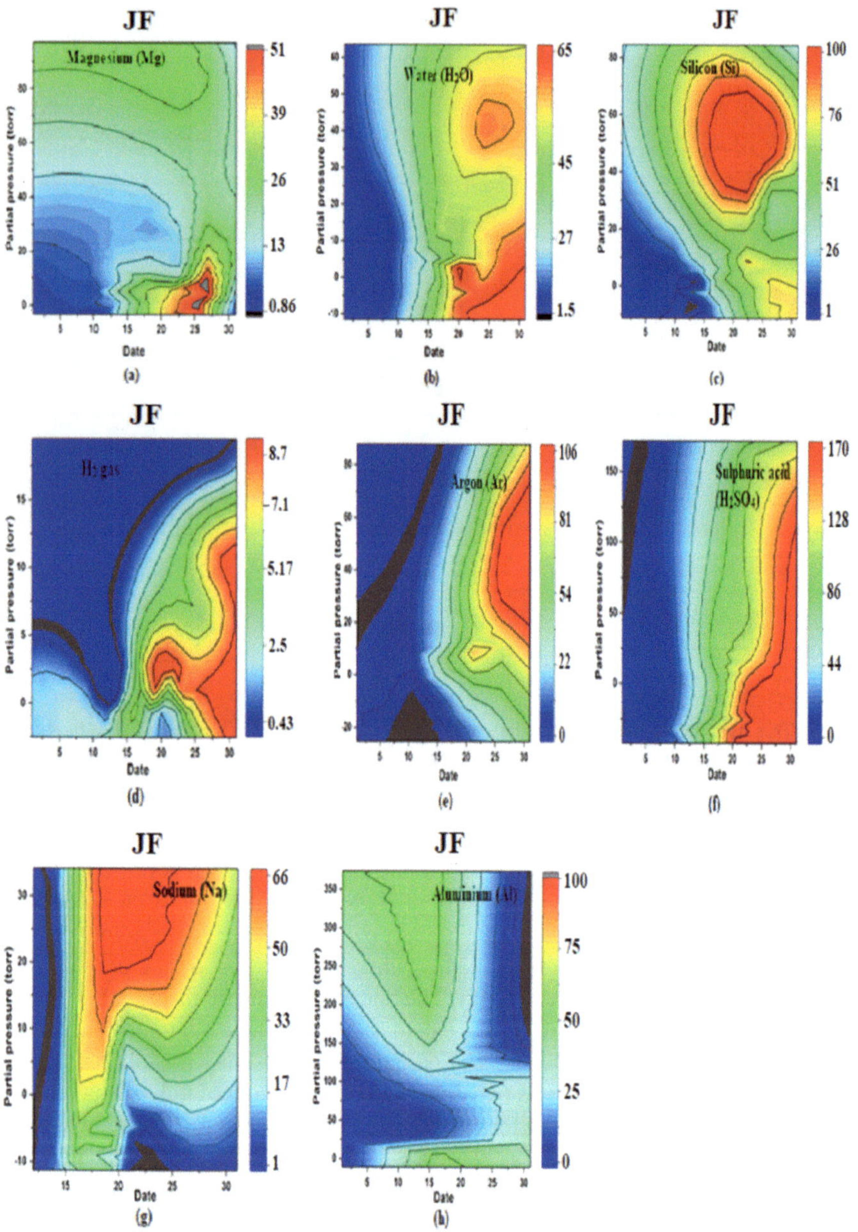

Fig. 6 The figure shows the contour plot of the partial pressure variation in torr of different gases for JF season during 2020

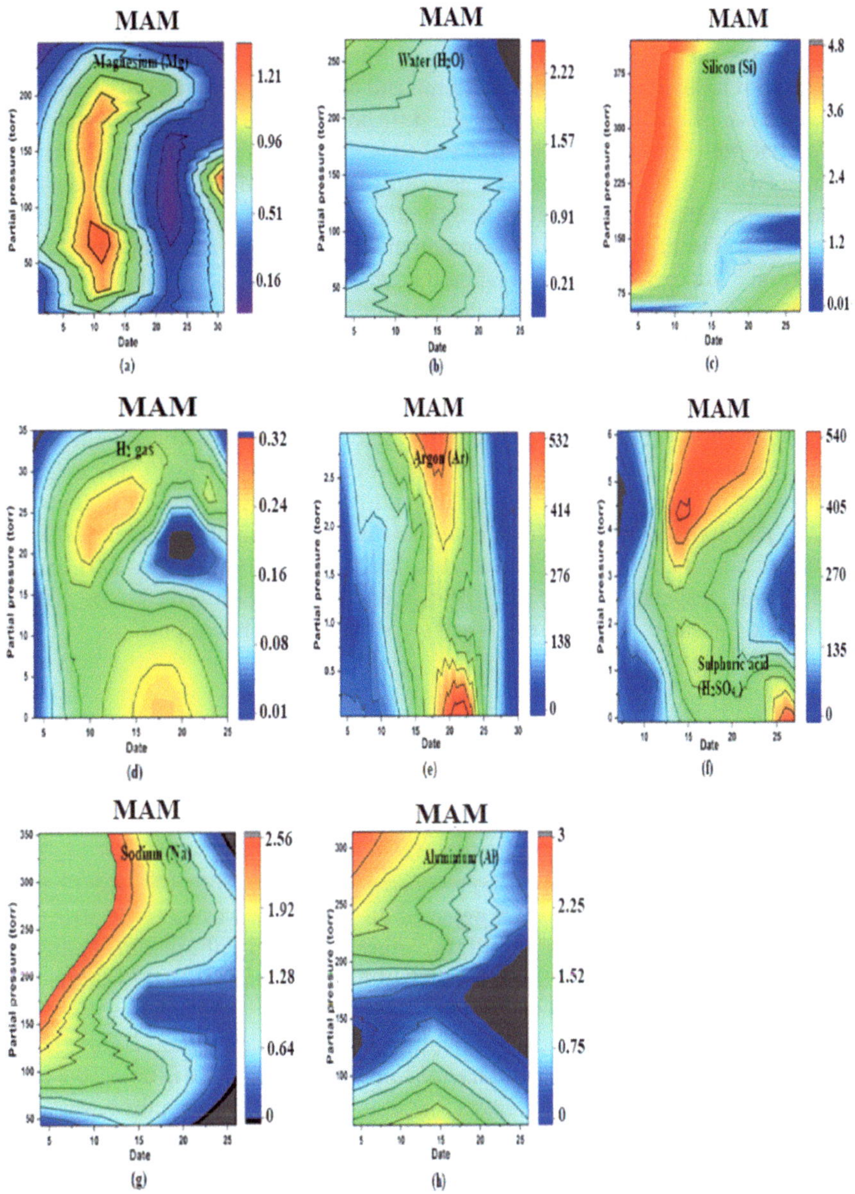

Fig. 7 The figure shows the contour plot of the partial pressure variation in torr of different gases for MAM season during 2020

highest partial pressure in the month of May is found to be 0.93 $\times$ 10^{-9} torr on 3 May, whereas the lowest partial pressure was found at 0.007 $\times$ 10^{-9} torr on 4 May. The linear trend line for the month of March and May was drawn, and the slope was calculated which came out to be 4.049 $\pm$ 4.07 and $-$ 0.029 $\pm$ 0.045, respectively. The partial pressure in May was observed to be almost similar, and the slope was negative, while the slope in March was positive. During MAM season, a peak of partial pressure for sodium (Na) (Figs. 3g and 7g) was found to be 35.14 $\times$ 10^{-9} torr on 16 March. The highest partial pressure in the month of May is found to be 0.25 $\times$ 10^{-9} torr on 3 May, whereas the lowest partial pressure was found at 0.001 $\times$ 10^{-9} torr on 23 May. The linear trend line for the month of March and May was drawn, and the slope was calculated which came out to be 6.66 $\pm$ 2.22 and $-$ 0.018 $\pm$ 0.013, respectively. The partial pressure in May was observed to be almost similar, and the slope was negative, while the slope in March was positive. In the MAM season, a peak for partial pressure of aluminium (Al) (Figs. 3h and 7h) was found to be 31.41 $\times$ 10^{-9} torr on 16 March. The highest partial pressure in the month of May is found to be 0.31 $\times$ 10^{-9} torr on 3 May, whereas the lowest partial pressure was found at 0.002 $\times$ 10^{-9} torr on 23 May. The linear trend line for the month of March and May was drawn, and the slope was calculated which came out to be 5.24 $\pm$ 2.01 and $-$ 0.021 $\pm$ 0.02, respectively. The partial pressure in May was observed to be almost similar, and the slope was negative, while the slope in March was positive.

3.3 *Variation of Partial Pressure for JJAS Season*

The highest partial pressure of magnesium (Figs. 4a and 8a) was observed on 17 June, i.e. 0.109 $\times$ 10^{-9} torr, whereas the lowest partial pressure was 0.0052 $\times$ 10^{-9} torr on 27 June. In the month of July, the highest partial pressure was 0.0056 $\times$ 10^{-9} torr, whereas the highest partial pressure detected on 23 August was 0.063 $\times$ 10^{-9} torr. The highest amount of variation in partial pressure was found in the month of August. The negative slope was observed in the month of August. During the JJAS season, the highest partial pressure of H_2O (Figs. 4b and 8b) was observed on 27 June, i.e. 0.13 $\times$ 10^{-9} torr. In the month of August, the highest partial pressure was 0.22 $\times$ 10^{-9} torr, whereas the lowest partial pressure detected on 24 August was 0.001 $\times$ 10^{-9} torr. Sudden increase in partial pressure was found on 17 August while a drop was seen on the next day. The slope was calculated using the linear trend line as $-$ 0.0059 $\pm$ 0.0075 and $-$ 0.0026 $\pm$ 0.011 for the month of June and August, respectively. The negative slope was observed in both the month of June and August. In the JJAS season, the highest partial pressure of silicon (Figs. 4c and 8c) was observed on 22 June, i.e. 0.35 $\times$ 10^{-9} torr and the lowest on 30 June, i.e. 0.009 $\times$ 10^{-9} torr. In the month of July, the highest partial pressure was 0.05 $\times$ 10^{-9} torr, whereas in the month of August, the highest partial pressure was 0.084 $\times$ 10^{-9} torr; whereas the lowest partial pressure detected on 24 August was 0.003 $\times$ 10^{-9} torr. Sudden increase in partial pressure was found on 22 June while a drop was seen on the next day with the difference of 0.331 $\times$ 10^{-9} torr partial pressure. The slope was calculated using the linear trend line as 0.0059 ± 0.018, – 0.0023 ± 0.0054, and – 0.0022 ± 0.0038 for the month of June, July, and August, respectively. The positive slope was observed in the month of June while negative was observed in July and August. During the JJAS season, the highest partial pressure of H_2 (Figs. 4d and 8d) was observed on 27 June, i.e. 0.028 × 10^{-9} torr and the lowest on 5 June, i.e. 0.008 × 10^{-9} torr. In the month of August, the highest partial pressure was 0.009 × 10^{-9} torr, whereas the lowest partial pressure detected on 21 August was 0.00012 × 10^{-9} torr. The positive slope was observed in the month of June while negative was observed in August.

In the JJAS season, the highest partial pressure of argon (Ar) (Figs. 4e and 8e) was observed on 29 June, i.e. 0.21 × 10^{-9} torr and the lowest on 18 June, i.e. 0.01 × 10^{-9} torr. In the month of August, the highest partial pressure was 0.17 × 10^{-9} torr, whereas the lowest partial pressure detected on 17 August was 0.01 × 10^{-9} torr. Sudden increase in partial pressure was found on 22 June while a drop was seen on the next day with the difference of 0.18 × 10^{-9} torr partial pressure. The positive slope was observed in the month of June while negative was observed in August. In the JJAS season, the highest partial pressure of H_2SO_4 (Figs. 4f and 8f) was observed on 22 June, i.e. 0.41 × 10^{-9} torr. In the month of July, the highest partial pressure was 0.026 × 10^{-9} torr on 16 July, whereas for the month of August, the highest partial pressure was 0.16 × 10^{-9} torr on 19 August, whereas the lowest partial pressure detected on 24 August was 0.002 × 10^{-9} torr. Sudden increase in partial pressure was found on 22 June while a drop was seen on the next day with the difference of 0.38 × 10^{-9} torr partial pressure. The slope was calculated using the linear trend line as – 0.016 ± 0.020, – 0.0021 ± 0.0024, and – 0.019 ± 0.0095 for the month of June, July, and August, respectively. The negative slope was observed in all the months of June, July, and August.

It is to be noted that in the JJAS season, the highest partial pressure of (Na) (Figs. 4g and 8g) was observed on 22 June, i.e. 0.19 × 10^{-9} torr. In the month of August, the highest partial pressure was 0.088 × 10^{-9} torr on 19 August, whereas the lowest partial pressure detected on 24 August was 0.004 × 10^{-9} torr. Sudden increase in partial pressure was found on 22 June while a drop was seen on the next day with the difference of 0.18 × 10^{-9} torr partial pressure. The slope was calculated using the linear trend line as – 0.0052 ± 0.011 and – 0.012 ± 0.0057 for the month of June and August, respectively. The negative slope was observed in both, June and August. During the JJAS season, the highest partial pressure of (Al) (Figs. 4h and 8h) was observed on 29 June, i.e. 0.24 × 10^{-9} torr. In the month of August, the highest partial pressure was 0.092 × 10^{-9} torr on 5 August, whereas the lowest partial pressure detected on 24 August was 0.004 × 10^{-9} torr. Sudden increase in partial pressure was found on 29 June while a drop was seen on the next day with the difference of 0.006 × 10^{-9} torr partial pressure. The slope was calculated using the linear trend line as 0.007 ± 0.013, and – 0.011 ± 0.005 for the month of June and August, respectively. The positive slope was observed in the month of June while it was negative in August.

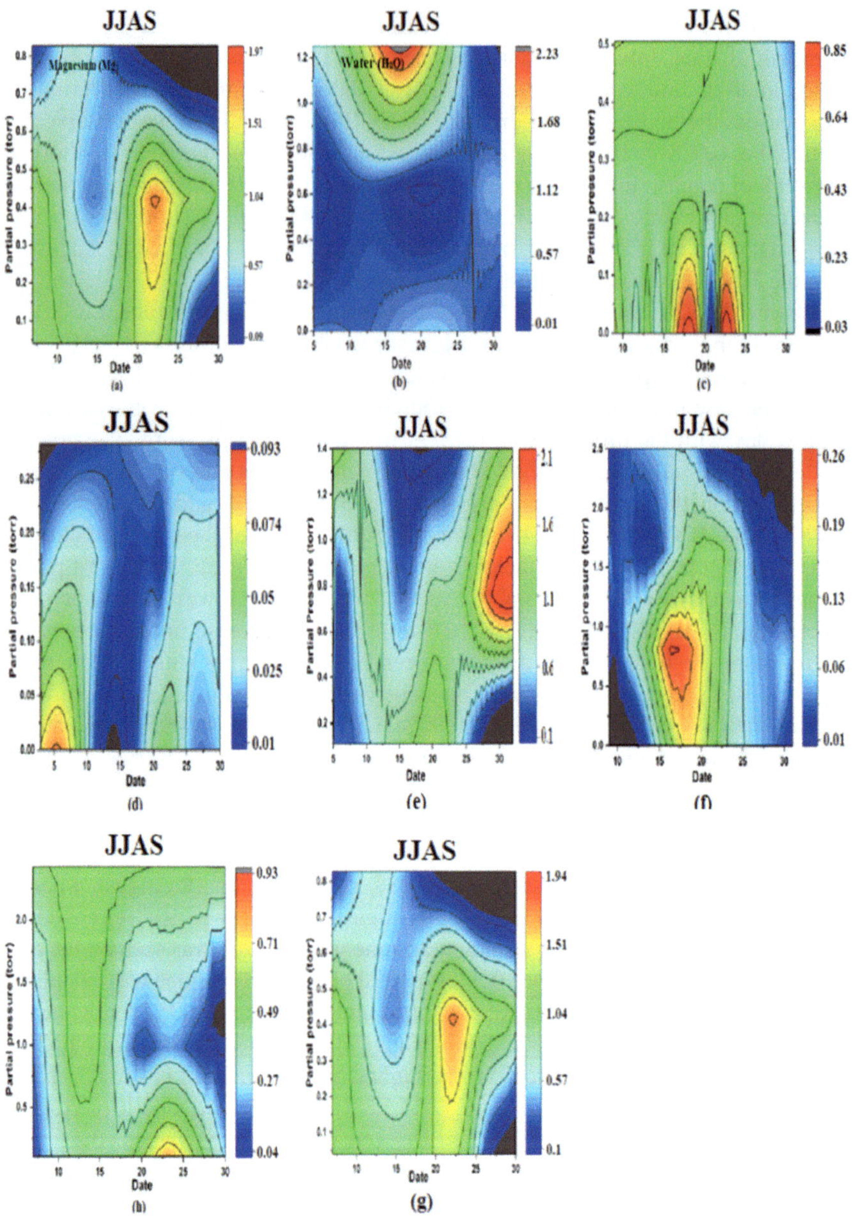

Fig. 8 The figure shows the contour plot of the partial pressure variation in torr of different gases for JJAS season during 2020

3.4 Variation of Partial Pressure for on Season

A peak for partial pressure of magnesium (Figs. 5a and 9a) was found to be 12.01 $\times 10^{-9}$ torr on 23 September, and the lowest value of partial pressure was observed as 0.006 $\times 10^{-9}$ torr on 1 September. The highest partial pressure in the month of October is found to be 23.09 $\times 10^{-9}$ torr, whereas the lowest partial pressure was found at 0.0041 $\times 10^{-9}$ torr. There was a peak observed with the highest value of partial pressure, i.e. 29.3 $\times 10^{-9}$ torr in the month of November. Using the linear trend line, the slope was calculated as 3.16 $\pm$ 0.6, – 5.03 $\pm$ 0.86, and 3.81 $\pm$ 1.62 for September, October, and November, respectively. The slope for October was observed to be negative, whereas for September and November, it was positive.

In the ON season, a peak for partial pressure of water (H_2O) (Figs 5b and 9b) was found to be 15.52 $\times 10^{-9}$ torr on 26 September, and the lowest value of partial pressure was observed as 0.007 $\times 10^{-9}$ torr on 1 September. The highest partial pressure in the month of October is found to be 21.7 $\times 10^{-9}$ torr, whereas the lowest partial pressure was found at 0.03 $\times 10^{-9}$ torr. Using the linear trend line, the slope was calculated as 3.44 $\pm$ 0.64, and – 5.74 $\pm$ 0.8 for September and October, respectively. The slope for October was observed to be negative, whereas for September, it was positive. During ON season, a peak for partial pressure of silicon (Figs. 5c and 9c) was found to be 25.8 $\times 10^{-9}$ torr on 28 September, and the lowest value of partial pressure was observed as 0.02 $\times 10^{-9}$ torr on 1 September. The highest partial pressure in the month of October is found to be 45.9 $\times 10^{-9}$ torr on 1 October, whereas the lowest partial pressure was found at 0.04 $\times 10^{-9}$ torr on 31 October. In the month of November, the highest partial pressure was observed to be 68.13 $\times 10^{-9}$ torr while the lowest was 2.25 $\times 10^{-9}$ torr on 15 November. Using the linear trend line, the slope was calculated as 5.36 $\pm$ 1.29, – 9.83 $\pm$ 1.54 and 7.5 $\pm$ 4.18 for September, October, and November, respectively. The slope for October was observed to be negative, whereas for September and November, it was positive.

In the ON season, a peak for partial pressure of H_2 gas (Figs. 5d and 9d) was found to be 2.5 $\times 10^{-9}$ torr on 20 September, and the lowest value of partial pressure was observed as 0.0015 $\times 10^{-9}$ torr on 1 September. The highest partial pressure in the month of October is found to be 4.26 $\times 10^{-9}$ torr on 1 October, whereas the lowest partial pressure was found at 0.03 $\times 10^{-9}$ torr on 16 October. In the month of November, the highest partial pressure was observed to be 5.4 $\times 10^{-9}$ torr while the lowest was 2.25 $\times 10^{-9}$ torr on 17 November. Using the linear trend line, the slope was calculated as 0.44 $\pm$ 0.14, – 0.9 $\pm$ 0.15 and 0.11 $\pm$ 0.33 for September, October, and November, respectively. The slope for October was observed to be negative, whereas for September and November, it was positive. During ON season, a peak for partial pressure of (Ar) (Figs. 5e and 9e) was found to be 24.79 $\times 10^{-9}$ torr on 23 September, and the lowest value of partial pressure was observed as 0.013 $\times 10^{-9}$ torr on 1 September. The highest partial pressure in the month of October is found to be 47.72 $\times 10^{-9}$ torr on 3 October, whereas the lowest partial pressure was found at 0.017 $\times 10^{-9}$ torr on 31 October. In the month of November, the highest partial pressure was observed to be 60.4 $\times 10^{-9}$ torr while the lowest was 2.4 $\times 10^{-9}$ torr on

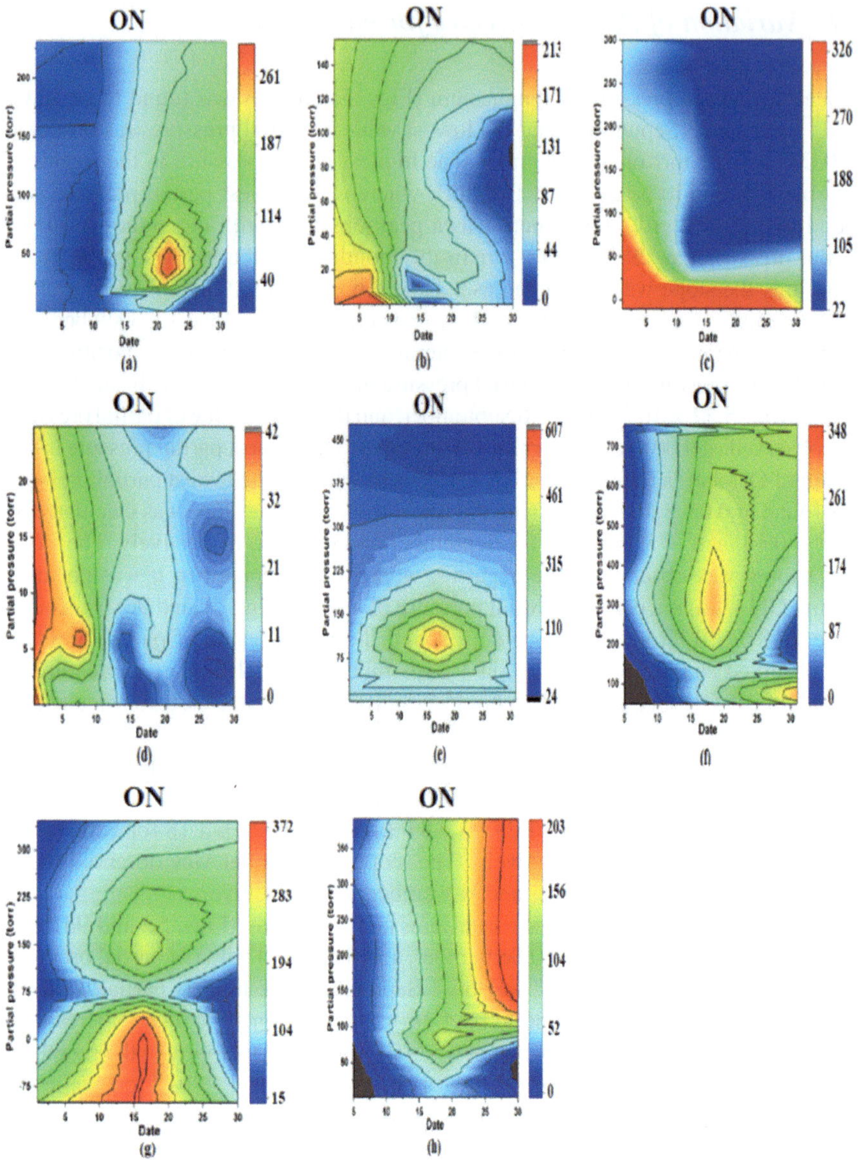

Fig. 9 The figure shows the contour plot of the partial pressure variation in torr of different gases for ON season during 2020

11 November. Using the linear trend line, the slope was calculated as 5.65 ± 1.44, − 9.98 ± 1.74 and 13.71 ± 4.67 for September, October, and November, respectively. The slope for October was observed to be negative, whereas for September and November, it was positive. The overall analysis of different gaseous elements for various months of a season is plotted in Figs. 10 and 11.

In the ON season, a peak for partial pressure of sulphuric acid (Figs. 5f and 9f) was found to be 34.6 × 10^{-9} torr on 26 September, and the lowest value of partial pressure was observed as 0.02 × 10^{-9} torr on 1 September. The highest partial pressure in the month of October is found to be 75.6 × 10^{-9} torr on 1 October, whereas the lowest partial pressure was found at 0.03 × 10^{-9} torr on 31 October. In the month of November, the highest partial pressure was observed to be 97.9 × 10^{-9} torr on 19 November while the lowest was 2.83 × 10^{-9} tor on 15 November. Using the linear trend line, the slope was calculated as 7.88 ± 1.84, − 17.50 ± 3.0, and 22.66 ± 7.01 for September, October, and November, respectively. The slope for October was observed to be negative, whereas for September and November, it was positive.

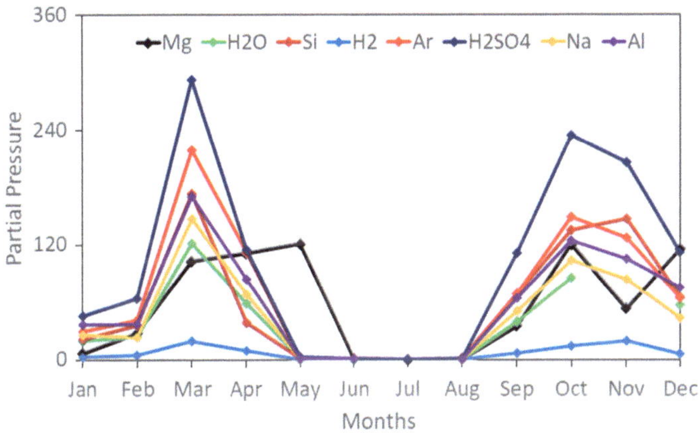

Fig. 10 Monthly variation of partial pressure of different trace gases over southern pole of Moon

Fig. 11 Seasonal variation of partial pressure of different gaseous elements over southern pole of Moon

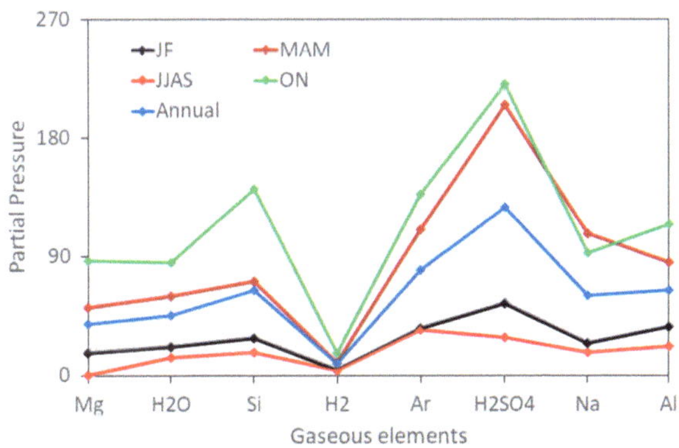

During ON season, a peak for partial pressure of sodium (Figs. 5g and 9g) was found to be 20.3×10^{-9} torr on 20 September, and the lowest value of partial pressure was observed as 0.009×10^{-9} torr on 6 September. The highest partial pressure in the month of October is found to be 34.5×10^{-9} torr on 5 October, whereas the lowest partial pressure was found at 0.01×10^{-9} torr on 31 October. In the month of November, the highest partial pressure was observed to be 37.1×10^{-9} torr on 16 November while the lowest was 1.5×10^{-9} torr on 7 November. Using the linear trend line, the slope was calculated as 3.98 ± 0.99, -7.57 ± 1.22, and 8.57 ± 3.19 for September, October, and November, respectively. The slope for October was observed to be negative, whereas for September and November, it was positive. In the ON season, a peak for partial pressure of aluminium (Figs. 5h and 9h) was found to be 20.6×10^{-9} torr on 25 September, and the lowest value of partial pressure was observed as 0.003×10^{-9} torr on 6 September. The highest partial pressure in the month of October is found to be 39.2×10^{-9} torr on 5 October, whereas the lowest partial pressure was found at 0.02×10^{-9} torr on 31 October. In the month of November, the highest partial pressure was observed to be 41.8×10^{-9} torr on 19 November while the lowest was 1.6×10^{-9} torr on 15 November. Using the linear trend line, the slope was calculated as 5.38 ± 1.038, -8.46 ± 1.51, and 10.58 ± 3.2 for September, October, and November, respectively. The slope for October was observed to be negative, whereas for September and November, it was positive.

4 Conclusions

The variation of partial pressure for different season like JF, MAM, JJAS, and ON for different gases was determined using CHACE-2 payload data of Chandrayan-2 mission. The highest and the lowest partial pressure was identified after which the linear trend analysis was done for each gas. The following conclusions were made. During JFs, the partial pressure was observed to be low while, it started increasing as soon as MAM approached. In the month of May, partial pressure was almost similar for each gas; hence, the trend line was constant and parallel to x-axis. It suggests that the partial pressure was depending on the temperature for different gases as per the observations. The semi-annual oscillation of inert gas Ar is found to be consistent with condensation loss in the seasonal cold traps at the poles due to the obliquity of the Moon. The results show that the exospheric morphology depends on the energetics of desorption of atoms from soil grains. In case of aluminium, during MAM season the highest partial pressure was seen in the month of May. The partial pressure in May was observed to be almost similar, and the slope was negative, while the slope in March was positive. For argon, there is a high variation in partial pressure observed during JJAS season, and the peak was observed in the month of June amongst June, July, and August of JJAS season. Slope of the all the three months was observed negative. The observed trend in the value of partial pressure versus date was calculated which cleared the fact about the negative and positive variation of partial pressure depending on the months and the seasons. Negative trend was observed at the end of JJAS while in the beginning of ON which depicts how the partial pressure changes with respect to temperature. On the other hand, a positive trend was seen at the end of JF and in the beginning of MAM. A similar variation was seen in left out gases such as sodium (Na), sulphuric acid (H_2SO_4), silicon (Si), and magnesium (Mg). In conclusion, this paper helped us to understand the concept of partial pressure and its variation on the lunar surface.

Acknowledgements The corresponding author is thankful to the different scientists of Department of Space (DOS) and ISRO, Bengaluru, for providing the partial pressure data through the Policy Based Data Retrieval, Analysis, Dissemination and Notification system (PRADAN) online portal. Authors are thankful to scholastic reviewers to provide the necessary suggestions to improve the manuscript of present work.

References

1. Agarwal N, Harias A, Khanna N, Srivastava P, Jain V (2019) Study of morphology and degradation of lunar craters using Chandrayaan–1 data. Planet Space Sci 167:42–53
2. Arya AS, Rajasekhar RP, Sur K, Gopala Krishna B, Suresh K, Srinivasan TP, Iyer KV, Chauhan P, Ajai, Kiran Kumar AS, Pandey AD, Khare A, Verma PK (2018) Morphometric and rheological study of lunar domes of Marius Hills volcanic complex region using Chandrayaan-1 and recent datasets. J Earth Syst Sc 127:70
3. Bhattacharya S, Chauhan P, Rajawat AS, Ajai, Kiran Kumar AS (2011) Lithological mapping of central part of Mare Moscoviense using Chandrayaan-1 Hyperspectral Imager (HySI) data. Icarus 212(2):470–479
4. Bhattacharya S, Lal D, Chauhan M, Chauhan P, Ajai, Kiran Kumar AS (2015) Detection of hydroxyl-bearing exposures of possible magmatic origin on the central peak of crater Theophilus using Chandrayaan-1 Moon Mineralogy Mapper (M3) data. Icarus 260:167–173
5. Chowdhury AR, Kumar V, Patel D, Joshi SR, Arya AS, Kumar A, Paul S, Shah D, Soni P, Karelia JC, Sampat M, Sharma S, Somani S, Bhagat HV, Sharma J, Amitabh Suresh K, Rajasekhar RP, Bokarwadia BB, Kumar M, Ghonia, DN (2020) Terrain Mapping Camera-2 onboard Chandrayaan-2 Orbiter. Curr Sci 118(4):566–572
6. Dhanya MB, Yadav CM, Thampi SV, Thampi RS, Das TP (2023) Peak filter algorithm for the calibration of mass spectra from CHACE-2 aboard Chandrayaan-2 orbiter. Int J Mass Spectrom 491:117098
7. Goswami JN, Annadurai M (2008) Chandrayaan-1 mission to the Moon. Acta Astronaut 11–12:1215–1220
8. Kamalakar JA, Bhaskar VS, Laxmi Prasad AS, Ranjith KA, Venketeswaran L, Alex K (2005) Lunar ranging instrument for Chandrayaan-1. J Earth Syst Sci 6:725–731
9. Kosambe S (2019) Chandrayaan-2: India's second lunar exploration mission. J Aircr Spacecraft Tech 3(1):221–236
10. Laxmiprasad AS, Sridhar Raja VLN, Menon S, Goswami A, Rao MVH, Lohar KH (2013) An in-situ laser induced breakdown spectroscope (LIBS) for Chandrayaan-2 rover: Ablation kinetics and emissivity estimations. Periodical: Adv Space Res 2:332–341
11. Narendranath S, Athiray PS, Sreekumar P, Radhakrishna V, Tyagi A, Kellett BJ et al (the CLASS team) (2014) Mapping lunar surface chemistry: new prospects with the Chandrayaan-2 large area soft X-ray spectrometer (CLASS). Adv Space Res 10:1993–1999
12. Narendranath S, Sreekumar P, Maddison BJ, Howe CJ, Kellett BJ, Wallner M, Erd C, Weider SZ (2010) Calibration of the C1XS instrument on Chandrayaan-1, Nuclear Inst. and Methods

in Phy. Res. Section A: Accelerators, Spectrometers, Detectors and Associated Equipment, 1–3, 344–353
13. Pillai NS, Narendranath S, Srikar KV, Tadepalli P, Radhakrishnan V, Tyagi A, Yadav R, Singh B, Sharan V, Athiray PS, Sreekar P, Sankarasubramanian K, Bhatt M, Sarbadhikari AB, Mithun NPS, Vadawale SV (2021) Chandrayaan-2 large area soft X-ray spectrometer (CLASS): calibration, In-flight performance and first results. Icarus 363:114436
14. Prabu T, Muthukkumaran K, Vnugopal I (2022) Assessment of dynamic properties of a new lunar highland soil simulant (LSS-ISAC-1) developed for Chandrayaan missions. Soil ynamis an Earthquak Eng 155:107178
15. Radhakrishna V, Tyagi A, Narendranath S, Vadodariya K, Yadav R, Singh S, Balaji G, Satya N, Shetty A, Suresha Kumar HN, Vaishali S, Pillai NS, Tadepalli S, Venkata RP, Sreekumar, Agarwal A, Valarmath N (2020) Chandrayaan-2 large area soft X-ray spectrometer. Curr Sci 118(2):219–225
16. Saha K, De Kumar B, Paul B, Guha A (2020) Satellite launch vehicle effect on the Earth's lower ionosphere: a case study. Periodical: Adv Space Res 11:2507–2514
17. Shanmugam M, Murty VS, Acharya YB, Goyal SK, Patel AR, Shah B, Hait AK, Patinge A, Subrahmanyam D (2014) Alpha particle X-ray spectrometer (APXS) on-board Chandrayaan-2 rover. Adv Space Res 54(10):1974–1984
18. Singh A, Sharma A, Kumar S, Chang L, Vashishtha A, Raj R, Agrawal S, Chauhan (2022) Dielectric characterization and polarimetric analysis of lunar north polar crater Hermite-A using Chandrayaan-1 Mini-SAR, Lunar Reconnaissance Orbiter (LRO) Mini-RF, and Chandrayaan-2 DFSAR data. Adv Space Res, Available online 10 May 2022
19. Sinha RK, Vijayan S, Bhatt M, Harish, Kumari N, Srivastava N, Varatharajan I, Ray, Wohlr, Bharwaj A (2020) Geological characterization of Chandrayaan-2 landing site in the southern high latitudes of the Moon. Icarus 337:113449
20. Sinha RK, Rani A, Ruj T, Bhardwaj A (2023) Geologic investigation of lobate scarps in the vicinity of Chandrayaan-3 landing site in the southern high latitudes of the moon. Icarus 402:115636
21. Sridharan R, Ahmed SM, Das TP, Sreelatha P, Kumar PP, Naik N, Supriya G (2010) Direct evidence for water (H_2O) in the sunlit lunar ambience from CHACE on MIP of Chandrayan I. Planet Space Sc 6:947–950
22. Sridharan R, Das TP, Ahmed SM, Supriya G, Bhardwaj A, Kamalakar JA (2013) Spatial heterogeneity in the radiogenic activity of the lunar interior: inferences from CHACE and LLRI on Chandrayaan-1. Adv Space Res 1:168–178
23. Sudhakar M, Sreekumar P (2012) Design optimization and background modeling of the HEX experiment on Chandrayan-I. Exp Astro 3:653–668
24. Vadawale SV, Shanmugam M, Acharya YB, Patel AR, Goyal SK, Shah B, Hait AK, Patinge A, Subrahmanyam D (2014) Solar X-ray monitor (XSM) on-board Chandrayaan-2 orbiter. Adv Space Res 10:2021–2028
25. Vashishtha A, Kumar S (2021) Characterization of geomorphological features of lunar surface using Chandrayaan-1 Mini-SAR and LRO Mini-RF data. Quatrnary Int 575–576:338–357
26. Verma PA, Chauhan M, Chauhan P (2022) Lunar surface temperature estimation and thermal emission correction using Chandrayaan-2 imaging infrared spectrometer data for H_2O & OH detection using 3 μm hydration feature. Icarus 383:115075
27. Vighnesam NV, Sonney A, Gopinath NS (2010) India's first lunar mission Chandrayaan-1 initial phase orbit determination. Acta Astronaut 7–8:784–792

Synthesis of Coumarins by Pechmann Condensation Using Nano-ferrites

Aaina Bansal, Kiran Bajaj, and Deepshikha Gupta

1 Introduction

Coumarin, a versatile natural pharmacophore, is widely available in plants like tonka beans and cinnamon and showcased widespread applications in pharmaceutical and material chemistry. Various coumarin scaffolds are distinctly known for their fungicidal, antimicrobial, anti-inflammatory, and anticancer profiles [1]. Due to its vast pharmacological profile, coumarins are also used as food additives [2]. In addition to drug and food industries, coumarins showcased vast applications in material chemistry, and due to their photophysical properties, coumarin derivatives are used in solar panels to store energy [3]. Structurally, coumarin is composed of two fused rings; the benzene ring is fused with the pyrone ring and has a 2H-1-benzopyrone core structure. Coumarin was isolated from tonka beans in 1820 for the first time. It was later synthesized in 1868. Warfarin was the first coumarin-based drug [4]. It is isolated from woodruff and lavender and acts as an anticoagulant. Umbelliferone, another coumarin derivative, has a chemical structure of 7-hydroxycoumarin and has many applications [5]. It is found naturally in parts of the plants such as carrots, coriander, etc., and is used as a sunscreen agent. Dicoumarol, which is another coumarin derivative, is naturally occurring and causes sweet clover diseases in cattle [6]. Imperatorin is a reversible acetylcholinesterase (AChE) inhibitor and acts in a dose-dependent manner [7]. Skin diseases in Egypt and India were treated by Psoralen (a photosensitizing drug) using sunlight exposure as early as 1200–2000 BC [8]. Structures of some of the naturally occurring important coumarins are given in Fig. 1. Coumarins play a significant role in the plant kingdom. Due to their widespread applications, extensive work has been reported for their synthetic

A. Bansal · K. Bajaj (✉) · D. Gupta (✉)
Amity Institute of Applied Sciences, Amity University Uttar Pradesh, Sector 125, Noida, India
e-mail: kbajaj@amity.edu

D. Gupta
e-mail: dgupta2@amity.edu

S. Rattan et al. (eds.), *Emerging Trends in Synthesis and Catalysis in Chemistry*, Springer Proceedings in Materials 51, https://doi.org/10.1007/978-981-97-5169-3_9

protocol. Coumarin and derivatives are known to be synthesized using several known organic reactions namely Wittig [9], Perkin [10], Knoevenagel [11], and Pechmann [12]. Among various conventional methods, Knoevenagel condensation and Pechmann reaction [13] are widely exploited by using simple reactants like phenols under mild reaction conditions utilizing a variety of acid catalysts from mineral acids to Lewis acids [14]. However, a variety of combination catalysts also complimented the synthetic protocol such as alumina-supported MoO_3 [15], silica-supported sulfuric acid [16], kegnin-type heteropoly compounds [17], and ion-exchange resins [18], and so on. Literature glance showcased that heterogenous catalysis ease the synthesis of coumarin with greater selectivity and economic viability [19]. Our reported magnetic nanocatalyst displays dual roles by providing an acidic medium for catalysis and a larger surface area for reaction. These magnetic nanoparticles can be regenerated after one catalytic cycle making this a cost-effective approach. In the present work, we explored nickel ferrite-based nanoparticles as a complementary catalyst for the synthesis of coumarins in user-friendly ways. Nickel ferrite finds its applications in various fields like reusable catalyst in the synthesis of heterocycles [20], carbon dioxide decomposition to carbon and oxygen [21], as a photocatalyst for dye degradation [22], catalyst for direct oxidation of glucose, NADH, and methanol [23], photocatalytic water oxidation using $[Ru(bpy)_3]^{2+}$ as photosensitizer and $S_2O_8{}^{2-}$ as sacrificial oxidant [24], and Fe–Ni alloy-based carbon nanotubes can also be used as urea oxidation catalysts [25, 26].

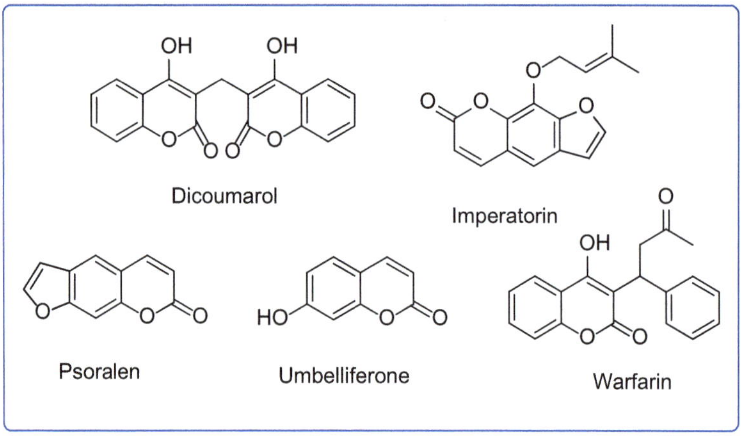

Fig. 1 Structure of some naturally occurring important coumarin-based molecules

2 Experimental and Methods

2.1 *Reagents and Analysis*

All chemicals required were procured from S.D. Fine chemicals Ltd., and SRL chemicals India. For XRD studies of the synthesized materials, Cu K radiation (λ = 1.5405 A°) in the range of 10°–80° at a scan rate of 2 per minute using Shimadzu XDD1 diffractometer were performed. Using a JEOL JSM-5300 microscope with an acceleration voltage of 20 kV, a scanning electron micrograph (SEM) was acquired. Before mounting on a sample holder, the sample powders were placed on carbon tape. The melting points were determined using an open capillary method. IR spectra were captured by Perkin–Elmer IR spectrophotometer using KBr pellets. ^{1}H-NMR spectra were obtained using a Bruker spectrometer at 400 MHz using TMS as internal standard. About 0.2-mm silica gel F-254 plates were used to monitor progress of reaction by TLC. All the coumarin products are known compounds and are characterized by comparing their IR, ^{1}H-NMR, and melting points.

2.2 *Preparation of Metal Ferrite Nanoparticles*

In this method, copper nitrate/manganese nitrate and ferric nitrate or nickel chloride hexahydrate ($NiCl_2.6H_2O$) and $FeCl_3$ were dissolved in 1:2 molar ratio in 100 ml of deionized water. Under vigorous continuous magnetic stirring, NaOH (2 M) was dropwise added until pH of 10–11 was reached. The precipitate of mixed metal hydroxide so obtained was filtered, centrifuged, washed with deionized water, and heated at 800 °C for 1 h for complete drying [27]. The synthesized nickel ferrite was characterized by SEM, XRD, EDAX, and magnetic coercivity measurement.

2.3 *Pechmann Condensation for Preparation of 4-Methyl-7-Hydroxy Coumarin*

To the solution of resorcinol **1a**(1 equiv, 1.8 mmol) in solvent (ethanol, 10 mL), added ethyl acetoacetate **2**(1.5 equiv, 2.7 mmol) dropwise with constant stirring. Metal ferrite nanoparticles (10 mol%, 0.18 mmol) were added to the reaction mass and refluxed for 16 h at room temperature. The reaction was monitored by TLC for completion. After completion, the reaction mass was filtered using a celite bed. Note catalytic nanoparticles were removed by putting a magnetic bar to the reaction mass. The filtrate obtained was quenched with ice and extracted with ethyl ethanoate. All organic layers were combined and washed with brine solution. It was then dried under reduced pressure over anhydrous sodium sulfate. The crude reaction mass obtained was further recrystallized by ethanol to afford pure coumarin derivative.

The recrystallized product **3a** was characterized by IR and ^{1}H-NMR spectroscopy as follows: IR (KBr) (ν, cm^{-1}): 3116, 2938, 1680, 1518, 1455. ^{1}H NMR (400 MHz, DMSO-d_6) δ 10.71 (s, 1H), 7.59 (d, J = 8.8 Hz, 1H), 6.81 (d, J = 8.4 Hz, 1H), 6.70 (s, 1H), 6.11 (s, 1H), 2.35 (s, 3H). ^{13}C NMR (101 MHz, DMSO-d_6) δ 161.11, 160.24, 154.79, 153.46, 126.53, 112.80, 111.97, 110.21, 18.05.

3 Result and Discussion

Our excursion starts with the synthesis of metal ferrite nanoparticles followed by their application in coumarin synthesis. Synthesis of ferrite nanoparticles was achieved by exploiting co-precipitation methods at room temperature by adding M^{2+} basic solution to the aqueous solution of Fe^{3+} ions. It was observed that at higher pH goethite (α-FeOOH) is formed as an impurity whereas at high temperature, concentration spherical nanoparticles are formed due to less selective crystallographic growth. The synthesis of nickel ferrite nanoparticles is discussed in the experimental section. The SEM micrograph and their corresponding EDX curves of $NiFe_2O_4$ are shown in Fig. 2a, b, respectively. SEM images indicated the formation of nanometre-sized quasi-spherical particles (with diameter of 150–210 nm) in the mixed ferrites, and EDAX images shows the presence of Ni, Fe, and oxygen. Nickel ferrite grows in inverse spinel-type crystal structures, which can be represented by $(Fe^{3+})[Ni^{2+}Fe^{3+}]O_4$. The cations inside the round brackets are in half of the tetrahedral (A) sites, and the cations inside square brackets occupy in octahedral (B) sites. M^{3+} occupy half of the tetrahedral holes while M^{2+} occupy octahedral holes. The spins of trivalent ions are canceled out since half of them belong to tetrahedral sites and the other half belong to octahedral sites. However, the spins of divalent ions are not cancelled and are responsible for the catalytic activity [28]. Hence, they show ferrimagnetism. The net magnetization in $NiFe_2O_4$ is the difference in the magnetizations of these two (A) and (B) sublattices [29]. All XRD peaks for $NiFe_2O_4$ MNPs are shown in Fig. 2c that match well with standard in JCPDS file 23–1119 [30]. Room temperature magnetic hysteresis loop (M-H) determined by VSM for $NiFe_2O_4$ MNPs is shown in Fig. 2d. Hysteresis loop depicts coercivity (G) and saturation magnetization (Ms). Ferromagnetic materials with high coercivity are called magnetically hard, while those with low coercivity are said to be magnetically soft. Coercivity (Hc) of nickel ferrite came out to be 229.46 mTesla, while saturation magnetization (Ms) came out to be 5.09 emu/g indicating it to be ferromagnetic material.

After the successful synthesis of nickel ferrite nanoparticles, we have explored their application in coumarin synthesis using Pechmann reaction. Model reaction was set up between resorcinol **1a** and ethyl acetoacetate **2** as a substrate. No reaction was observed between the starting materials in the absence of any acidic catalyst. However, with the addition of copper ferrite nanoparticles in 10 mol% using water as a solvent doesn't work (Table1, **entry 1**). Fortunately, by changing the solvent from water to ethanol yielded the target compounds in 30% yield at 60 °C for 16 h.

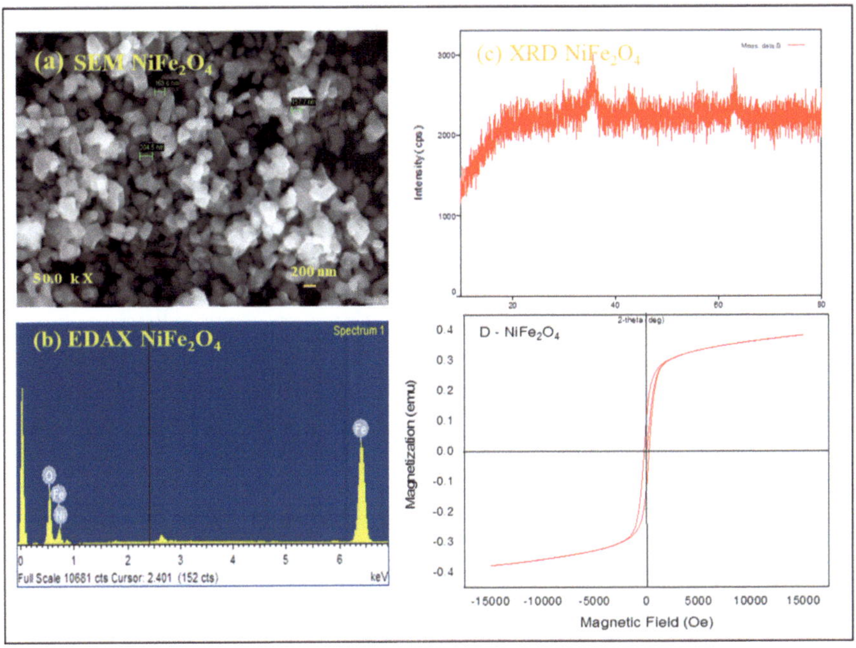

Fig. 2 **a** SEM image of $NiFe_2O_4$, **b** XRD of $NiFe_2O_4$, **c** EDAX curve of $NiFe_2O_4$ (O-51%, Fe-35.2%, Ni-13.8%), **d** magnetic hysteresis loop of $NiFe_2O_4$

However, change in temperature further enhances the reaction profile and the desired product **3a** is obtained with enhanced yield (Table 1, **entry 3**). To explore other catalysts, it was observed that replacement of copper ferrite by manganese ferrite nanoparticles lowers the product yields to 30% (Table 1, **entry 4**). Further, changing the catalyst from copper ferrite to nickel ferrite nanocatalyst reinforces the reaction profile, and desired coumarin was obtained in better yield (Table 1, **entry 5**). It was worth mentioning that further change in catalyst amount doesn't trigger the reaction profile. However, by lowering the catalyst loading, product is obtained in poor yields. Next, we have explored the role of temperature by fixing the catalyst and its amount. Unfortunately, by changing the reaction parameters, product was obtained in poor yields. (Table 1, **entry, 7–8**).

Having the optimized condition in hand, next we have explored the scrutinized catalyst for the synthesis of variety of coumarins as depicted in Table 2. Following the protocol, various substituted coumarins were synthesized in good yields under optimized reaction conditions.

Table 1 Optimization of coumarin synthesis

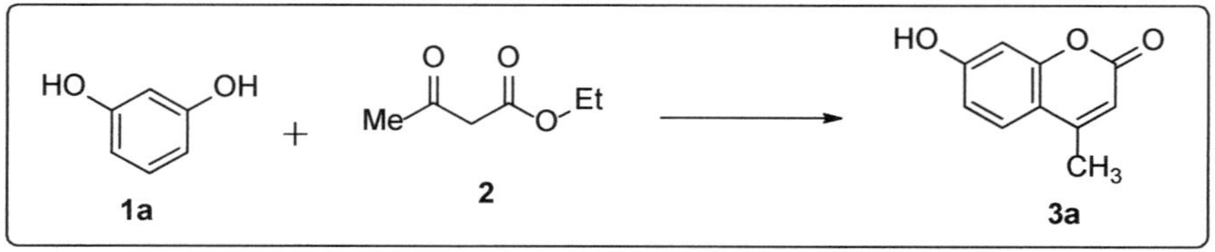

S. No.	Catalyst used	Solvent used	Temperature (°C)	Yield (%)
1	$CuFe_2O_4$ (10 mol %)	H_2O	90	–
2	$CuFe_2O_4$ (10 mol %)	Ethanol	60	30
3	$CuFe_2O_4$ (10 mol %)	Ethanol	80	50
4	$MnFe_2O_4$ (10 mol %)	Ethanol	80	30
5	**$NiFe_2O_4$ (10 mol %)**	**Ethanol**	**80**	**60**
6	$NiFe_2O_4$ (5 mol %)	Ethanol	80	30
7	$NiFe_2O_4$ (5 mol %)	Ethanol	60	55
8	$NiFe_2O_4$ (5 mol %)	Methanol	60	50

Reaction Conditions: The reactions were carried out with **1a** (1.8 mmol) and **2a** (2.7 mmol) in the presence of metal ferrite catalyst in ethanol (10 mL) at 80 °C for 16 h.

Bold signifies the optimized reaction conditions

Table 2 Synthesis of coumarins using nickel ferrite nanoparticles

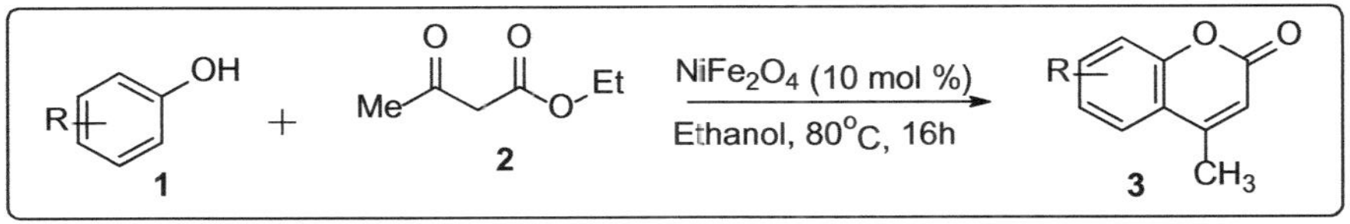

S. No.	Starting 1	Compound 3	Yield (%)	λ_{max} (in nm)	M. pt (Lit M. Pt)	References
1	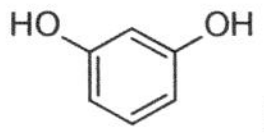 1a	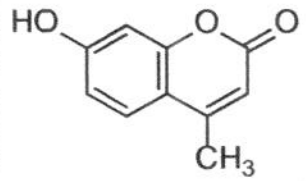 3a	60	357.2	185 °C (188 °C)	[31]
2	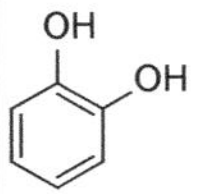 1b	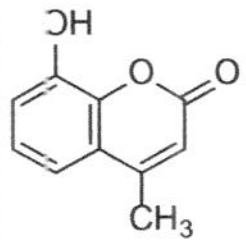 3b	65	362.8	168 °C (168–171 °C)	[32]
3	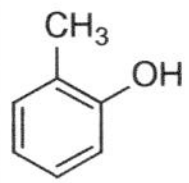 1c	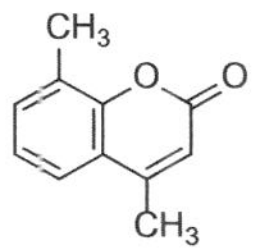 3c	68	579.2	110 °C (114–115 °C)	[33]

(continued)

Table 2 (continued)

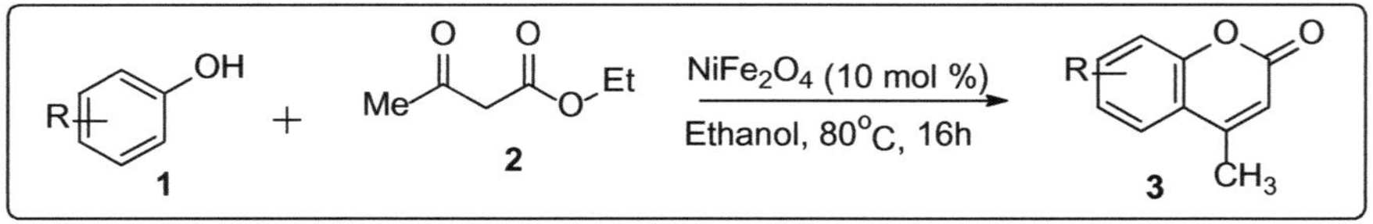

S. No.	Starting 1	Compound 3	Yield (%)	λ_{max} (in nm)	M. pt (Lit M. Pt)	References
4	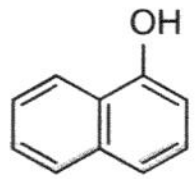 **1d**	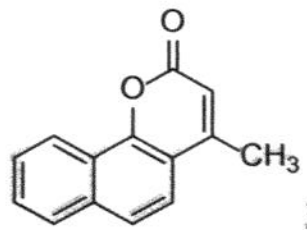 **3d**	72	370.6	155 °C (155–156 °C)	[34]
5	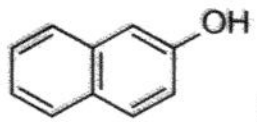 **1e**	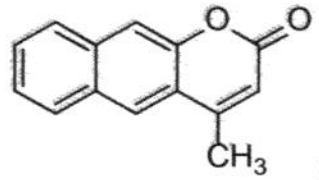 **3e**	62	351.2	282 °C (285 °C)	[35]
6	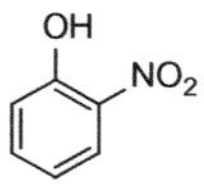 **1f**	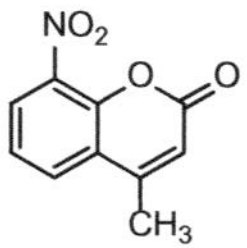 **3f**	50	379.4	180 °C (184–185 °C)	[32]

4 Conclusions

In this work, we have reported a novel application of synthesized magneto ferrite nanoparticles for the synthesis of structurally diverse coumarins under mild conditions. Among various metal ferrite nanoparticles, nickel-based magnetic nanoparticles were found to catalyze Pechmann condensation efficiently under mild reaction conditions. Synthesis of nickel ferrite nanoparticles was also discussed. The generality of reaction was also showcased to various phenols having electron withdrawing as well as electron donating groups. Electron donating groups present on phenol reinforce the reaction and afforded the products in high yields. The present protocol showcased the application of nickel metal ferrite nanoparticles to Pechmann-type coumarin synthesis.

References

1. Tsivileva OM, Koftin OV, NV E (2022) Coumarins as fungal metabolites with potential medicinal properties. Antibiotics (Basel) 11(9):1156. https://doi.org/10.3390/antibiotics11091156
2. Lončar M, Jakovljević M, Šubarić D, Pavlić M, Buzjak Služek V, Cindrić I, Molnar M (2020) Coumarins in food and methods of their determination. Foods 9(5):645
3. Sánchez-de-Armas R, San Miguel MA, Oviedo J, Sanz JF (2012) Coumarin derivatives for dye sensitized solar cells: a TD-DFT study. Phys Chem Chem Phys 14(1):225–233
4. Kontogiorgis C, Detsi A, Hadjipavlou-Litina D (2012) Coumarin-based drugs: a patent review (2008–present). Expert Opin Ther Pat 22(4):437–454
5. Singh D, Rahman M (2020) Umbelliferone loaded nanocarriers for healthcare applications. Curr Biochem Eng 6(1):25–33
6. Rashed K (2021) Biological evidences of dicoumarol: a review. Plantae Scientia 4(2):121–124
7. Kozioł E, Skalicka-Woźniak K (2016) Imperatorin–pharmacological meaning and analytical clues: profound investigation. Phytochem Rev 15:627–649
8. Pathak MA, Fitzpatrick TB (1992) The evolution of photochemotherapy with psoralens and UVA (PUVA): 2000 BC to 1992 AD. J Photochem Photobiol, B 14(1–2):3–22
9. Desai VG, Shet JB, Tilve SG, Mali RS (2003) Intramolecular Wittig reactions. A new synthesis of coumarins and 2-quinolones. J Chem Res 2003(10):628–629
10. Johnson JR (2004) The Perkin reaction and related reactions. Org React 1:210–265
11. Vekariya RH, Patel HD (2014) Recent advances in the synthesis of coumarin derivatives via Knoevenagel condensation: a review. Synth Commun 44(19):2756–2788
12. Heravi MM, Khaghaninejad S, Mostofi M (2014) Pechmann reaction in the synthesis of coumarin derivatives. In: Advances in heterocyclic chemistry, vol 112. Elsevier, pp 1–50
13. Jadhav NH, Sakate SS, Rasal NK, Shinde DR, Pawar RA (2019) Heterogeneously catalyzed Pechmann condensation employing the tailored Zn_{0.925}Ti_{0.075}O NPs: synthesis of Coumarin. ACS Omega 4(5):8522–8527. https://doi.org/10.1021/acsomega.9b00257
14. Nikoofar K, Khademi Z (2016) A review on green Lewis acids: zirconium (IV) oxydichloride octahydrate ($ZrOCl_2\cdot 8H_2O$) and zirconium (IV) tetrachloride ($ZrCl_4$) in organic chemistry. Res Chem Intermed 42:3929–3977
15. Rajabi F, Feiz A, Luque R (2015) An efficient synthesis of coumarin derivatives using a SBA-15 supported Cobalt (II) nanocatalyst. Catal Lett 145:1621–1625
16. Rad-Moghadam K, Montazeri N (2009) Coumarin synthesis via Pechmann condensation on silica-supported sulfuric acid under microwave irradiation. Asian J Chem 21(1):499–503

17. Heravi MM, Khorasani M, Derikvand F, Oskooie HA, Bamoharram FF (2007) Highly efficient synthesis of coumarin derivatives in the presence of H14 [NaP5W30O110] as a green and reusable catalyst. Catal Commun 8(12):1886–1890
18. Sabou R, Hoelderich W, Ramprasad D, Weinand R (2005) Synthesis of 7-hydroxy-4-methylcoumarin via the Pechmann reaction with Amberlyst ion-exchange resins as catalysts. J Catal 232(1):34–37
19. Adimule VM, Nandi SS, Kerur SS, Khadapure SA, Chinnam S (2022) Recent advances in the one-pot synthesis of coumarin derivatives from different starting materials using nanoparticles: a review. Top Catal. https://doi.org/10.1007/s11244-022-01571-z
20. Moghaddam FM, Koushki Foroushani B, Rezvani HR (2015) Nickel ferrite nanoparticles: an efficient and reusable nanocatalyst for a neat, one-pot and four-component synthesis of pyrroles. RSC Adv 5(23):18092–18096. https://doi.org/10.1039/c4ra09348h
21. Lin K-S, Adhikari AK, Tsai Z-Y, Chen Y-P, Chien T-T, Tsai H-B (2011) Synthesis and characterization of nickel ferrite nanocatalysts for CO_2 decomposition. Catal Today 174(1):88–96. https://doi.org/10.1016/j.cattod.2011.02.013
22. Jadhav SA, Khedkar MV, Somvanshi SB, Jadhav K (2021) Magnetically retrievable nanoscale nickel ferrites: an active photocatalyst for toxic dye removal applications. Ceram Int 47(20):28623–28633
23. Galindo R, Gutiérrez S, Menéndez N, Herrasti P (2014) Catalytic properties of nickel ferrites for oxidation of glucose, β-nicotiamide adenine dinucleotide (NADH) and methanol. J Alloy Compd 586:S511–S515. https://doi.org/10.1016/j.jallcom.2013.01.049
24. Hong D, Yamada Y, Nagatomi T, Takai Y, Fukuzumi S (2012) Catalysis of nickel ferrite for photocatalytic water oxidation using $[Ru(bpy)3]^{2+}$ and $S2O8^{2-}$. J Am Chem Soc 134(48):19572–19575. https://doi.org/10.1021/ja309771h
25. Modak A, Mohan R, Rajavelu K, Cahan R, Bendikov T, Schechter A (2021) Metal-organic polymer-derived interconnected Fe–Ni alloy by carbon nanotubes as an advanced design of urea oxidation catalysts. ACS Appl Mater Interfaces 13(7):8461–8473. https://doi.org/10.1021/acsami.0c22148
26. Modak A, Mankar AR, Sonde RR, Pant KK (2023) One-pot conversion of glucose to 5-hydroxymethylfurfural under aqueous conditions using acid/base bifunctional mesoporous silica catalyst. Renew Energy 212:97–110. https://doi.org/10.1016/j.renene.2023.05.058
27. Gupta D, Rishi V, Gupta TK (2021) Synthesis of MFe_2O_4 (M: Cu, Mn Co, Ni) magnetic nanoparticles and their efficient catalytic role in nitrophenol reduction. Mater Res Innov 25(7):393–398. https://doi.org/10.1080/14328917.2020.1831145
28. Anandan S, Selvamani T, Prasad GG, M. Asiri A, J. Wu J (2017) Magnetic and catalytic properties of inverse spinel $CuFe_2O_4$ nanoparticles. J Magnetism Magnetic Mater 432:437–443. https://doi.org/10.1016/j.jmmm.2017.02.026
29. Ortiz-Quiñonez J-L, Pal U, Villanueva MS (2018) Structural, magnetic, and catalytic evaluation of spinel Co, Ni, and Co–Ni ferrite nanoparticles fabricated by low-temperature solution combustion process. ACS Omega 3(11):14986–15001. https://doi.org/10.1021/acsomega.8b02229
30. Gao Q, Sun Z (2018) Facile fabrication of uniform MFe_2O_4 (M=Co, Ni, Cu) hollow spheres and their recyclable superior catalytic activity towards 4-nitrophenol reduction. J Nanosci Nanotechnol 18(8):5645–5653
31. Palaniappan S, Chandra Shekhar R (2004) Synthesis of 7-hydroxy-4-methyl coumarin using polyaniline supported acid catalyst. J Mol Catal A: Chem 209(1):117–124. https://doi.org/10.1016/j.molcata.2003.08.006
32. Rafiee E, Fakhri A, Joshaghani M (2013) Coumarins: facile and expeditious synthesis via Keggin-type heteropolycompounds under solvent-free condition. J Heterocycl Chem 50(5):1121–1128
33. Wienhold M, Molloy JJ, Daniliuc CG, Gilmour R (2021) Coumarins by direct annulation: β-borylacrylates as ambiphilic C3-synthons. Angew Chem Int Ed 60(2):685–689. https://doi.org/10.1002/anie.202012099

34. Pandey V, Tusi Z, Tusi S, Joshi M (2012) 4-Methyl-2h-benzo [h] chromen-2-one as focal intermediates for the synthesis of 4-methyl-naphtho [1, 2-b] pyridin-2-one-1-(2′, 3′-diphenyl isoquinolinyl [1, 5-c] azoles. Chem Sci J 3:1
35. Kalaria PN, Satasia SP, Purohit VB, Avalani JR, Sapariya NH, Raval DK (2021) Microwave-assisted solvent-free rapid synthesis of 4-methyl coumarin derivatives by using silica functionalized-so3h as an effective heterogeneous catalyst. Polycyclic Aromat Compd 41(6):1212–1222. https://doi.org/10.1080/10406638.2019.1666887

Deployment of Solar Energy Concentrators Across the Globe

Roshni, Sanjyotpote, Anita Gupta, Parul Khurana, and Sheenam Thatai

1 Introduction

Due to its environmentally sustainable nature and absence of detrimental emissions, renewable energy has gained significant appeal as a viable solution to global energy. Solar energy stands out among the diverse array of renewable energy sources for its remarkable versatility, as it can be harnessed through a wide range of solar collectors, including solar concentrators. This makes solar power an excellent option for reducing environmental impact while meeting energy needs [1]. Solar energy concentrators are a form of solar technology that employs mirrors or lenses to focus sunlight onto a small area, creating a concentrated amount of solar energy that can be utilized for generating electricity. By reflecting or refracting sunlight onto a single point or line, these concentrators can generate more solar energy than traditional solar panels. As a free and abundant source of energy, solar power presents an excellent choice for industrial applications, making it a popular option across the globe. However, the adoption of solar energy concentrators depends on a range of factors, including economic feasibility, policy regulations, market demand, production capacity, and financial assistance.

To effectively compete with other energy sources, cost-reducing strategies can be adopted to make solar energy concentrators more affordable. However, despite their many benefits, the widespread adoption of solar energy concentrators faces significant challenges, including high upfront costs, technical intricacies, and manufacturing limitations. As a result, it is imperative to establish suitable policies, financial aid, and market incentives to encourage their large-scale deployment in countries

Roshni · Sanjyotpote · A. Gupta · S. Thatai (✉)
Amity Institute of Applied Sciences, Amity University, Noida 201313, India
e-mail: sthatai@amity.edu

P. Khurana
G.N. Khalsa College of Arts, Science and Commerce, University of Mumbai, Mumbai 400019, India

S. Rattan et al. (eds.), *Emerging Trends in Synthesis and Catalysis in Chemistry*, Springer Proceedings in Materials 51, https://doi.org/10.1007/978-981-97-5169-3_10

like India and elsewhere across the globe [2]. This chapter aims to provide insights into solar energy concentrators, exploring their features, methods of deployment, global usage rates, as well as the challenges they encounter. Through this analysis, we will develop a comprehensive understanding of solar concentrators, including their benefits and limitations of deployment.

2 Solar Energy Concentrators

Using mirrors or lenses, solar energy concentrators concentrate sunlight onto a small area, resulting in a highly concentrated amount of solar energy that can be used to produce electricity. This technology enhances the amount of solar energy that can be harnessed by reflecting or bending sunlight onto a single point or line. Compared to conventional solar panels, solar energy concentrators offer superior power generation efficiency due to their capacity to create higher temperatures. Concentrated solar power (CSP) uses a fluid-filled receiver to collect direct sunlight reflected through mirrors, which then produces heat for a steam turbine to generate electricity [1].

Moreover, their ability to concentrate sunlight makes them ideal for deployment in regions with inadequate sunlight. Another advantage of solar energy concentrators is their scalability, which allows for their application in a range of settings, from small residential units to large solar power plants as shown in Fig. 1.

Their versatility makes them a suitable choice for various energy requirements. Despite these benefits, the cost of solar energy concentrators remains a significant challenge, though their price is declining with increased market competition and technological advancement. Solar concentrators provide numerous advantages, leading to their wide range of applications.

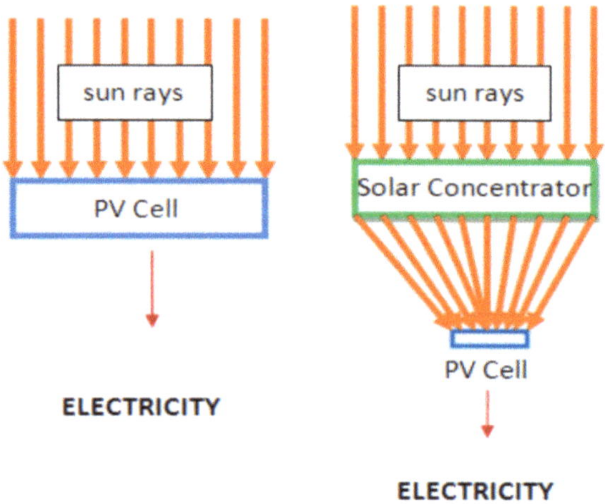

Fig. 1 Producing solar electricity using sunlight, both with and without the utilization of solar energy concentrators [3]

Benefits of Using Solar Energy Concentrators

- By focusing the energy that is available over a vast area on a smaller surface, it has the capacity to improve the intensity of solar energy.
- It helps to lower the cost of a solar power generation system by switching out a large, expensive receiver for a smaller reflecting area.
- There is a significant reduction in heat loss.
- Lower solar energy production costs and greater solar energy intensity.
- Given that the receiving area's temperature is relatively high, it is possible to obtain a thermodynamic match, which is encouraging for solar power.

Applications of Solar Energy Concentrators

- Increasing the energy on photovoltaic modules
- Heating fluids for huge electrical power plants
- Heating fluids for other uses, such as domestic hot water, food-processing facilities, hospitals, and other industrial uses.

Solar energy concentrators can be sorted into distinct categories based on varying parameters. We will explore two primary types of categorizations, one based on whether sunlight is concentrated onto a point or a line, and the other based on their optical principles.

3 Classification Based on Point or Line Concentration of Sunlight

3.1 Point Solar Concentrators

Point solar concentrators are a type of solar energy concentrator that uses mirrors or lenses to focus sunlight onto a singular point or a confined region as seen in Fig. 2. By focusing sunlight onto a small region, they are made to reach high temperatures that can be utilized to produce electricity or power a variety of industrial activities. They include heliostat field collectors (HFCs) and parabolic dish collectors (PDCs) [1].

3.1.1 Heliostat Field Collectors (HFCs)

Heliostat field collectors (HFCs) are an array of reflective mirrors that are capable of tracking the movement of the sun on two axes as shown in Fig. 3. Arranged in a specific configuration, these mirrors concentrate sunlight onto a central receiver positioned at the pinnacle of a tower. This design allows for maximum solar energy

Fig. 2 Point solar concentrator

collection and enables the production of electricity or other useful forms of energy. This technology offers numerous notable benefits, one of which is its capacity to attain exceptional conversion efficiency due to the elevated temperatures generated at the focal point of the HFC. Moreover, employing heliostats with restricted reflective surfaces results in a substantial decrease in thermal losses to the surroundings. Another advantage is the ability to couple solar energy capture with a sensible heat energy storage device [4].

Regarded as one of the most promising methods for generating solar power in the medium-scale power sector, this technology holds immense potential. With a

Fig. 3 Heliostat field collectors (HFCs)

100 MWe capacity SPT plant under construction in the USA, such systems have previously demonstrated their capability to generate 20 MWe of clean energy in Spain. The literature shows that the heat transfer fluid utilized in this procedure ranges from 150 to 2000 °C [1].

3.1.2 Parabolic Dish Collectors (PDCs)

Parabolic dish collectors (PDCs) are small-scale solar energy conversion devices that utilize a reflective surface in the shape of a parabola to concentrate and focus sunlight onto a receiver positioned at the dish's focal point as shown in Fig. 4. They are one of the concentrating solar power (CSP) systems and are typically smaller in size, ranging from 5 to 25 kWe. The largest PDC with a 489 m^2 reflective surface and a 13.4-m focal length was created and utilized in Australia [5].

Parabolic dish collectors (PDCs) offer several advantages, including the ability to be installed even on uneven ground, making them suitable for a wide range of locations. They are also well-suited for standalone and off-grid applications and can be easily integrated with thermal storage for increased efficiency. Additionally, PDCs have a higher conversion efficiency compared to other point and linear solar energy concentrator technologies. Although PDCs offer several advantages, they have some drawbacks. Compared to other solar concentrators, the initial investment and operational expenses associated with parabolic dish collectors are comparatively higher, and ensuring accurate positioning and regulation is essential to sustain optimal efficiency. The PDC system's overall thermal performance is greatly influenced by the PDC local concentration ratio (LCR) in the absorber point. Because of this,

Fig. 4 Parabolic dish collectors in Arizona, USA

the right layout of this focal point and the wise selection of its position are crucial elements. Even a slight failure in positioning and regulation can cause a significant reduction in PDC efficiency. According to information found in the literature, the temperature range for the heat fluid used in point receiver systems is reported to be between 100 and 1500 °C [1].

3.2 Line Solar Concentrators

3.2.1 Linear Fresnel Solar Reflectors (LFRs)

Linear Fresnel solar reflectors (LFRs) are a form of solar thermal technology that employs flat mirrors to reflect and concentrate sunlight onto a receiver pipe. The mirrors are arranged in a linear pattern and can be adjusted to track the movement of the sun throughout the day as shown in Fig. 5. The concentrated sunlight heats a working fluid that flows through the receiver pipe, which can then be used to generate electricity or provide thermal energy for various applications [6]. The technology is named after Augustin-Jean Fresnel, a French scientist renowned for his invention of the Fresnel lens, widely utilized in lighthouses during the eighteenth century [7]. The lens was designed to be lighter and less expensive than traditional lenses while still accurately directing light rays. Compared to parabolic mirrors, the reflective flat mirrors used in LFR solar reflectors are more economical and experience less wind-related mechanical stresses due to their flat configuration. Nonetheless, this technology suffers from a comparatively lower efficiency when compared to alternative solar concentrator types. The available literature suggests that the heat fluid used in this solar concentrator typically operates within the temperature range of 60–250 °C [1].

Fig. 5 Linear solar concentrator

Fig. 6 Parabolic trough collector

3.2.2 Parabolic Trough Collectors (PTCs)

Parabolic trough collectors (PTCs) consist of series of curved glass mirrors arranged in rows, which concentrate beam radiation onto a receiver pipe. Within the receiver pipe, a heat transfer fluid is present, which is heated by the concentrated sunlight as shown in Fig. 6. This heated fluid is then utilized to generate steam, which powers a turbine, ultimately producing electricity [8]. The cost of designing PTCs is higher than LFRs because it requires shaping glass to create the parabolic shape of the mirrors. The heat fluid used in PTCs ranges from 60 to 300 °C, according to studies in the literature [1].

4 Classification Based on Optical Principle

4.1 Reflector

Reflecting type of concentrating solar power (CSP) systems use mirrors or reflectors to focus sunlight onto a receiver. These systems are also known as mirror-based CSP systems. The reflector is designed to capture solar radiation and redirect it toward a receiver, which can then convert it into either heat or electricity.

Various types of reflector-based CSP systems exist, such as the hyperbolic concentrator, parabolic trough, parabolic dish, and compound parabolic concentrator (CPC) trough. The selection of the reflector relies on the specific purpose and the desired degree of concentration for the intended use [3].

One of the main advantages of reflector-based CSP systems is their ability to concentrate sunlight, which leads to high energy efficiency. Additionally, these systems can store thermal energy and generate electricity on demand.

4.2 *Refractor*

Refracting type of concentrating solar power systems use lenses or other transparent materials to bend or refract sunlight and focus it onto a photovoltaic (PV) cell. The PV cell is typically placed at the focal point of the lens or other refractive element, where it receives the concentrated sunlight and converts it into electricity. A commonly encountered refracting solar concentrator is the Fresnel lens, which consists of concentric rings that refract sunlight toward a central focal point. Other types of refracting solar concentrators include lenses made from acrylic or other transparent materials, as well as compound lenses that use a combination of refractive elements to achieve high levels of concentration [9].

Refracting solar concentrators have the advantage of being able to achieve high-concentration ratios with relatively simple and low-cost components. However, they are generally less efficient than reflecting concentrators, as some of the sunlight is lost due to absorption and scattering within the refractive element. Additionally, refractive elements are typically larger and heavier than reflective elements, making them less suitable for some applications [9].

4.3 *Hybrid*

Hybrid concentrating solar power (CSP) systems use a combination of both reflecting and refracting technologies to maximize the efficiency of solar energy conversion. These systems utilize both mirrors and lenses to concentrate sunlight onto a small area, which is then directed toward the PV cell. One example of a hybrid CSP system is the dielectric totally internally reflecting concentrator (DTIRC), which uses a combination of parabolic trough mirrors and refractive lenses to concentrate and direct sunlight [10]. Another example is the flat high-concentration device, which uses a combination of reflective and refractive elements to focus sunlight onto a limited region.

The advantage of hybrid CSP systems is that they can achieve higher concentration ratios and overall efficiency compared to systems that use only reflecting or refracting technology. However, the cost and complexity of these systems can be higher due to the need for both types of concentrators and the associated tracking and control systems [3].

4.4 Luminescent

Luminescent concentrating solar power systems utilize total internal reflection to direct photons toward the photovoltaic cell. In this system, a luminescent material is placed between the concentrator and the PV cell. The incoming light is absorbed by the luminescent material, which then emits light of a longer wavelength. This light is trapped within the material due to total internal reflection and directed toward the PV cell. This technology holds the promise of enhancing the efficiency of solar cells by harnessing a broader spectrum of solar energy. An example of a luminescent concentrator is the quantum dot concentrator (QDC) [3, 11].

5 Deployment of Solar Energy Concentrators

Concentrating solar power (CSP) is well-suited for regions abundant in direct solar radiation, which refers to sunlight directly emitted by the sun and reaching the Earth's surface. While non-directional radiation cannot be concentrated, it can still be harnessed for energy conversion through solar photovoltaic (PV) technology. The plant's effectiveness relies heavily on the amount of direct sunlight, known as direct normal irradiance (DNI), present at a particular location. Developers usually seek an annual direct normal irradiance (DNI) range of 1900 to 2100 kWh/m^2 as a threshold. Although CSP technology can be utilized in areas with lower values, the associated energy production costs are typically very high to justify the investment. CSP plants now come equipped with thermal energy storage to reduce energy costs and provide more flexibility in power generation. Thermal energy storage is accomplished by heating thermal oil or molten salt with the help of the solar field and storing it in tanks for several hours or even days. This allows for the generation of electricity for up to 24 h a day, provided that the solar field and storage capacity are large enough [12].

Currently, there are approximately about 6 GW of operational CSP plants worldwide, with the majority located in China, Spain, the United States, Morocco, India, and South Africa. According to reports, concentrating solar power (CSP) has the potential to produce a large amount of energy, with an estimated annual capacity of 885,000,000 terawatt-hours (TWh). In comparison, the world consumed 23,400 TWh of energy in 2018. If CSP capacity reaches 300 GW worldwide, it could potentially fulfill over 5% of the current global demand [12].

The construction of new CSP plants is moving away from traditional markets like the USA and Spain toward countries like China, Morocco, and United Arab Emirates. These markets possess the potential to leverage CSP technology and, upon achieving competitive prices, can engage in international power markets to export surplus energy supply. China has the highest installed capacity of solar energy concentrators globally, with 1651 MW as of 2021. The Chinese government has been actively promoting renewable energy sources and investing heavily in solar

Fig. 7 The largest operational concentrated solar power (CSP) plant: Noor Ouarzazate

technology research and development, resulting in a highly competitive solar energy industry.

The United States ranks second in installed capacity, with 1096 MW as of 2021, and has been promoting the use of solar power through federal incentives and state policies. Spain ranks third with 532 MW of installed capacity and has been a leader in solar energy development due to its ideal location for solar energy deployment and policies promoting renewable energy sources. India ranks fourth with 510 MW of installed capacity and has implemented policies to promote solar energy adoption, recognizing the importance of renewable energy sources to meet increasing energy demand. The largest operating concentrated solar power (CSP) complex globally, with a capacity of 510 MW of CSP and a 72 MW photovoltaic (PV) plant, is located in Morocco and is called Noor Ouarzazate as shown in Fig. 7.

Furthermore, Morocco has been granted the Noor Midelt 1 project, an 800 MW CSP-PV hybrid that will store the output from both CSP and PV as heat in molten salt tanks. The development of additional phases in the Midelt solar complex suggests the potential for additional CSP installations in Morocco [12].

In the upcoming years, it is anticipated that the global deployment of solar energy concentrators would expand as nations seek to diversify their energy portfolios and reduce their dependence on fossil fuels. Solar energy technology is anticipated to become more affordable over time, making it a more attractive choice for generating electricity. Solar energy concentrators will make a big contribution to lowering greenhouse gas emissions and battling climate change as they become more widely used.

6 SWOT Analysis of Deployment of Solar Energy Concentrators

Solar energy is becoming increasingly popular and cost-effective as countries and companies invest in advancing technologies. To promote awareness among policy makers, governments, and companies, it is important to analyze the strengths, weaknesses, opportunities, and obstacles of solar energy concentrators. SWOT analysis is a commonly used tool in strategic planning to evaluate internal and external factors that affect the success of a business or project. This analysis assists in recognizing the project's strengths and weaknesses, as well as identifying opportunities and threats in the external environment that may affect its objectives. Only specific and verifiable factors are evaluated, prioritized, and used for decision-making [13].

6.1 Strengths

(a) **High Efficiency**: Solar energy concentrators are more efficient in generating power compared to traditional solar panels. By focusing sunlight onto a small area, they can produce higher temperatures than traditional solar panels, resulting in more effective energy production.

(b) **Viable Option for Areas with Limited Sunlight Availability**: In locations with little access to sunshine, solar energy concentrators can be used. They can nevertheless produce a sizable quantity of energy by focusing the available sunlight onto a small region. They can produce more energy with less sunshine, which makes them a feasible option for locations with low sunlight availability. Additionally, thermal energy storage systems can be utilized in conjunction with solar concentrators to store and use the energy produced when there is no sunlight. This makes them a more appealing option for locations with little access to sunlight because they may still serve as a consistent source of energy during times of low sunlight.

(c) **Scalability**: Solar energy concentrators are a highly adaptable technology that can be used for both small- and large-scale energy production, making them incredibly versatile. These concentrators can be utilized in a vast array of applications, from large-scale power plants for utility companies to off-grid and portable power systems for remote locations. Their adaptability and flexibility allow solar energy concentrators to be used in many different industries, including agriculture, telecommunications, and transportation.

(d) **Renewable and Clean**: Solar energy concentrators are a renewable and environmentally friendly source of energy. To create energy, they use the sun's power, an abundant and natural resource. Solar energy concentrators generate clean energy that doesn't contaminate the environment or contribute to climate change, in contrast to fossil fuels, which when burned emit dangerous greenhouse gases into the atmosphere. They are therefore an ethical and sustainable option for supplying our energy demands while safeguarding the environment for future generations [14].

6.2 Weaknesses

(a) **High Initial Cost**: Solar energy concentrators can be more expensive to manufacture and install compared to traditional solar panels, this may limit their adoption in some areas where cost is a primary concern. The concentrator system is more complex than traditional solar panels and requires more materials, resulting in higher costs. However, as technology advances and economies of scale are achieved, costs are expected to decrease over time.

(b) **Maintenance Requirements**: Solar energy concentrators require regular cleaning and maintenance to function properly, which can be time-consuming and expensive. Solar energy concentrators require regular maintenance to keep them operating at peak efficiency. Components such as mirrors or lenses must be kept clean and aligned properly to ensure optimal performance. This can add to the overall cost of the system over time.

(c) **Reliance on Direct Sunlight**: While solar concentrators can still generate energy in areas with limited sunlight availability, their efficiency and power output are dependent on direct sunlight. This means that their performance may be affected by factors such as cloud cover, rain or shading, which can reduce their efficiency [15].

6.3 Opportunities

(a) **Government Incentives**: Governments worldwide are encouraging the use of renewable energy technologies such as solar energy concentrators by providing financial incentives in the form of tax credits, grants, and rebates, among others. These incentives make solar concentrators more affordable, making them more appealing to both households and commercial users. Furthermore, the availability of these incentives creates a favorable market environment for solar concentrator manufacturers and suppliers. They can increase production and reduce manufacturing costs by achieving economies of scale. Adopting solar concentrators can help governments achieve their renewable energy goals, reduce their reliance on fossil fuels, and minimize greenhouse gas emissions.

(b) **Job Creation**: The deployment of solar energy concentrators not only helps to produce clean and renewable energy but can also create job opportunities in various sectors like manufacturing, installation, and maintenance. As the demand for solar energy concentrators increases, the need for skilled workers in these fields is also expected to rise. According to the International Renewable Energy Agency (IRENA) report in 2018, around 34,000 people were employed in the concentrated solar power (CSP) sector globally. Additionally, during the construction phase of CSP projects, up to 20 jobs per MW are created. This indicates that the deployment of solar energy concentrators has the potential to not only support the growth of renewable energy but also boost the global job market [12].

(c) **Increasing Demand for Renewable Energy**: The increasing demand for renewable energy presents a major opportunity for solar energy concentrators. The global shift away from fossil fuels toward cleaner and sustainable energy sources due to concerns about climate change and environmental impact has created a need for solar concentrators. With the ability to generate energy from a renewable source, solar energy concentrators can play a significant role in meeting this demand. As more people and businesses seek out renewable energy options, the demand for solar concentrators is expected to increase, making them more affordable and appealing [12].

(d) **Technological Advancements**: Solar energy concentrators hold immense potential due to the ongoing technological advancements in this area. These improvements are expected to lower the manufacturing and installation costs of solar energy concentrators, making them a more affordable and accessible option for a wider range of consumers. Furthermore, as technology continues to evolve, solar concentrators will become more efficient, reliable, and long-lasting, making them even more attractive to potential customers. With reduced costs, the investment in this technology will likely increase, leading to further technological developments and improved efficiency. As result, solar energy concentrators are poised to play a crucial role in creating a cleaner and more sustainable energy future [2].

6.4 *Threats*

(a) **Competition from Other Renewable Energy Technologies**: Solar energy concentrators face competition from other renewable energy technologies such as wind and hydro power, which can limit their widespread use. Wind turbines and hydroelectric dams generate electricity by utilizing the kinetic energy of water and wind, respectively. These renewable energy sources are often installed in areas with strong water flow or wind, which could reduce the deployment of solar energy concentrators in those regions. Furthermore, because wind and hydro power have been available for a longer time, they have a track record of being efficient and dependable, making it difficult for solar energy concentrators to compete with these technologies, especially where they are well-established.

(b) **Lack of Public Awareness**: The lack of public knowledge about the advantages of solar energy concentrators poses a considerable risk to their success. Many consumers may be unfamiliar with the potential benefits of utilizing this technology, such as its ability to provide clean and sustainable energy, decrease greenhouse gas emissions, and reduce energy expenses. This lack of awareness could restrict the implementation of solar energy concentrators, especially in regions where people do not recognize the significance of renewable energy sources.

(c) **Regulatory Challenges**: Regulatory obstacles are a major threat to the widespread adoption of solar energy concentrators. Governments may implement zoning laws that restrict the construction of solar energy concentrators in certain locations or require additional permits, making their installation more difficult and time-consuming. In addition, grid connection requirements can pose a challenge, especially in areas with inadequate infrastructure to support the integration of solar energy concentrators into the power grid. These regulations can result in additional expenses and barriers for the installation and use of solar energy concentrators, limiting their implementation in certain regions [16].

In summary, the SWOT analysis of solar energy concentrators reveals their potential strengths, weaknesses, opportunities, and threats. Although solar energy concentrators offer clean and renewable energy and are becoming more affordable, challenges like energy storage and dependence on sunlight remain. Opportunities exist due to increasing demand and technological advancements, but competition from other renewable energy technologies, lack of awareness, and regulatory hurdles pose potential threats. Addressing these issues can help solar energy concentrators play an essential role in our shift toward sustainable energy.

7 Economics of Solar Energy Concentrators

The economics of solar energy concentrators are heavily dependent on factors such as the cost of manufacturing, installation, and maintenance, as well as the efficiency of the technology in converting sunlight into usable energy. While solar energy concentrators can be more expensive to manufacture and install compared to traditional solar panels, they offer higher efficiency and can generate more energy per unit area, making them cost-effective in the long term [12].

Another economic factor to consider is the potential for government incentives and subsidies. Many governments offer tax credits, rebates, and other incentives for the adoption of renewable energy technologies, including solar energy concentrators. These incentives can help to reduce the cost of installation and increase the adoption of solar energy concentrators in a wide range of settings [17].

8 Policies and Regulations

Government policies and regulations can also play a significant role in the deployment of solar energy concentrators. For example, zoning restrictions may limit the placement of solar energy concentrators in certain areas, while grid connection requirements may impose additional costs on consumers who wish to connect their solar energy concentrators to the grid [16].

But numerous countries are making efforts to encourage the use of solar energy concentrators through regulations like renewable portfolio standards, which mandate that a specific proportion of energy be produced from renewable sources. Additionally, some governments are implementing net metering policies, which allow consumers to sell excess energy generated by their solar energy concentrators back to the grid, further incentivizing their adoption [17].

9 Market Outlook of Solar Concentrators

The desire for renewable and clean energy sources is anticipated to fuel further growth in the market for solar energy concentrators in the upcoming years. As the technology behind solar energy concentrators continues to improve, their cost-effectiveness is also expected to increase, making them a more attractive option for energy production in a wide range of settings.

Additionally, as the global focus on sustainability and reducing greenhouse gas emissions continues to grow, there is likely to be increased investment in renewable energy technologies, including solar energy concentrators. This could further drive growth in the market and increase the adoption of solar energy concentrators in a wide range of applications.

Since solar concentrators reduce the carbon footprint of oil production without increasing oil consumption, they are becoming a new source of income for oil corporations. At the Coalinga Oil Field, businesses like Chevron Corp. and Bright Source Energy developed a 29-megawatt solar-to-steam facility. Other oil firms are considering turning decommissioned coal sites into concentrated solar power plants. Solar concentrators are used to harness solar energy at high temperatures in both thermal and photovoltaic applications. Solar concentrators provide thermal energy, or energy at a high temperature, which powers an engine to produce electricity.

Developing nations around the world represent a lucrative potential for solar concentrator manufacturers since the world is adopting the "Go Green" mentality and lowering the utilization of traditional energy sources, which have been affecting the environment for a very long time. Rapid technology advancements, notably in the off-grid energy storage sector, and increased consumer demand for reasonably priced thermal energy storage solutions are expected to fuel the solar concentrator market.

The market for solar energy concentrators is growing rapidly, driven by increasing demand for renewable and clean energy sources as shown in Fig. 8. While there are economic, regulatory, and policy barriers to their deployment, these challenges can be overcome through a combination of government policies, public education and outreach, and technological advancements. As the technology behind solar energy concentrators continues to improve, they are likely to become an increasingly cost-effective and attractive option for energy production in a wide range of settings.

The deployment of solar concentrators needed a significant upfront investment, and the cost of electricity per unit is often expensive, which is anticipated to restrain the market's expansion.

The global solar concentrators market is projected to experience a compound annual growth rate (CAGR) of 14.3% during the forecast period. This growth is expected to drive the market value from USD 22.28 billion in 2023 to USD 56.79 billion by 2029 [18].

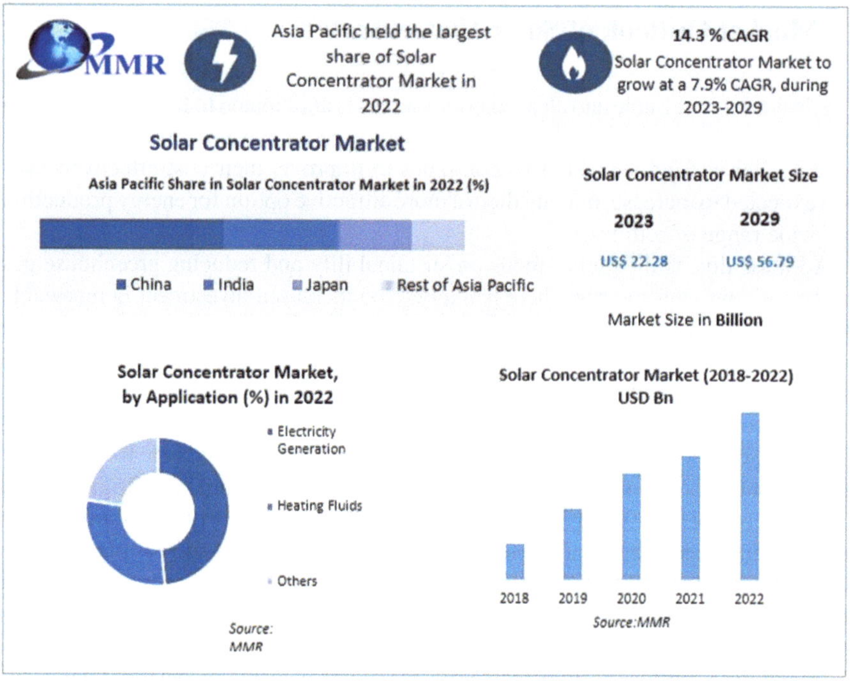

Fig. 8 Solar concentrator market overview [18]

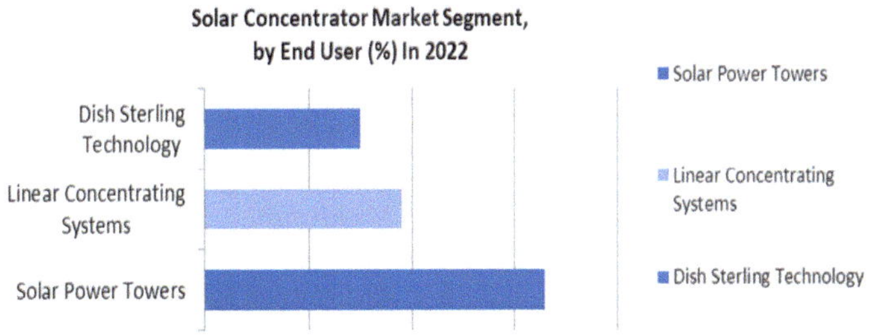

Fig. 9 Solar concentrator market segment analysis chart by end user (%) in 2022

10 Competitive Environment for Solar Concentrators

The analysis includes information about mergers and acquisitions, current developments, and strategies for several major competitors in the solar concentrator market, including Luminescent, BrightSource, Solar Reserve, Siemens AG, Reliance New Energy, and Abors Green GmbH as shown in Fig. 9.

11 Market Segmentation Research for Solar Concentrators

Solar Power Towers

A market sector with a market share of 60.1% in 2022 is anticipated to expand rapidly over the next five years. Heliostats, sometimes referred to as flat, sun-tracking mirrors, are used in solar power towers. The entire energy production process is dependent on sunshine, which emits no carbon. Molten salt energy storage, a recent trend, is useful for creating power without sunshine. The most dependable energy source today is solar power towers.

Based on End Use

Based on end use, it is predicted that the industrial category would hold the most revenue share in the market. Different businesses are encouraged to use renewable energy sources by shifting governmental laws. The market for solar concentrators is anticipated to grow more rapidly as a result. Solar concentrators find application across various industries such as electricity generation, chemical production, and paper manufacturing.

Based on Application

With a market share of over 72.1%, electricity generation is unquestionably the most prominent application-based segment. The expanding industrial area is looking for less disturbing and more affordable electrical supply options as discussed in Table 1. Due to the increased use of alternative energy sources for the generation of electricity in commercial, residential, and industrial areas, the category has been growing in the market.

Table 1 Solar concentrator market

Coverage of report	Details		
Base year	2022	Period of forecast	2023–2029
Historical data	2017–2022	2022 Market size	USD 22.28 Bn
Forecast period 2023–2029 CAGR	14.3%	2029 Market size	USD 56.79 Bn
Covered segments	By type By technology By end use By application	1. Cylindrical parabolic (2D Hubs) 2. Parabolic revolution (3D Hubs) 1. Solar power towers 2. Linear concentrating systems 1. Residential 2. Commercial 3. Industrial 1. Electricity generation 2. Heating fluids 3. Others	

12 Conclusion

Solar energy concentrators have the remarkable potential to revolutionize the global energy industry by making solar technology more affordable without compromising its performance. There is still opportunity for improvement because there are several available designs, each of which has pros and cons. For concentrated solar thermal energy, the Middle East and North Africa (MENA) region—known for its outstanding solar fields—is the best place to look. By leveraging advanced technologies, large-scale production, economies of scale, and improved operations, the cost of using solar concentrators can be significantly reduced. Solar energy, being abundant, clean, and suitable for industrial applications emerges as the optimal choice among renewable energy sources. Through the widespread deployment of solar energy concentrators, the vast potential of solar energy can be harnessed worldwide, transforming the energy landscape, and paving the way for a sustainable future.

References

1. Ghodbane M, Benmenine D, Khechekhouche A, Boumeddane B (2020) Brief on solar concentrators: differences and applications. Instrum Mesure Metrologie 19(5):371–378. https://doi.org/10.18280/i2m.190507
2. Liang D, Zhao C, Challenge and research trends of solar concentrators. [Online]. Available: www.mdpi.com/journal/energies
3. Muhammad-Sukki F, Ramirez-Iniguez R, Mcmeekin SG, Stewart B, Stewart BG, Clive B (2010) Solar concentrators LED based indoor localization view project low energy consumption

illumination sources based on novel beam shaping optics view project solar concentrators, 2010. Available: https://www.researchgate.net/publication/228348640
4. Saghafifar M, Gadalla M, Mohammadi K (2019) Thermo-economic analysis and optimization of heliostat fields using AINEH code: analysis of implementation of non-equal heliostats (AINEH). Renew Energy 135:920–935. https://doi.org/10.1016/j.renene.2018.12.060
5. Kalogirou SA (2004) Solar thermal collectors and applications. Prog Energy Combust Sci 30(3):231–295. https://doi.org/10.1016/j.pecs.2004.02.001
6. Mokhtar G, Boussad B, Noureddine S (2016) A linear Fresnel reflector as a solar system for heating water: theoretical and experimental study. Case Stud Therm Eng 8:176–186. https://doi.org/10.1016/j.csite.2016.06.006
7. "Silvi C (2009) The pioneering work on linear Fresnel reflector concentrators (LFCs) in Italy. In: Paces S (ed) 15th solar paces conference, solar power and chemical energy systems, Berlin, Germany
8. Ghodbane M, Boumeddane B (2017) Optical modeling and thermal behavior of a parabolic trough solar collector in the Algerian Sahara. Model, Measur Control B 86(2):406–426. https://doi.org/10.18280/mmc_b.860207
9. Lorenzo E, Luque A (1982) Comparison of Fresnel lenses and parabolic mirrors as solar energy concentrators
10. Ning X, Winston R, O'Gallagher J (1987) 'Dielectric totally internally reflecting concentrators'. Appl Opt 26(2):300–305
11. Gallagher SJ, Norton B, Eames PC (2007) Quantum dot solar concentrators: electrical conversion efficiencies and comparative concentrating factors of fabricated devices. Solar Energy 81:813–821
12. Concentrating solar power clean power on demand 24/7, 2020. Available: www.worldbank.org
13. Guangul FM, Chala GT (2019) Solar energy as renewable energy source: SWOT analysis. In: 2019 4th MEC International conference on Big Data and smart city, ICBDSC 2019, Institute of Electrical and Electronics Engineers Inc. https://doi.org/10.1109/ICBDSC.2019.8645580
14. Foroudastan SD, Dees O, Solar power and sustainability in developing countries
15. Bethea RM, Barriger MT, Williams PF, Chin S (1981) Environmental effects on solar concentrator mirrors
16. Sharma NK, Tiwari PK, Sood YR (2012) Solar energy in India: strategies, policies, perspectives and future potential. Renew Sustain Energy Rev 16(1):933–941. https://doi.org/10.1016/j.rser.2011.09.014
17. Timilsina GR, Kurdgelashvili L, Narbel PA (2012) Solar energy: markets, economics and policies. Renew Sustain Energy Rev 16(1):449–465. https://doi.org/10.1016/j.rser.2011.08.009
18. Solar Concentrator Market: Global market growth by type, technology, end use, application and region forecast (2023–2029)

Analyzing of 6-APA for Synthesis of Amoxicillin

Ankita Sharma, Sanjeev Thakur, Tejendra K. Gupta, Parul Khurana, and Sheenam Thatai

1 Introduction

The 6-APA (6-Aminopenicillanic acid) is preserved in tightly closed containers in pharmaceutical industry, stored at a temperature of 10 °C. It appears almost white crystalline powder. Research shows that the maximum percentage of penicillin is produced directed toward 6-APA (6-Aminopenicillanic acid) production [1, 2]. The manufacturing of 6-APA (6-Aminopenicillanic Acid) can take place with the help of enzymatic hydrolysis of penicillin as well as acylases, which leads to a more systematic process than the chemical process of 6-APA (6-Aminopenicillanic Acid). It is a pharmaceutical product that uses a source of intermediates so that it can be used as semisynthetic antibiotics, for example, amoxicillin, ampicillin, and many more. It has been found that raw material of 6-Aminopenicillanic acid is now just easily conversion of penicillin into raw 6-APA (6-Aminopenicillanic acid) in a simple way by just recirculate an aqueous penicillin solution through a shallow bed which comprises penicillin amylase catalyst in certain temperature and pH. The three main methods to modify Penicillin G to 6-aminopenicillanic acid are—enzymatic hydrolysis, biosynthesis, and chemical hydrolysis. The Penicillin acylase was first initiated from chrysogenum Q176 penicillin as mentioned by Sakaguchi and Marao in 1950 [3–5]. It is an important industrial enzyme used to produce 6-APA (6-Aminopenicillanic acid). It is also an intracellular enzyme which is produced by several micro-organisms such as yeast, fungi, and bacteria. Penicillin acylase is kept under some conditions where it

A. Sharma · T. K. Gupta · S. Thatai (✉)
Department of Applied Chemistry, Amity University, Noida 201303, India
e-mail: sthatai@amity.edu

P. Khurana
G. N. Khalsa College of Arts, Science and Commerce, University of Mumbai, Mumbai 400019, India

S. Thakur
Vardhman Ltd, Derabassi, India

S. Rattan et al. (eds.), *Emerging Trends in Synthesis and Catalysis in Chemistry*, Springer Proceedings in Materials 51, https://doi.org/10.1007/978-981-97-5169-3_11

acquires soluble form so that conversion of penicillin to 6-APA (6-Aminopenicillanic acid) takes place [6, 7]. The active acylases against penicillin are grouped into three such as Penicillin V acylases, Penicillin G acylases, and Ampicillin acylase which are further used in the process. The most used acylases for commercial use are Penicillin G and Penicillin V. Some biologically inactive products were detected when the fermentation of penicillin chrysogenum was carried out by Kato in 1953 and hydrolysis took place with the help of penicillinase. There is a direct method of fermentation of 6-APA (6-Aminopenicillanic acid) where penicillin strain grown is defined and it was originally adopted for the production phenethicillin [8–10]. The semi-synthetic penicillin conversion takes place by the isolation of 6-APA pure followed by the mechanism of extraction complex procedures from the crude broth. The biosynthetic route to 6-APA (6-Aminopenicillanic acid) is antiquated from a commercial viewpoint, but their overall fermentation was low and cost productions were high. Most common fermentation contrast of over last thirty years are Penicillin G and Penicillin V which were successfully developed. There are some necessary conditions for efficient fermentation to take place such as balance of true nutrients and growth factors with a sufficient supply to precursor side chain in the cultural medium. Other conditions are to provide efficient aeration to support reactions such as biosynthetic as well as respiratory and to regulate proper pH and temperature for growth in the medium culture [9, 10]. Initial methods were done for 6-APA (6-Aminopenicillanic acid) where it transformed directly into natural penicillin by means of hole plate method in which mechanism is carried out within holes in an agar plate. During the formation of Penicillin G, phenyl acetic acid (PAA) is added into the fermentation result and it has been discovered that it also being accepted as precursor by the mold to fit in the side chain of Penicillin G. The availability of natural Penicillin G and Penicillin V is provided at very low rates and are reliable sources as raw material for conversion into 6-APA (6-Aminopenicillanic acid) which further is required for the synthesis of semi-synthetic penicillin [11–13]. Recently, there has been increased stress on impurity profiling of APIs (active pharmaceutical ingredients). According to ICH (International council for Harmonisation of technical requirements for pharmaceuticals for human use) guidelines, an impurity of drug substance is not a part of chemical entity and impurities are unavoidable and will be present in minor amounts in any kind of drug to follow permissible limits of impurities, to launch a drug product into the market [14].

2 Experimental Procedure

The solution was prepared for testing of impurities in 6-APA by adding 25 g raw material of 6-APA and 100 mL water in round bottom flask. Then keep the stirrer on for about 2–3 min. The pH should be maintained at 7.5 by adding liquid ammonia into the solution. Now the solution was turned out from white to colorless, and the temperature must be at 10–15 °C throughout the process by keeping ice outside of the round bottom flask. Again, pH should be maintained at 7.53 with the help of ammonia. Then, add 30 g enzyme into the solution while keeping the stirrer on. Immediately, add ester of about 24 mL into the solution. There was a sudden increase in the temperature, but pH started decreasing. After 45 min, ester is again added into the solution of about 24 mL. Maintained the temperature at 20 °C by removing the ice. The solution was slowly converted into powder. The last step was to add ester (24 mL) again into the solution. It takes 2–4 h to prepare the solution for filtration. Then, the enzyme was separated with the help of filtration. The powder was kept in oven for 6–7 h for drying completely at 35 °C.

3 Results and Discussion

Carryout the identification test (by HPLC) to follow the procedure (assay test by HPLC) and maintain the retention time of the test sample and the standard sample. To perform mobile phase, add 6.8gms KH_2PO_4 in 1000 mL water into the beaker and maintained the pH 6.0 with the help of NaOH. Keep the beaker for stirring and it had been known as buffer solution. Add 800 mL of buffer solution and 200 mL of Acetonitrile in the ratio of 4:1. Sonicate the solution for about 5 min.

Chromatographic Conditions:

Mobile phase—0.05 M Phosphate, Buffer: Acetonitrile

Flow—1.0 ml/min

Detection—UV at 210 nm

Injection volume—20 μl

Run time—20 min

6-APA STD RT 2.60 min.

Preparation of Standard Solution:

- Weighed accurately about 10 mg each of phenyl acetic acid (PAA) and Penicillin G. Potassium working standard and then mix up into the volumetric flask.
- Dissolve the upper solution with mobile phase and finally make up the volume up to the mark.
- Dilute the solution with 5 to 50 mL solution of 100 mL volumetric flask and add the volume with the help of mobile phase.

For Sample Solution:

Weighed 40 mg of sample accurately and transferred into a volumetric flask of 100 mL. Then, added mobile phase into the solution and finally make up the volume up to mark with mobile phase.

Mobile Phase as Blank:

Table 1 Specification of PAA and PEN-G-related substance by HPLC test

Test	Specifications
Individual unknown impurity	Not more than 1.00% w/w
Phenyl acetic acid	Not more than 0.50% w/w
Penicillin G potassium	Not more than 0.50% w/w

- Equilibrate the column by running the mobile phase for about 30 min. Inject 20 μL of blank solution in chromatographic system and record the chromatogram.
- Inject 20 μL each of standard solution and test solution in chromatographic system and record the chromatogram.
- The resolution between 6-APA Peak and phenyl acetic acid peak is not less than 4.0. Record the peak area of individual impurity. Table 1 indicates specification of PAA and PEN-G-related substance by HPLC test.

Only impurity greater than 0.1% to be prepared. If any peak is observed in the blank chromatogram, then subtract the area of peak from corresponding area of peak in the sample chromatogram and calculate percentage area of the impurity.

All the samples were tested on HPLC to check the impurities. Purging is to be done for 5 min. Integration is done to get the major peaks as shown in Figs. 1, 2 and 3.

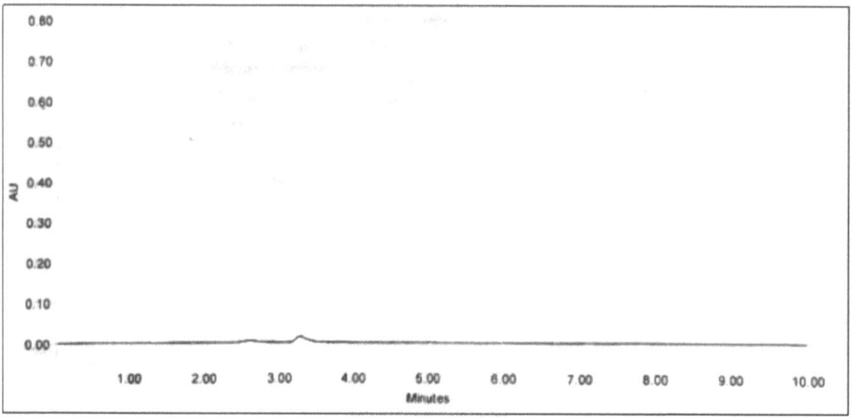

Fig. 1 Indicates blank sample solution tested in high performance liquid chromatography

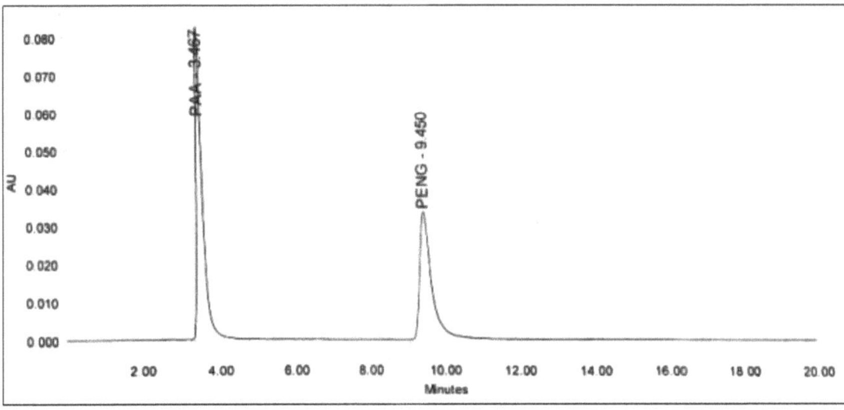

Fig. 2 Indicates standard sample solution tested in high performance liquid chromatography

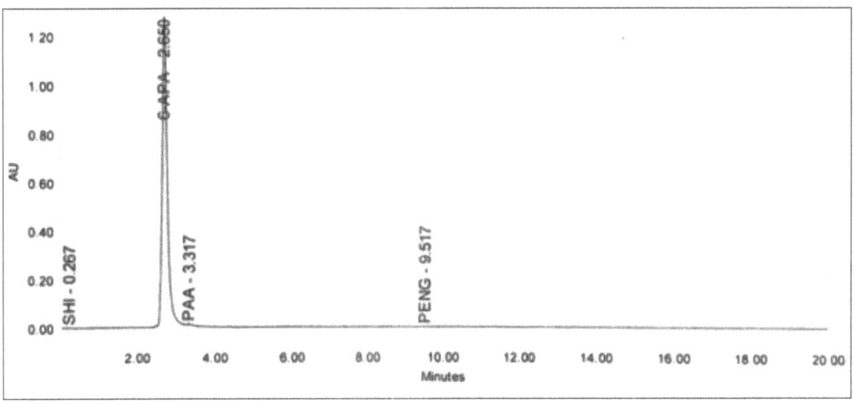

Fig. 3 Indicates an unknown sample solution tested in high performance liquid chromatography

4 Calculations

$$\text{PAA \& PEN-G Impurity (\% w/w)} = \frac{\text{AT} \times \text{DS} \times P}{\text{AS} \times \text{DT}}$$

where

AT Area count of PAA & PEN-G impurity peak in the chromatogram
AS Area count of phenyl acetic acid or Penicillin G potassium peak in chromatogram of STD preparation
DS Dlution factor of standard solution

Table 2 Calculation of final sample solution where two impurities PAA and PEN-G % area is not more than 0.50%

	Peak name	RT (retention time)	Area	% area	Total area
1	6-APA	2.650	11,522,743	99.67	11,560,462
2	PAA	3.317	32,922	0.28	11,560,462
3	PEN-G	9.517	1997	0.02	11,560,462

DT Dilution factor of test solution
P Percentage purity of PAA and Penicillin G potassium standard

Table 2 shows the result (sample solution) calculations where both impurities PAA and PEN-G % Area is not more than 0.50%. Hence, it is proved that impurities are in its minor form and 6-APA can be used further for the production of antibiotics such as amoxicillin.

5 Conclusion

The study is based on the impurities shown in the 6-Aminopenicillanic acid commonly known as 6-APA. The highly unwanted chemicals found, which are commonly known as impurities in the raw material which must be minimized otherwise can cause huge impact in the pharmaceutical industry by lowering the shelf life of related substances, can cause problems in formulations as well as can cause health issues. Furthermore, 6-Aminopenicillanic acid (6-APA) can be used further for the formation of antibiotics in the pharma industry. Testing of 6-APA in HPLC (high performance liquid chromatography) method experimentally, two peaks of PEN-G and PAA are generated where their specifications should not more than 0.50% w/w to achieve the safety measure of a drug in pharmaceutical industry and hence can further used for the synthesis of amoxicillin.

References

1. Anupama P, Harish K, Marwaha SS, Kennedy JF (2000) Advances in enzymatic transformation of penicillins to 6-aminopenicillanic acid (6-APA). Biotech Adv, 289–301
2. Avnish K, Monika A, Krishn G, Vinod S (2017) Microbial production of enzymes. Phys Sci Rev 7
3. Fatih E, Hasniye Y (2022) Determination of amocixillin trihydrate impurities 4-HPG and 6-APA by means of ultraviolet spectroscopy, 7037
4. Rolinson GN, Geddes AM (2007) The 50th anniversary of the discovery of 6-aminopenicillanic acid (6-APA). Int J Antimicrob Agents 29:3–8
5. Uri J, Sztaricskai F (1961) Simple quick methods for detection of 6-aminopenicillanic acid. Nature 4794

6. Carrington TR (1971) The development of commercial processes to produce 6-aminopenicillanic acid, 321–333
7. Geddes A, Wilcox RML (1978) Cefoxitin in abdominal sepsis. J Antimicrob Chemother, 151–153
8. Rolinson G (1998) Forty years of-lactam research. J Antimicrob Chemother, 589–603
9. Batchelor F, Doyle F, Nayler J, Rolinson G (1959) Synthesis of penicillin 6-aminopenicillanic acid in penicillin fermentations. Nature, 257–258
10. Evans M, Wilson P, Leung T, Williams J (1978) Pharmacokinetics of piperacillin following intravenous administration. J Antimicrob Chemother, 255–261
11. Ellis C, Geddes A, Davey P (1979) Mezlocillin and azlocillin: an evaluation of two new-lactam antibiotics. J Antimicrob Chemother, 517–525
12. Gang X, Qiang Z, Bin H, Jinghui Z, Fuxiang C (2018), Directed evolution of a penicillin V acylase from *Bacillus sphaericus* to improve its catalytic efficiency for 6-APA production. Enzyme Micro Tech, 1010–1016
13. Gozzard D, Geddes A, Farrell I (1982) Ceftazidime—a new extended-spectrum cephalosporin. Lancet, 1152–1156
14. Gaylor DW, Gold LS (1995) Quick estimate of the regulatory virtually safe dose based on the maximum tolerated dose for rodent bioassays. Regul Toxicol Pharmacol, 57–63

An Innovative Cation Exchange Water Treatment Method Using Zeolite Synthesized from Industrial Waste

Karan Gehlot, Richa Kasana, Lipika Sehgal, and Sangeeta Tiwari

1 Introduction

The necessity to find new and long-lasting methods of treating water has become critical due to water constraints and the increase in global demand for clean water. Recent times have seen a lot of interest in the conversion of industrial wastes into value-added products as a possible approach to solving both environmental and economical problems [1]. Zeolites have played a significant role in the development of water treatment techniques thanks to their remarkable adsorption and ion-exchange capabilities. Particularly for cation exchange in water treatment applications, zeolites made from industrial wastes like fly ash and red mud present a creative and environmentally beneficial option [2, 3].

Zeolites are acknowledged for their unique structural characteristics that allow them to selectively exchange cations. They are aluminosilicate minerals with crystalline, tiny holes. Adsorption, gas separation, and catalysis are just a few of the many businesses that can make use of these materials. Zeolites are created from industrial waste materials, which reduce their negative environmental effects while also making them valuable water filtration resources [4–7].

Red mud, a waste generated during the production of aluminum, and fly ash, a consequence of coal combustion, are two typical industrial wastes that pose significant disposal challenges [8]. There may be environmental problems from the regular landfilling of these wastes, such as soil and groundwater contamination. However, by employing inventive techniques, they can be converted into zeolites with exceptional cation exchange properties, making them a long-term choice for water treatment methods [9–12].

K. Gehlot · R. Kasana · L. Sehgal · S. Tiwari (✉)
Department of Chemistry, Amity Institute of Applied Sciences, Amity University, Noida, Uttar Pradesh, India
e-mail: stiwari2@amity.edu

S. Rattan et al. (eds.), *Emerging Trends in Synthesis and Catalysis in Chemistry*, Springer Proceedings in Materials 51, https://doi.org/10.1007/978-981-97-5169-3_12

Using zeolites generated from industrial wastes for cation exchange in water filtration has a number of important advantages [13]. First off, it encourages a circular economy while also controlling waste and purifying water by converting trash into usable items. Second, the availability of red mud and fly ash ensures an affordable and convenient supply for the manufacturing of zeolite. This makes water treatment processes more cheaply feasible, particularly in locations where traditional water purification methods could be pricy [14–19].

In the current investigation, a novel approach is being presented, to using industrial waste-derived zeolite as a potent cation exchange material in water treatment applications. Synthesis procedures to produce zeolites from red mud and fly ash will be suggested while examining the effects of various variables on the cation exchange capacity. The efficacy of these zeolites in removing cations of environmental concern will also be evaluated.

By understanding the significant function of zeolite made from industrial wastes in cation exchange for water treatment, this research work aims to improve ecologically acceptable and financially sustainable water purification systems. The findings of this study have the potential to fundamentally modify how water is treated, offering a ground-breaking solution to the global issues of water pollution and scarcity.

2 Experimental

2.1 Materials

Fly Ash

The fly ash sample used in this study was obtained from National Thermal Power Station (NTPC), Badarpur, New Delhi, India. NaOH, fumed silica (FS) and sodium aluminate (SA) from Sigma-Aldrich were used for the study. All other chemicals used were of analytical grade. Fly ash (FA) is a waste by-product containing aluminosilicate, it is produced in large amount during the coal combustion in the coal-fired power plants. Fly ash contains minerals such as quartz, mullite, and amorphous aluminosilicate. The use of coal fly ash for zeolite synthesis has environmental as well as industrial significance, and hence it can operate as a disposal technique with remarkable commercial advantages. FA-derived zeolites demonstrated good ammonium adsorption capacity and a high magnetic property.

Red Mud

The red mud sample used in this study was also obtained from National Thermal Power Station (NTPC), Badarpur, New Delhi, India. Red mud (RM) is a waste product created during the alumina production process when bauxite ores are subjected to caustic leaching. It is mineralogically distinguished by the presence of iron oxides (mainly hematite and goethite) and a trace of aluminum hydroxides (boehmite).

Table 1 Various compositions for the catalysts

S. no	Ratio of raw materials used	Sample code
1	2:3 (RM + FA)	Z1
2	1.0:0.5 (RM + FA:SA)	Z2
3	0.5:1:0.5 (FS:RM + FA:SA)	Z3

It is expected that the red mud, which contains iron-based oxides with magnetic properties, will induce magnetic behavior in the final synthetic zeolites.

Other Chemicals

Sodium aluminate and fumed silica were used in the preparation of zeolite. NaOH pellets were used for the alkaline fusion. For determining the cation exchange capacity, O.1 M HCl solution and 0.1 M NaOH were used. A phenolphthalein indicator was used for titration.

Synthesis

The numerous literature research offered a clear understanding of the synthesis step to be followed for zeolite synthesis. Fly ash (FA) and red mud (RM) were used for this synthesis. The synthesis of zeolite involves two steps, alkaline fusion of fly ash and red mud followed by hydrothermal treatment.

Alkaline Fusion of FA and RM

The first step in zeolite synthesis includes fusion of fly ash and red mud with NaOH pellets. Fly ash and red mud were grounded with NaOH pellets to a fine powder. Fumed silica and sodium aluminate were added in different proportions and thoroughly mixed to homogeneous mixture. One sample was prepared with only red mud and fly ash mixture without the addition of FS or SA. The various combinations of FA, RM, FS, and SA prepared are detailed as below in Table 1.

The above mixtures were fused at 600 °C for 3 h in a Muffle furnace.

Hydrothermal Treatment

The fumed mass obtained from alkaline fusion step was allowed to cool and it was further blended to produce a fine powder. Distilled water is added to the fumed mass in different solid-to-water ratios. The purpose of adding different solid-to-water ratio is to analyze the effect of water in determining the size of crystals of zeolite during aging. The fused fly ash and red mud mixture was allowed to age with constant stirring in a magnetic stirrer. The resulting slurry was then transferred to an autoclave lined with Teflon for hydrothermal treatment. The resulting reactive mass was filtered, rinsed several times with distilled water to remove excess alkali, and then dried at room temperature. All three samples were prepared under the same hydrothermal conditions. Zeolite samples were prepared without any structure-directing template.

Iron-Coated Zeolite

Iron-coated zeolite was synthesized by the following method. In total, 10 g zeolite was measured and transferred into a beaker and distilled water was added to it, and the mixture was stirred for 15 min to wet the zeolite and obtain a homogeneous suspension. To this, $FeCl_3$ solution was added, and it was again stirred for 40 min minutes on magnetic stirrer. After this 2.5 M NaOH was added and constantly stirred for 30 min. Now, the mixture was filtered, and the synthesized product was washed with distilled water and air-dried for 2–3 days.

3 Results and Discussion

XRD analysis was performed for the two samples Z2 and Z3. As shown in Fig. 1, both samples were identical to Zeolite Y. The distinctive diffraction pattern with particular values that indicated Zeolite Y's crystalline planes, especially (1 1 1) and (2 2 2), was used to confirm the presence of the mineral. The strong, intense peaks demonstrated the great crystallinity of Zeolite Y, demonstrating the substance's potential for water treatment and catalysis. In general, the presence of zeolite Y in the XRD analysis is indicated by the (2 2 0) and (1 1 0) peaks at roughly 6.3 degrees and 7.3 (2 theta), respectively. And, average zeolite size calculated using Scherrer equation was found to be ~ 600 nm for Z2.

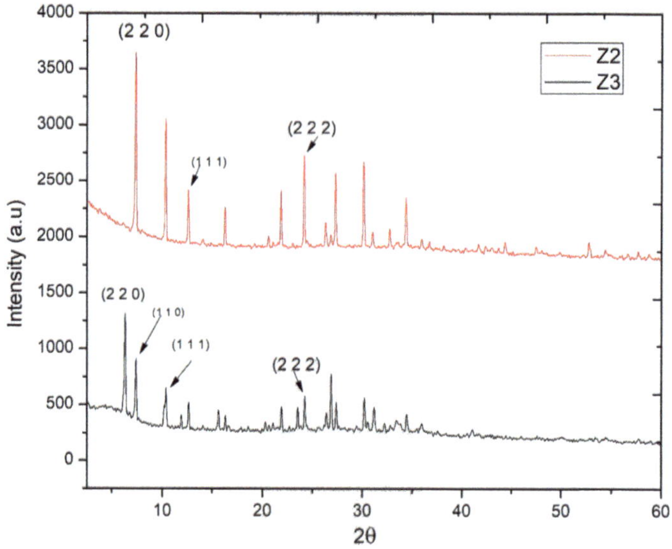

Fig. 1 XRD pattern for zeolite synthesized from RM + FA + SA (Z2) and RM + FA + SA + FS (Z3)

Three different samples were synthesized by using industrial wastes by varying the Si/Al ratio in each of them. The synthesized zeolites from fly ash and red mud are cubical shaped and not spherical particles which is the shape of raw fly ash particles. Figure 2 indicates SEM images of Z1, Z2, and Z3 samples reveals ultra-fine particles having typical cubic morphology which is associated with Zeolite type A. The size of the Z1 particles is observed to be ranging from 1.2 to 1.5 μm. SEM results confirm that alkali fusion prior to the hydrothermal treatment helps in the dissolution of the quartz and mullite phases present in Fly ash. SEM images for Z1 at different resolutions are illustrated below.

The SEM images of zeolite A synthesized from fly ash and red mud is shown in Fig. 3. The SEM images of each sample synthesized contained lumps of cubic shape. The morphology is type of zeolite A which is reported in the literature. The spherical fly ash particles were able to dissociate in the presence of NaOH and recrystallized to appear as regular cubic-shaped zeolite particles during hydrothermal method. The SEM images in Fig. 3. Below illustrated shows particle size ranging from 300 to 800 nm.

In Fig. 4, SEM images of sample Z3 synthesized also exhibit cubic morphology and the type of zeolite formed is Zeolite A. The particle size is ranging between 1 and 3 μm.

Fig. 2 SEM images of zeolite prepared from FA + RM (Z1)

Fig. 3 SEM images of zeolites prepared from FA + RM + SA (Z2)

Fig. 4 SEM images of Zeolite prepared from FA + RM + FS + SA (Z3)

SEM images of iron-coated zeolite. From SEM images, it could be interpreted that iron is deposited on the cubic structure of zeolite.

The fly ash consists of SiO_2, Al_2O_3, and the main constituent of red mud is Fe2O3. The percentage composition of each element present in Z1, Z2, and Z3 are given below (Table 2).

Investigation gave vital proof of zeolite's effectiveness in chromium absorption for water treatment applications. The favorable adsorption isotherms and observed decrease in absorbance in the UV–Vis spectra depicted in Fig. 5. It show that zeolite has the potential to be a successful adsorbent for chromium removal. After including zeolite as the adsorbent, we observed a significant reduction in the intensity of the chromium absorption peaks in the UV–Vis spectra [20–23]. the fact that there was a sizable decrease in absorbance following the adsorption process shows that zeolite was successful in removing chromium ions from the aqueous solution. The findings emphasize zeolite's potential as an effective instrument for chromium removal in water treatment and demonstrate its capacity for adsorption.

The specific surface area identified by the BET analysis provides invaluable insights into the quality and porosity of the produced zeolite. A porous material has a higher surface area, which indicates a greater potential for adsorbing cations and pollutants from water. Furthermore, a detailed BET analysis enables researchers to accurately change the preparation conditions to produce the optimal surface area for

Table 2 Elemental composition of Z1, Z2, and Z3

Element	Z1		Z2		Z3	
	Weight %	Atomic %	Weight %	Atomic %	Weight %	Atomic %
O	50.15	64.52	51.08	64.86	51.14	65.03
Na	12.39	11.10	13.54	11.97	11.84	10.47
Al	12.46	9.51	13.56	10.21	11.57	8.72
Si	15.54	11.39	13.98	10.11	18.08	13.09
Fe	9.45	3.48	7.48	2.85	7.37	2.68
Total	100	100	100	100	100	100

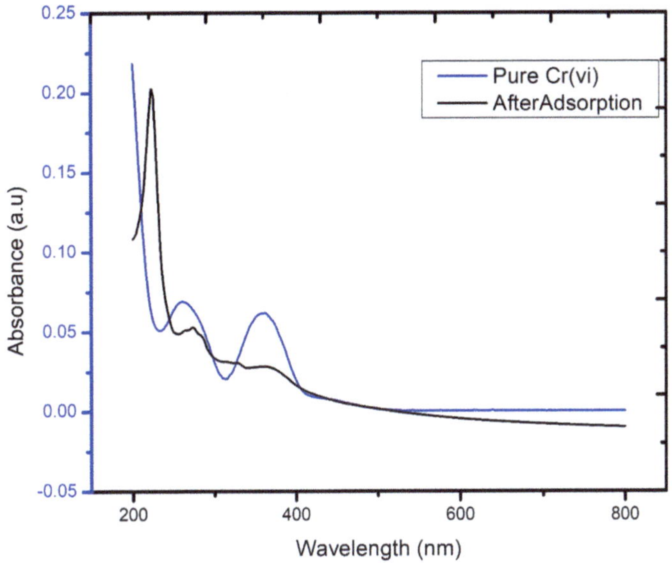

Fig. 5 UV–Vis spectroscopy for removal of Cr metal

the particular application, aiding in the optimization of the zeolite synthesis process [24, 25]. Figures 6 and 7 provide BET adsorption curves following isotherm-IV.

The particular surface area numbers from the two BET assessments could vary because of variations in the zeolite manufacturing process. BET surface area of sample Z3 was 43.2405 m^2/g. The inclusion of sodium aluminate and fumed silica may encourage the formation of zeolite crystals and produce a more porous structure when compared to zeolite made simply from red mud and fly ash. This can produce a bigger specific surface area [26].

Cation Exchange Capacity (CEC)

Cation exchange capacity is an important property which makes zeolites useful for various applications. The table depicts the cation exchange capacity of various zeolites synthesized (Table 3). Gradual increase and then decrease is predictable as Si/Al ratio leads to reorganization of crystals of Z1 offering more exchangeable sites. The CEC of Z1 was 2.38 meq/g which was decreased to 2.15 meq/g on increase of silica in Z3 [27, 28].

All zeolites should have acceptable cation exchange capabilities, however, due to its increased surface area and greater porosity, the zeolite produced from the mixture of precursors may function more efficiently. The inclusion of chemical agents may help cations enter the zeolite framework more successfully.

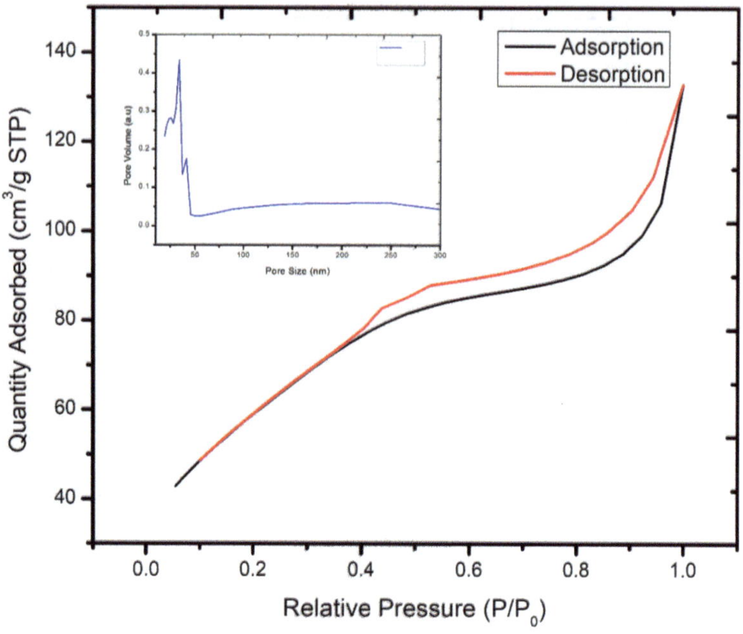

Fig. 6 BET of zeolites prepared from RM + FA + SA

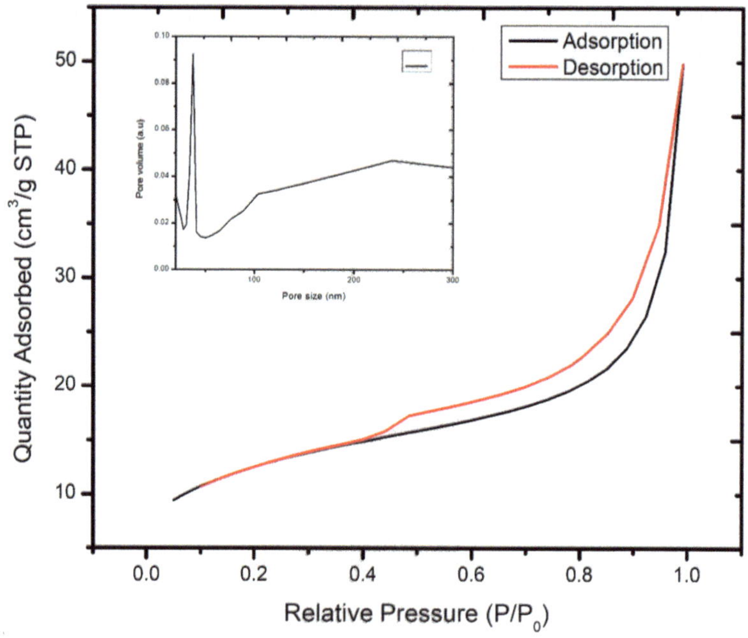

Fig. 7 BET of zeolite prepared from RM + FA + SA + FS (Z3)

Table 3 CEC for Z1, Z2, and Z3

Sample	Si/Al ratio	CEC (mEq/g)
Z1 (RM + FA)	1.24	2.38
Z2 (RM + FA + SA)	1.03	3.23
Z3 (RM + FA + FS + SA)	1.56	2.15

4 Conclusion

In conclusion, this work suggests an innovative and sustainable technique for treating water with cations utilizing zeolite produced from industrial waste. The innovative strategy demonstrated potential for lowering issues with water pollution while simultaneously enhancing waste management and environmental stability. The use of red mud and fly ash in the synthesis of zeolite, which reduces the demand for virgin raw materials by employing waste as a productive input, is one of the best examples of embracing the circular economy notion. The efficient and integrated use of resources that this strategy fosters develops the industrial ecosystem and raises environmental consciousness. A detailed investigation of the zeolite made from red mud and fly ash also reveals that it has the structural integrity, ion-exchange ability, and chemical stability that are necessary to judge its effectiveness as a water treatment medium. The performance test results show that zeolite has a remarkable ability to selectively extract cations from contaminated water, making it a feasible option for large-scale water treatment applications. As a result of using red mud and fly ash as precursors for the synthesis of zeolites, a novel approach to simultaneously tackling waste management and water pollution has been discovered. This research creates a precedent for industrial waste-based water filtration methods that are ethical, feasible, and profitable. By accepting the promise of zeolite created from red mud and fly ash, we get one step closer to creating a future where garbage is accepted as a valuable resource and water treatment becomes an integral part of a circular and ecologically conscious industrial landscape. As we advance in this sector, it is impossible to overestimate the positive effects of these discoveries on raising global sustainability standards and water quality.

References

1. Mallapur VP, Oubagaranadin JUK (2017) A brief review on the synthesis of zeolites from hazardous wastes. Trans Indian Ceram Soc 76(1):1–13. https://doi.org/10.1080/0371750X.2016.1231086
2. Ríos RCA, Williams CD, Roberts CL (2009) A comparative study of two methods for the synthesis of fly ash-based sodium and potassium type zeolites. Fuel 88(8):1403–1416. https://doi.org/10.1016/j.fuel.2009.02.012.
3. Ozdemir OD, Piskin S (2019) A novel synthesis method of zeolite X from coal fly ash: alkaline fusion followed by ultrasonic-assisted synthesis method. Waste Biomass Valorizat 10(1):143–154. https://doi.org/10.1007/s12649-017-0050-7

4. Jha B, Singh DN (2011) A review on synthesis, characterization and industrial applications of flyash zeolites. J Mater Educ 33(1–2):65–132
5. Ahmaruzzaman M (2010) A review on the utilization of fly ash. Prog Energy Combust Sci 36(3):327–363. https://doi.org/10.1016/j.pecs.2009.11.003
6. Szerement J, Szatanik-Kloc A, Jarosz R, Bajda T, Mierzwa-Hersztek M (2021) Contemporary applications of natural and synthetic zeolites from fly ash in agriculture and environmental protection. J Clean Prod 311:127461. https://doi.org/10.1016/j.jclepro.2021.127461
7. Sushil S, Batra VS (2008) Catalytic applications of red mud, an aluminium industry waste: a review. Appl Catal B Environ 81(1–2):64–77. https://doi.org/10.1016/j.apcatb.2007.12.002
8. Li Y et al (2021) Conversion of biomass ash to different types of zeolites: a review. Energy Sources, Part A Recover Util Environ Eff 43(14):1745–1758. https://doi.org/10.1080/15567036.2019.1640316
9. Tanaka H, Fujii A (2009) Effect of stirring on the dissolution of coal fly ash and synthesis of pure-form Na-A and -X zeolites by two-step process. Adv Powder Technol 20(5):473–479. https://doi.org/10.1016/j.apt.2009.05.004
10. Yang L et al (2019) Green synthesis of zeolite 4A using fly ash fused with synergism of NaOH and Na_2CO_3. J Clean Prod 212:250–260. https://doi.org/10.1016/j.jclepro.2018.11.259
11. Cundy CS, Cox PA (2005) The hydrothermal synthesis of zeolites: Precursors, intermediates and reaction mechanism. Microp Mesop Mater 82(1–2):1–78. https://doi.org/10.1016/j.micromeso.2005.02.016
12. Wang CF, Li JS, Wang LJ, Sun XY (2008) Influence of NaOH concentrations onsynthesis of pure-form zeolite A from fly ash using two-stage method. J Hazard Mater 155(1–2):58–64. https://doi.org/10.1016/j.jhazmat.2007.11.028
13. Król M (2020) Natural vs. synthetic zeolites. Crystals 10(7):1–8. https://doi.org/10.3390/cryst10070622
14. Schirmer W (1995) Molecular transport and reaction in zeolites—design and application of shape selective catalysis 191(2)
15. Belviso C et al (2018) Red mud as aluminium source for the synthesis of magnetic zeolite. Microp Mesop Mater 270:24–29. https://doi.org/10.1016/j.micromeso.2018.04.038
16. Zhao Y et al (2016) Removal of ammonium ions from aqueous solutions using zeolite synthesized from red mud. Desalin Water Treat 57(10):4720–4731. https://doi.org/10.1080/19443994.2014.1000382
17. Yuna Z (2016) Review of the natural, modified, and synthetic zeolites for heavy metals removal from wastewater. Environ Eng Sci 33(7):443–454. https://doi.org/10.1089/ees.2015.0166
18. Belviso C (2018) State-of-the-art applications of fly ash from coal and biomass: a focus on zeolite synthesis processes and issues. Prog Energy Combust Sci 65:109–135. https://doi.org/10.1016/j.pecs.2017.10.004
19. Soni R, Shukla DP (2019) Synthesis of fly ash based zeolite-reduced grapheme oxide composite and its evaluation as an adsorbent for arsenic removal. Chemos 219:504–509. https://doi.org/10.1016/j.chemosphere.2018.11.203
20. Belviso C et al (2015) Synthesis of magnetic zeolite at low temperature using a waste material mixture: fly ash and red mud. Microp Mesop Mater 202(C):208–216. https://doi.org/10.1016/j.micromeso.2014.09.059
21. Aldahri T, Behin J, Kazemian H, Rohani S (2016) Synthesis of zeolite Na-P from coal fly ash by thermo-sonochemical treatment. Fuel 182:494–501. https://doi.org/10.1016/j.fuel.2016.06.019
22. Ren X et al (2020) Synthesis of zeolites from coal fly ash for the removal of harmful gaseous pollutants: a review. Aerosol Air Qual Res 20(5):1127–1144. https://doi.org/10.4209/aaqr.2019.12.0651
23. Querol X et al (2002) Synthesis of zeolites from coal fly ash: an overview. Int J Coal Geol 50(1–4):413–423. https://doi.org/10.1016/S0166-5162(02)00124-6
24. Hu T, Gao W, Liu X, Zhang Y, Meng C (2017) Synthesis of zeolites Na-A and Na-X from tablet compressed and calcinated coal fly ash. R Soc Open Sci 4(10). https://doi.org/10.1098/rsos.170921

25. Rath PC, Misra VN, Paramguru RK (2004) Mineral processing and extractive metallurgy review: an international journal. Trends Red Mud Utilizat—A Rev, May 2013
26. Gamero-Vega KY, Medina-Ramírez A, Khamkure S, Orozco-Núñez SI, Aguilera-González EN, Gamero-Melo P (2019) Upscaling of W zeolite direct synthesis from coal fly ash and its water adsorption capacity. J Chem Technol Biotechnol 94(11):3479–3487. https://doi.org/10.1002/jctb.5922
27. Mgbemere HE, Ekpe IC, Lawal GI (2017) Zeolite synthesis, characterization and application areas: a review. Int. Res. J. Environ. Sci. 6(10):45–59
28. Yoldi M, Fuentes-Ordoñez EG, Korili SA, Gil A (2019) Zeolite synthesis from industrial wastes. Microp Mesop Mater 287:183–191. https://doi.org/10.1016/j.micromeso.2019.06.009

Chlor-Alkali Waste in the Production of Wall Tiles: A Sustainable Approach

Charu Tyagi, Tejendra K. Gupta, Parul Khurana, Aarti Sharma, Arvind Singh, and Sheenam Thatai

1 Introduction

The production of wall tiles plays a crucial role in the construction and interior design industries, providing aesthetically pleasing and functional surfaces for homes, offices, and public spaces. As the demand for tiles continues to grow, manufacturers face the challenge of meeting this demand while simultaneously addressing environmental concerns. One particular issue is the generation and management of waste during the tile production process. Among the various waste materials encountered in this industry, chlor-alkali waste stands out as a significant by-product that requires effective handling and disposal. However, by exploring innovative approaches and utilizing this waste, manufacturers can not only mitigate their environmental impact but also contribute to a more sustainable and resource-efficient tile production process. A by-product of the manufacturing of chlorine and caustic soda, two essential chemicals used in numerous industrial applications, is chlor-alkali waste. This garbage has often been handled as a hazardous substance and is subject to strict disposal rules. However, as businesses everywhere work to adopt more sustainable practises, a paradigm change has taken place. This change has made it possible to use waste chlor-alkali as a valuable resource in a variety of manufacturing processes, including the manufacture of wall tiles [1, 2].

There are various environmental advantages to using chlor-alkali waste in the production of tiles. The first benefit is that it lessens the amount of garbage that the

C. Tyagi · T. K. Gupta · S. Thatai (✉)
Amity Institute of Applied Sciences, Amity University, Sector-125, Noida 201313, India
e-mail: sthatai@amity.edu

P. Khurana · A. Singh
G. N. Khalsa College of Arts, Science and Commerce, University of Mumbai, Mumbai 400019, India

A. Sharma
Shri Ram Institute of Industrial Research, Delhi, India

S. Rattan et al. (eds.), *Emerging Trends in Synthesis and Catalysis in Chemistry*, Springer Proceedings in Materials 51, https://doi.org/10.1007/978-981-97-5169-3_13

chlor-alkali sector produces overall, easing the strain on landfills and waste management facilities. Manufacturers can help preserve natural resources and lessen the impact of waste management on the environment by diverting this garbage from traditional disposal techniques [3].

Second, the use of leftover chlor-alkali in the manufacture of wall tiles promotes raw material conservation. The usage of virgin raw materials like clay, silica, and feldspar can be partially replaced by the incorporation of this waste into the tile manufacturing process. As the recycled waste can partially replace them, this not only lowers the demand for these resources but also increases their availability. By encouraging resource efficiency and minimizing the reliance on the extraction and manufacturing of new raw materials, this strategy promotes the tenets of a circular economy.

Additionally, the utilization of chlor-alkali waste in the production of tiles can help cut down on energy use and greenhouse gas emissions. In general, producing tiles from recycled waste materials uses less energy than mining and processing raw materials from virgin sources. Manufacturers can drastically reduce their energy needs by using chlor-alkali waste, which results in a lower carbon footprint and a more sustainable production process. This is in line with efforts being made around the world to tackle climate change and move towards a low-carbon economy. In addition to the environmental advantages, the incorporation of chlor-alkali waste in wall tile production can also have economic benefits. By utilizing this waste, manufacturers can potentially reduce their production costs, as recycled materials are often more cost-effective compared to virgin raw materials. This cost savings can be passed on to consumers, making sustainable wall tiles more accessible and appealing in the market. Moreover, the adoption of sustainable practices can enhance a company's reputation, attracting environmentally conscious customers and boosting brand value.

The "brine" is a solution with an extraordinarily high concentration of salts, such as sodium chloride, which can occur either naturally (as with saltwater, deep-water ocean pools, salt lakes, etc.), artificially (as with industrial processes, etc.), or as an industrial by-product. The by-products and residues created during the electrolysis of brine (sodium chloride solution) in chlor-alkali facilities are referred to as chlor-alkali waste. The main waste streams include wasted electrolyte, mercury waste, and chlor-alkali waste (in older facilities). These waste products are potentially detrimental to ecosystems and human health because they contain a variety of contaminants, including salts, heavy metals, and chlorinated organic compounds. Improper disposal or release of chlor-alkali waste can lead to severe environmental consequences. When discharged into water bodies, the high salt content of the waste can disrupt the aquatic ecosystem, affecting aquatic organisms and plant life. Additionally, the release of chlorinated organic compounds may contribute to water pollution, posing risks to human health and aquatic biodiversity. Furthermore, mercury waste, a significant concern in older chlor-alkali plants, can lead to widespread contamination if not handled correctly. Mercury is a toxic heavy metal that bio accumulates in the food chain, posing risks to humans through the consumption of contaminated seafood. It can also evaporate into the atmosphere, contributing to air pollution and further contamination [2–4].

An inventive and sustainable option for the tile manufacturing business is the use of chlor-alkali waste in the creation of wall tiles. Manufacturers can lessen their environmental effect, conserve natural resources, and generate less trash by diverting this material from conventional disposal techniques and incorporating it into the production process. The use of chlor-alkali waste also has financial advantages, such as cost savings and enhanced brand recognition. The incorporation of chlor-alkali waste in the production of wall tiles is a significant step towards a more resource-aware and ecologically conscious sector as the demand for sustainable construction materials rises.

Sustainable management techniques are essential to reduce the hazards that chlor-alkali waste poses to the environment and human health. This problem has been addressed using a variety of strategies, waste minimization, treatment and neutralization, resource recovery, secure storage and disposal, and regulatory framework.

Chlor-alkali waste poses significant environmental and health hazards due to its complex composition of pollutants (Al_2O_3: 5.22%, Fe_2O_3:5.5%, CuO: 0.05%, ZnO: 0.03%, K_2O_3: 0.31%, MgO: 8.757%, BaO: 40.3%, SiO_2: 9.16%, Cl: 5.3%, SO_3: 12.32%, CaO: 12.32%). However, through sustainable management practices, these risks can be mitigated. Waste minimization, treatment and neutralization, resource recovery, secure storage, and a robust regulatory framework are key elements in ensuring the safe and sustainable management of chlor-alkali waste. By adopting these practices, the chlor-alkali industry can reduce its environmental footprint, protect ecosystems, and safeguard human health for a more sustainable future.

Walls can be enhanced with texture and visual interest by using decorative wall tiles. These tiles can be used to make a wide range of beautiful patterns and designs because they are available in a variety of forms, sizes, and materials. Decorative wall tiles are frequently made of ceramic, porcelain, glass, stone, and metal. Each of these materials offers distinct functional and aesthetic qualities like toughness, moisture resistance, and cleaning simplicity.

Decorative wall tiles can be used in a variety of interior and exterior settings, such as bathrooms, kitchens, living rooms, and outdoor spaces. They can be used as accents, borders, or full-wall installations, and can be arranged in various patterns and layouts to create different design effects. When selecting decorative wall tiles, it is important to consider the size and layout of the space, as well as the style and colour scheme of the surrounding decor. It is also important to choose tiles that are appropriate for the specific application, such as tiles that are rated for use in wet areas for bathroom installations. Decorative wall tiles are an attractive and adaptable way to add texture and visual interest to walls in any setting.

Wall tiles come in a variety of styles and varieties. Several of the most common varieties of wall tiles include ceramic tiles, porcelain tiles, glass tiles, stone tiles, metal tiles, vinyl tiles, and mosaic tiles.

Overall, the type of wall tile that is best for you will depend on your personal style, the type of space you are designing, and the durability and practicality of the tile. Chlor-alkali waste is a waste material generated by the desalination process. It is a by-product of the treatment of saline water to remove dissolved salts and other impurities. The disposal of chlor-alkali waste is a major environmental concern, as it can lead to soil and water pollution. However, chlor-alkali waste can be utilized effectively in various applications, including the making of wall tiles.

It is possible to make decorative wall tiles from chlor-alkali waste, but it would require a specialized process and equipment. Chlor-alkali waste is a by-product of the chlor-alkali industry, which produces chemicals like chlorine and caustic soda. The waste contains a high concentration of salt and other minerals that can be harmful to the environment if not properly disposed of.

Izidoro et al. prepared brine sludge from chlor-alkali waste and used in the cement materials for construction material products [5]. Hossain et al. has discussed about the use of harmful industrial waste materials for ceramic marketable value added products [6]. A study published in the Journal of Building Engineering examined the use of chlor-alkali waste in the production of self-compacting concrete (SCC) tiles. The study found that the waste could be used as a partial replacement for cement in SCC production. The resulting tiles had a compressive strength comparable to traditional SCC tiles and had a lower environmental impact due to the reduced use of cement. The study concluded that using chlor-alkali waste in SCC production could be a sustainable and cost-effective approach to waste management [7]. Mojapelo et al. has discussed about the waste water sludge for light weight concrete and the use of waste water as a curing medium [4].

Despite the potential benefits of using chlor-alkali waste as a raw material for tile production, there are also challenges associated with this approach. One challenge is the variability in the composition of chlor-alkali waste, which can make it difficult to produce consistent and high-quality tiles. Another challenge is the potential for the waste to contain contaminants that could impact the safety and quality of the resulting tiles.

To address these challenges, several studies have proposed different production methods and quality control measures for making tiles from chlor-alkali waste and other industrial wastes. In a study, researchers has used the wet milling process to produce a homogeneous mixture of chlor-alkali waste and other raw materials. The resulting mixture was then dried and fired in a kiln to produce tiles with a consistent texture and colour. Another study published in the Journal of Hazardous Materials proposed using a high-temperature incineration process to remove contaminants from the waste prior to use in tile production [6, 8, 9].

The utilization of chlor-alkali waste in the production of wall tiles is a relatively new and emerging field, and research and studies on this specific topic may be limited. However, the concept of incorporating industrial waste materials into tile production processes and other construction materials has gained attention in recent years due to the growing importance of sustainable and environmentally friendly practices.

In general, the literature on waste utilization in tile production highlights the potential benefits of incorporating recycled materials, including chlor-alkali waste. Some studies have explored the possibility of using waste materials as partial replacements for traditional raw materials, such as clay, silica, and feldspar. By doing so, the waste materials can serve as alternatives that reduce the demand for virgin resources and decrease the environmental impact associated with their extraction and processing.

One key aspect that researchers have focussed on is the characterization and evaluation of chlor-alkali waste and its suitability for tile production. This involves assessing the physical, chemical, and mineralogical properties of the waste material to determine its compatibility and performance in the tile manufacturing process. The waste material needs to meet certain quality standards and should not adversely affect the final product's properties or aesthetics.

Studies have also investigated the effects of incorporating chlor-alkali waste on the properties of the tiles, such as their mechanical strength, water absorption, and resistance to chemical and thermal stress. These investigations are crucial to ensure that the tiles produced from the waste material meet the required standards and performance criteria for various applications.

Furthermore, the environmental implications and sustainability aspects of utilizing chlor-alkali waste in tile production have been examined. Life cycle assessments (LCAs) have been conducted to evaluate the overall environmental impact and resource efficiency of incorporating waste materials. LCAs consider factors such as energy consumption, greenhouse gas emissions, waste generation, and potential risks to human health and the environment.

While specific literature focussing solely on the production of wall tiles from chlor-alkali waste may be limited, related studies on waste utilization in tile manufacturing and other construction materials can provide valuable insights and guidance. It is recommended to search academic databases, journals, and conference proceedings related to sustainable construction, waste management, and ceramic materials to find more specific and up-to-date literature on the topic.

Please note that the information provided here is a general overview and may not cover the latest research developments or specific studies conducted on the production of wall tiles from chlor-alkali waste.

2 Experimental Methods

2.1 Materials Used

Chlor-Alkali Waste

Chlor-alkali waste refers to the by-products generated during the production of chlorine and caustic soda, which are essential chemicals used in various industrial processes. The chlor-alkali industry utilizes an electrolysis process, where a solution of brine (sodium chloride or saltwater) is passed through an electrolytic cell to produce chlorine gas and sodium hydroxide (caustic soda).

During this electrolysis process, chlor-alkali waste is generated as a result of the side reactions and impurities present in the brine solution. The composition of chlor-alkali waste can vary depending on factors such as the quality of the brine, operating conditions, and purification processes. Common components found in chlor-alkali waste include heavy metals, such as mercury, as well as various salts and organic compounds.

Historically, chlor-alkali waste was considered a hazardous material due to its potential environmental and health risks. Therefore, it required careful management and disposal to prevent contamination of soil, water, and air. However, with growing concerns about environmental sustainability and resource management, there has been increasing interest in finding beneficial uses for this waste rather than simply disposing of it.

The use of chlor-alkali waste into various manufacturing procedures, such as the creation of wall tiles, is a key application. While chlor-alkali waste is used in place of or in addition to typical raw materials while making tiles, it helps conserve resources and cut down on waste production. By encouraging the reuse and recycling of waste resources, this strategy adheres to the ideals of a circular economy.

There are various advantages to using chlor-alkali waste in the manufacture of tiles. The first benefit is that it lessens the environmental impact of trash management and disposal. The load on waste management systems is reduced by avoiding traditional treatment techniques like landfilling or incineration with this material. Additionally, it is possible to substitute or scale back the extraction and processing of virgin raw materials.

Secondly, incorporating chlor-alkali waste in tile production can contribute to energy savings and lower greenhouse gas emissions. Manufacturing tiles from recycled materials generally requires less energy compared to the production of tiles from virgin raw materials. By utilizing chlor-alkali waste, manufacturers can reduce their energy consumption and carbon footprint, contributing to climate change mitigation efforts.

Moreover, the utilization of chlor-alkali waste (as shown in Fig. 1) in tile production can have economic advantages. By using recycled materials, manufacturers can potentially reduce their production costs, making their products more competitive in the market. This cost-saving benefit can be passed on to consumers, making sustainable wall tiles more accessible and appealing. It's important to note that the incorporation of chlor-alkali waste in tile production requires proper testing, quality control, and adherence to environmental regulations to ensure the safety and performance of the end products. However, when implemented responsibly, the utilization of chlor-alkali waste presents an opportunity for the tile manufacturing industry to reduce waste, conserve resources, and promote a more sustainable production process.

In addition to wall tiles, chlor-alkali waste, a by-product of the chlor-alkali business, can be used to make a variety of other goods. Several instances include paper and pulp, textiles and fabrics, cleaning and disinfecting agents, pharmaceuticals and chemical manufacturing, water treatment chemicals, polyvinyl chloride (PVC), and epoxy resins.

Fillers Used in Making of Wall Tiles

Fillers are substances that are added to the base material for wall tiles to enhance its qualities, such as strength, durability, or fire resistance. The type of tile being made

Fig. 1 Chlor-alkali waste

and its intended usage will determine the filler to employ. Here are some examples of fillers that are frequently used in the manufacture of tiles:

1. **Gypsum**: A naturally occurring mineral known as gypsum is frequently utilized as a filler in the manufacture of tiles. Gypsum-based fillers are a preferred option for tiles in commercial structures, institutions of higher learning, and health-care facilities because of their superior fire resistance and acoustic insulation qualities.
2. **Mineral Wool**: Made of natural or synthetic minerals, mineral wool is a form of insulation material. It frequently serves as a filler in tiles to insulate sound.
3. **Expanded Polystyrene (EPS)**: EPS is a lightweight, rigid, and durable foam material that is commonly used as filler in tiles. EPS fillers provide excellent thermal insulation and sound absorption properties, making them ideal for use in residential and commercial buildings.
4. **Perlite**: Perlite is a type of volcanic glass that is used as filler in tiles to provide lightweight insulation properties. Perlite fillers are often used in combination with gypsum or

Fillers play an important role in the production of false ceilings, providing enhanced properties such as fire resistance, thermal insulation, and acoustic insulation. The choice of filler depends on the type of tiles being produced and its intended use. Gypsum, mineral wool, EPS, perlite, and chlor-alkali waste are some commonly used fillers in the production of tiles.

The Chlor-Alkali Waste Processing

Semi-dry chlor-alkali waste was procured from M/s. Shriram Alkali & Chemicals, Bharuch (Gujarat). Sludge of semi-dry brine was brown in colour and had coarser particles, yet moisture is present. Before being used for development activities, chlor-alkali waste underwent processing. To get rid of the moisture, it was oven dried at 110 °C for 24–48 h. Due to drying, chlor-alkali waste produced lumps that must be ground into fine powder using a pulverizer.

Making and Designing of Mould

Various moulds were designed and made in order to develop the material into the proper form, size, and thickness. For the purpose of creating false ceilings with various sizes, four moulds were created.

M/s. Amar Engineering & Plastic Industries, New Delhi, made the moulds and dies. Mould tests were conducted in a lab, and samples were made using chlor-alkali waste with a binder and other additives. Moulds were discovered to be satisfactory.

Process for Making Wall Titles Using Chlor-Alkali Waste

Chlor-alkali waste (dried powder) was mixed with Plaster of Paris (POP), cement, and filler in certain ratio, and then, water was added to prepare a homogeneous mixture. The prepared mixture was poured in the mould of desired shape and thickness. It was kept at room temperature for 74 h (air-dry). Tiles material was made by varying the quantity of chlor-alkali waste (40, 45, 50)% along with binder and filler. The complete process for formation of tiles is shown in the Fig. 2.

Fig. 2 Flow diagram for the steps taken, and the procedure for the formation of tiles

2.2 Characterization

Density Measurement

The density of wall tiles can vary depending on the material used to make them. Here are some general ranges of densities for common types of wall tiles:

1. Ceramic Tiles: The density of ceramic tiles can range from about 1500 to 2000 kg/m^3.
2. Porcelain Tiles: Porcelain tiles are denser than ceramic tiles and can have a density of 2000–2500 kg/m^3.
3. Glass Tiles: Glass tiles are typically less dense than ceramic and porcelain tiles, with a density range of around 1000–1500 kg/m^3.
4. Stone Tiles: Stone tiles can have a wide range of densities depending on the type of stone used, but generally range from 2500 to 3000 kg/m^3.
5. Metal Tiles: Metal tiles can also vary in density depending on the material used, but generally range from 5000 to 8000 kg/m^3.
6. Vinyl Tiles: Vinyl tiles are generally less dense than other types of wall tiles, with a density range of around 1000–1500 kg/m^3.
7. Mosaic Tiles: Mosaic tiles are typically made of small pieces of material, so their density can vary depending on the type of material used.

It is important to note that the density of a wall tile can affect its durability, strength, and ability to resist wear and tear. For example, denser tiles tend to be stronger and more durable than less dense tiles. However, denser tiles can also be more difficult to install and may require stronger adhesives and support structures.

Ultimately, the density of a wall tile is one factor to consider when choosing the right type of tile for a specific application, along with factors such as the size, shape, colour, and texture of the tile, as well as its intended use and the overall design aesthetic of the space.

The density can be calculated as follows:

$$\textbf{Density}(\rho) = \textbf{Mass (m)}/\textbf{Volume (V)},$$

where

Density (ρ) is measured in units such as kg/m^3, g/cm^3, or lbs/ft^3

Mass (m) is measured in units such as kilograms (kg), grams (g), or pounds (lb).

Volume (V) is measured in units such as cubic metres (m^3), cubic centimetres (cm^3), or cubic feet (ft^3).

From the above results as shown in Table 1, it is found that the wall tiles created from chlor-alkali waste have slight lower density (1.5 gm/cc) as compared to the conventional tiles (1.9 gm/cc).

Water Absorption

The ability of a material to absorb water or moisture from its environment is referred to as water absorption. The amount of water that a material absorbs can vary based

Table 1 Density measured for the given formulations

Batch number	Formulation	Density (g/cm^{-3})
W1	Pop; cement	1.368
W2	Waste; pop; cement	1.422
W3	Waste; pop; cement	1.299

on a number of variables, including the material's type, porosity, surface area, and time in contact with water.

Materials can be affected by water absorption in both positive and negative ways. Water absorption can occasionally improve a material's characteristics, such as boosting its flexibility, toughness, or durability. Some varieties of concrete and ceramic tiles, for instance, are made to absorb water, which might lessen the risk of cracking and other damage brought on by freeze–thaw cycles. However, excessive water absorption can also have negative effects on materials. When materials absorb water beyond their capacity, they may become weakened, brittle, or susceptible to damage. Water absorption can also lead to the growth of mould, bacteria, and other microorganisms, which can cause health problems and degrade the quality of the material.

Materials can be coated with coatings, sealants, or other protective measures to stop or reduce water absorption. Water infiltration and damage can be avoided with proper installation and maintenance. When choosing materials for certain applications, it is crucial to take water absorption capabilities into account. It is also crucial to keep an eye on and maintain the materials' condition over time.

Water absorption percentage can be calculated as follows:

$$\text{Water absorption percentage} = \frac{W_f - W_i}{W_i} \times 100,$$

where

W_f is the final weight of the material after immersion in water, and

W_i is the initial weight of the material before immersion in water

In water absorption experiment, which is shown in Fig. 3, the weight of the sample has been taken before placing in the water. Thereafter, the samples were placed in the water for atleast 24 h and the weight is taken again. The difference in the weight can confirm that volume of water is absorbed by the sample.

It was found that the wall tiles made from chlor-alkali waste absorbed water more quickly than conventional tiles made of feldspar and kaolin. The sample as a whole absorbs water on average at a rate of 21.74% which is shown in Table 2.

Flexural Strength

The ability of a material to resist bending or breaking under a stress applied perpendicular to its longitudinal axis is measured by flexural strength, also known as modulus of rupture. It is a crucial mechanical characteristic for building and engineering materials since it can indicate which applications are best suited for them.

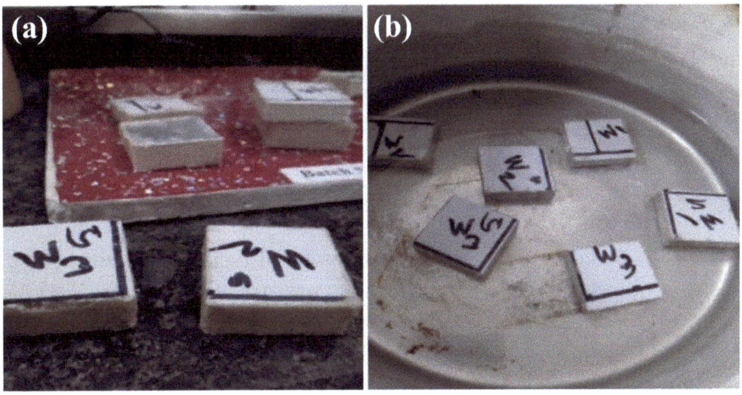

Fig. 3 Sample placed in water for 24 h, and the weight of sample before and after placing in water has been taken

Table 2 Water absorption percentages of the given formulations

Batch number	Composition	Water absorption (%)		
		W1	W2	W3
W1	Pop; cement	29.46	33.55	30.048
W2	Waste; cement; pop	17.31	16.79	16.23
W3	Waste; pop; cement	17.188	17.98	17.142

A three-point bending test is commonly used to gauge a material's flexural strength. The material is supported at two spots during this test, and a load is delivered to the middle until the material fractures. The material's maximum load before it breaks is noted and used to determine the flexural strength. The formula for flexural strength is:

$$\text{Flexural Strength}(\sigma) = (3 \times \text{Load} \times \text{Length})/\left(2 \times \text{Width} \times \text{Height}^2\right)$$

where

- Load is the maximum load applied to the material during the test.
- Length is the distance between the supports.
- Width is the width of the material.
- Height is the thickness of the material.

Flexural strength is typically reported in units of stress, such as mega Pascal's (MPa) or pounds per square inch (psi).

A material's composition, production method, and environmental circumstances are only a few examples of the variables that might affect a material's flexural strength. For applications requiring strong structural integrity and load-bearing capacity, materials with higher flexural strengths may be a preferable choice because they are often

Table 3 The average value of the flexural strength of the given formulations

Batch number	Composition	Average value of the flexural strength (MPa)
W1	Pop; cement	0.3943
W2	Waste; cement; pop	0.3146
W3	Pop; cement; waste	0.2248

more resistant to bending and breaking. An essential quality that affects wall tiles' resistance to bending and breaking when subjected to external forces is their flexural strength. By exerting a force perpendicular to a tile's surface until it bends or breaks, flexural strength is evaluated. The kind of material used, the thickness of the tile, and the manufacturing method is only a few of the variables that affect the flexural strength of wall tiles. For instance, due to their composition and production method, ceramic tiles typically have lower flexural strength than porcelain tiles. Due to their greater density and hardness, porcelain tiles have a higher flexural strength.

Flexural strength is also influenced by the thickness of the tile. Compared to thinner tiles, thicker tiles are often tougher and more resistant to bending and breaking. However, heavier and more challenging to install thicker tiles. For ceramic tiles and porcelain tiles, the minimum flexural strength required to meet industry standards is 35 M Pa (mega Pascals), and 50 M Pa, respectively. These requirements make certain that tiles can withstand the pressures put on them during installation and use, such as foot traffic or the weight of things set onto them. When selecting wall tiles, you should also take into account flexural strength, water absorption, abrasion resistance, and slip resistance as shown in Table 3. To ensure longevity and safety, it's critical to pick tiles that are suitable for the particular area and purpose.

Overall, a key factor in determining a wall tile's resistance to external forces is its flexural strength. Varying amounts can be produced by various materials and manufacturing processes.

3 Conclusion

Waste from the chlor-alkali industry, which is produced during the production of chemicals like chlorine and caustic soda, is known as chlor-alkali waste. This trash has a high concentration of salt and other minerals that, if improperly handled and disposed of, could harm the ecosystem. The manufacture of ornamental wall tiles using this trash as a raw material is one possible use for it. Chlor-alkali waste must first be cleaned to remove any potentially dangerous impurities before being combined with other materials like sand and clay and fired at high temperatures to create wall tiles. The produced tiles can be used as decorative tiles because of their distinctive texture and colour.

Using chlor-alkali waste as a source for wall tiles has a number of potential advantages. One benefit is that it allows for the recycling of materials that would otherwise be disposed of as hazardous trash. By developing a more sustainable and circular approach to waste management, it also has the potential to lessen the environmental effect of the chlor-alkali sector. Making wall tiles from chlor-alkali waste presents certain difficulties, though. For starters, the procedure might be costly because it calls for specialized tools and knowledge. When dealing with this kind of waste material, there can also be environmental and legal factors that need to be taken into mind. Additionally, the final tiles could not have the same degree of consistency and quality as those made from more traditional raw materials.

Despite these difficulties, it may be possible to use waste chlor-alkali as a source for beautiful wall tiles. This waste-to-wall tile manufacturing method is still in its infancy and needs more investigation and development. However, if it is successful, it might present a brand-new method of garbage management that is both ecologically beneficial and sustainable.

Overall, it is a good idea that might potentially minimize waste and provide a more sustainable method of waste management to use chlor-alkali waste as a raw material for beautiful wall tiles. To assess this strategy's viability and deal with any potential technological, economic and regulatory difficulties, more study and development are required.

References

1. Ferreira WM et al (2022) Perspective of the application of ash from the ceramic industry in the development of alkali-activated roof tiles. Ceram Int 48(5):6250–6257
2. Zanelli C et al (2021) Waste recycling in ceramic tiles: a technological outlook. Resour Conserv Recycl 168:105289
3. Bignozzi MC, Saccani A (2012) Ceramic waste as aggregate and supplementary cementing material: a combined action to contrast alkali silica reaction (ASR). Cement Concr Compos 34(10):1141–1148
4. Mojapelo KS et al (2021) Utilization of wastewater sludge for lightweight concrete and the use of wastewater as curing medium. Case Stud Construct Mat 15:e00667
5. Izidoro JDC et al (2021) Brine sludge waste from a Chlor-alkali industry: characterization and its application for non-structural and structural construction materials. J Appl Mat Tech 3(1):1–7
6. Hossain SS, Roy PK (2020) Sustainable ceramics derived from solid wastes: a review. J Asian Ceramic Soc 8(4):984–1009
7. Qu F et al (2021) Durability deterioration of concrete under marine environment from material to structure: a critical review. J Build Eng 35:102074
8. Acchar W, Avelino KA, Segadães AM (2016) Granite waste and coffee husk ash synergistic effect on clay-based ceramics. Adv Appl Ceram 115(4):236–242
9. Acchar W, Silva JE, Segadães AM (2013) Increased added value reuse of construction waste in clay based building ceramics. Adv Appl Ceram 112(8):487–493

MWCNT/TPU Nanocomposites-Based UV Photodetector

Shivraj Singh, Arpit Verma, Srishtee Chauhan, Sheenam Thatai, Christine Jeyaseelan, Ravi Kant Triapthi, Sidhharth Sirohi, B. C. Yadav, and Tejendra K. Gupta

1 Introduction

Due to their wide range of applications in a variety of fields, such as optoelectronic devices [1, 2], light-emitting diodes (LEDs) [3], electroluminescence [4], flat panel displays [5–7], infrared windows, sensors [6, 8, 9], lasers and bio-devices [10, 11], solid state lightning [12, 13], etc., wide band gap inorganic semiconducting materials have drawn significant attention in the last two decades [14, 15]. In these above-mentioned applications, optoelectronic devices, in particular UV photodetectors, are one of the most crucial classes used in both the civilian and military fields for missile tracking, space communication, astronomical studies, flame sensing, biological and chemical analysis, etc. [16–18]. High surface-to-volume ratio, high transparency, and anisotropic transmission properties are necessary for this class of optoelectronic devices. Due to their shape, size, and surface structure dependent physical and chemical properties, functional nanomaterials are easily able to satisfy these fundamental requirements of optoelectronic devices [19]. In the region of the electromagnetic spectrum, UV photodetectors have found their uses in wavelengths below 400 nm. This region of the electromagnetic spectrum is typically divided into three groups:

S. Singh · S. Chauhan · S. Thatai · C. Jeyaseelan · T. K. Gupta (✉)
Amity Institute of Applied Sciences, Amity University Uttar Pradesh, Noida 201313, India
e-mail: tejendra.amu@gmail.com

A. Verma · B. C. Yadav
Nanomaterials and Sensors Research Laboratory, Department of Physics, Babasaheb Bhimrao Ambedkar University, Lucknow, Uttar Pradesh 226025, India

R. K. Triapthi
Department of Physics, Lal Bahadur Shastri Smarak Post Graduate College, Maharajganj, Uttar Pradesh, India

S. Sirohi
Bhaskaracharya College of Applied Sciences, University of Delhi, Delhi 110075, India

S. Rattan et al. (eds.), *Emerging Trends in Synthesis and Catalysis in Chemistry*, Springer Proceedings in Materials 51, https://doi.org/10.1007/978-981-97-5169-3_14

UV-A (320–400 nm), UV-B (290–320 nm), and UV-C (200–290 nm). These radiations are extremely dangerous for humans because due to their harmful nature. However, the radiation of UV-C is totally absorbed by the earth's environment, while UV-B radiation can be easily absorbed by sunscreen lotions. The longer wavelengths (UV-A), however, can easily pass through the earth's atmosphere and sunscreen, according to current study, and may even cause skin cancer [20].

ZnO is one of the most suitable candidate materials for UV photodetector among many wide-bandgap semiconductor materials due to its outstanding properties, which includes a direct bandgap of 3.3 eV, and high exciton binding energy of 60 meV, as well as excellent photoelectric conversion characteristics, nontoxicity, thermal stability, ease of preparation, and low cost. ZnO-based UV photodetectors still have issues with weak light response and prolonged reaction times, which restricts the use of these devices in UV photodetectors. Therefore, increasing the carrier life and decreasing the carrier composite rate are crucial for bettering ZnO-based UV photodetectors. Recent research has demonstrated that nanostructured ZnO is advantageous for photogenic carrier transport, p-n heterojunction nanocomposite materials improve photoconductivity, and the formation of nanostructured ZnO p-n heterojunction is anticipated to be the key to solving the issue in these UV photodetector issues.

Polymer-based flexible UV photodetectors have gained considerable attention due to their unique properties, such as flexibility, lightweight, low cost, and ease of processing. These photodetectors have a range of applications, including wearable electronics, biomedical sensors, and flexible displays. Polymer-based photodetectors can be fabricated using a variety of techniques, such as spin-coating, inkjet printing, and solution casting. The photodetectors consist of a thin film of a semiconducting polymer that absorbs UV radiation and generates a photocurrent, which is measured by electrodes. One of the key advantages of polymer-based photodetectors is their flexibility, which allows them to conform to various shapes and surfaces, making them suitable for wearable and implantable devices. Additionally, polymer-based photodetectors can be fabricated on flexible substrates, such as plastics, paper, or textiles, which reduces their weight and enhances their mechanical robustness. Polymer-based photodetectors also have the potential for low-cost mass production, which is critical for applications that require large-scale deployment, such as environmental monitoring or security systems.

Thermoplastic polyurethane (TPU) nanocomposites have been investigated for their potential as UV photodetectors due to their unique combination of properties, including flexibility, transparency, and good mechanical and thermal stability. The incorporation of nanoparticles into polyurethane matrices can further enhance the sensitivity and performance of the photodetector.

2 Experimental Part

2.1 Materials

TPU was purchased from COIM group, Laripur, India (LPR grade). MWCNT were synthesized via chemical vapor deposition method and the discussed elsewhere [21, 22].

Preparation of MWCNT/TPU Nanocomposite Films

TPU nanocomposite films reinforced with 0.1 and 0.2 wt.% of MWCNTs loadings have been prepared via solvent mixing and casting method as shown in scheme (Fig. 1) and are abbreviated as PUCNT0.1 and PUCNT0.2, respectively. In this process, both MWCNTs and TPU were dissolved in N, N-Dimethylformamide (DMF) using method as described earlier [22]. The solution of MWCNT/TPU nanocomposites was finally transferred in to the petri dish and dry in an oven at around 80 °C.

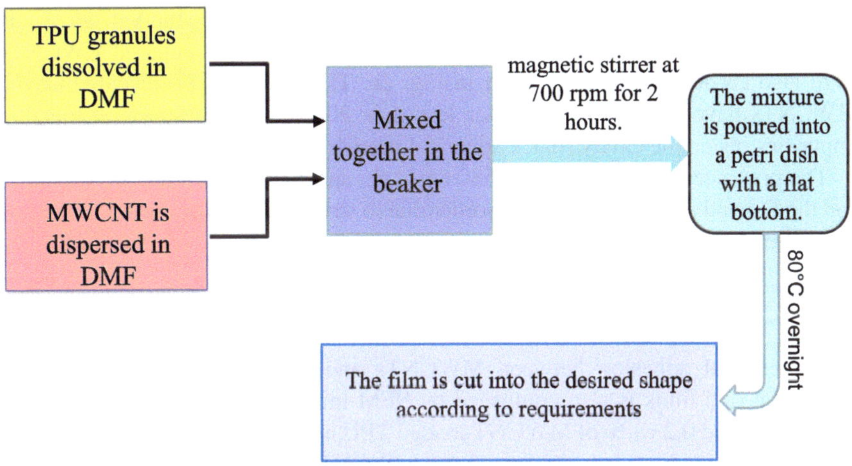

Fig. 1 Schematic diagram of MWCNT/TPU nanocomposite films

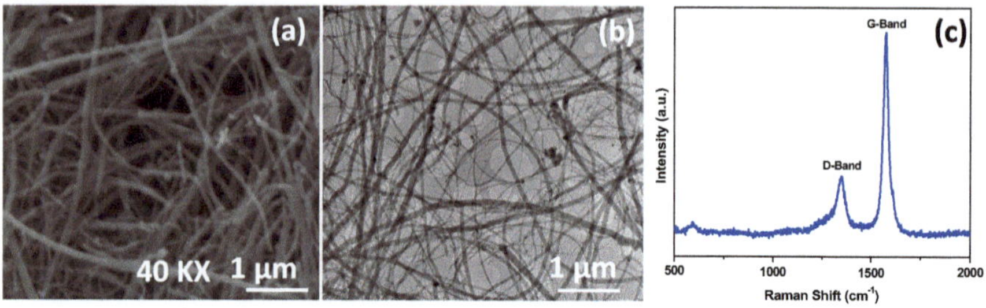

Fig. 2 SEM micrograph (**a**), TEM micrograph (**b**), and Raman spectra (**c**) of as grown MWCNTs

3 Results and Discussion

3.1 Characteristics of MWCNTs

As shown in Fig. 2a, the individual as prepared MWCNTs is shown in the form of tubular shape with small amount of catalyst (white particles on the tubes). Before SEM imaging, these MWCNTs were dispersed in DMF to reduce the Van der Waals force of attraction and then dried in oven. TEM micrograph of as prepared MWCNTs were also carried out and is shown in Fig. 2b. TEM micrograph reveals that the MWCNTs have an average diameter less than 30 nm with presence of metallic impurities inside the cavity (black spherical particles).

The presence of the G-band at 1580 cm^{-1} for sp^2 hybridized graphitic structure, and the D-band at 1350 cm^{-1} for disorder due to structural defects during folding of graphene sheet in Raman spectroscopic study of as prepared MWCNTs in Fig. 2c confirm the presence of MWCNTs [23].

Microscopic Study of MWCNT/TPU Nanocomposite Films

The interfacial adhesion between MWCNTs and TPU matrix in MWCT/TPU nanocomposite films is also confirmed by SEM imaging of the fractured surfaces of 0.1 wt.% and 0.2 wt.% of MWCNT-loaded TPU nanocomposite films as shown in Fig. 3. From Fig. 3a, b, it is confirmed that MWCNTs are distributed and embedded uniformly inside the TU matrix.

UV Photodetection Performance

A depletion layer forms between the oxygen species and the detection material when oxygen molecules are adsorbed on the surface of the nanomaterial in the presence of the ambient air conditions [24]. These oxygen species reacted with the photon after being exposed to the radiation to remove O2 from the device's surface. The materials' photoresponse is also a result of this gas sensing process [25]. Under illumination, electron–hole pairs are formed in the detector material and transported towards the appropriate electrodes under the applied potential drift. Upon

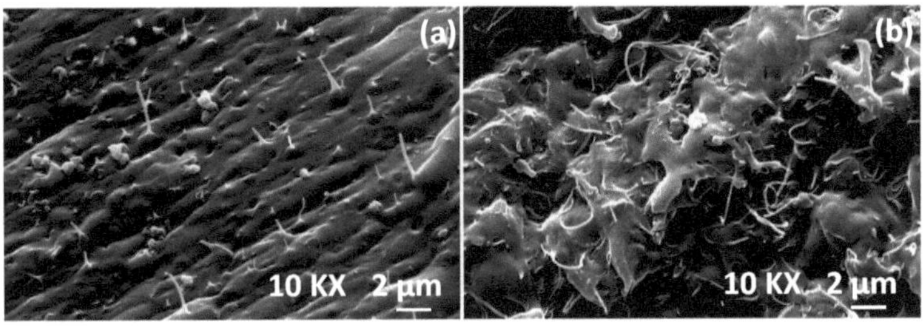

Fig. 3 SEM micrographs of the fracture surface of MWCNTs/TPU nanocomposite films for **a** 0.1 wt.%, and **b** 0.2 wt.% of MWCNTs loaded TPU nanocomposite films

Table 1 Photodetection parameters of MWCNT/TPU nanocomposite films

Samples	Responsivity (mA/W)	Detectivity (Jones)
PUCNT0.1	7.18	4.32×10^{10}
PUCNT0.2	5.26	7.82×10^{10}

illumination, oxygenated species are eliminated, and high order conductivity is seen [26–29].

The responsivity (*R*) and detectivity (*D*) are the important parameters, which defines the performance of the photodetector device [30, 31]. All these parameters are calculated and are presented in Table 1. According to Relation (1), the photogenerated current upon each incident photon of each unit area determines the photodetector device's responsiveness (*R*).

$$R = \frac{I_p - I_d}{AP_{in}} \text{ A/W} \tag{1}$$

Responsivity typically refers to the ability of a system, device, or organism to respond appropriately and sensitively to stimuli, inputs, or changes in the environment. It is a measure of how effectively and accurately a system reacts or adjusts to external factors. The highest responsivity was observed to 5.26 mA/W for the device made from 0.2 wt.% of MWCNTs loaded TPU nanocomposite films. This is because of the introduction of more conduction channels into the TPU matrix as the CNT's conductivity increases [32].

Detectivity (*D*) is another important parameter of the photodetector device which is shown in relation (2).

$$D = \sqrt{\frac{A}{2eI_d}} R \text{ Jones} \tag{2}$$

where, Id is the dark current, 'A' the area of device, and 'e' is electronic charge. When very weak light signals need to be detected, this detectivity parameter is important. The general definition of the term is the amount of weak signal that a photodetector devices can detect [25]. The detectivity of 0.1–0.2 wt.% of MWCNTs loaded TPU nanocomposite films are shown in Table 1. The data suggests that the maximum value of detectivity of 7.82 × 10^{10} Jones was observed for photodetector device prepared from 0.2 wt.% loaded MWCNT/TPU nanocomposite film. Detectivity is directly proportional to the concentration of the MWCNTs in the TPU matrix.

4 Conclusion

TPU nanocomposites filled with 0.1–0.2 wt.% of MWCNTs have been prepared successfully via solution mixing method and UV photodetector performance has been observed. SEM, TEM, and Raman spectroscopic studies were carried out to confirm that the CNTs are multiwalled in nature. SEM of the fractured surface of the TPU nanocomposite films confirms that the MWCNTs are distributed uniformly inside the TPU matrix. The photodetector device made from MWCNTs/TPU nanocomposite films show better photoresponse due to higher absorption in the UV range. 0.2 wt.% of MWCNTs loaded TPU nanocomposite film-based UV photodetection device showed best performance with responsivity, 5.26 mA/W, and detectivity, 7.82 × 10^{10} Jones. With an increase in MWCNT concentration in the TPU matrix, the detectability increases. The results show that MWCNTS/TPU nanocomposites can be used to fabricate highly sensitive, and flexible visible blind UV photodetectors.

Acknowledgements The authors (SS, SC, ST, CJ, and TKG) are grateful to the Director of Amity Institute of Applied Sciences, Amity University Uttar Pradesh, India for providing the necessary infrastructure and facilities to carry out this work.

Declaration of Competing Interest The authors declare no known competing financial interests.

References

1. Pisula W (2023) Inorganic semiconductors in electronic applications. Elect Mat 4(3):136–138
2. Li H, Cao Y, Wang Z, Feng X (2019) Flexible and stretchable inorganic optoelectronics. Opt Mater Express 9(10):4023–4049
3. Cao YB, Zhang D, Zhang Q, Qiu X, Zhou Y, Poddar S, Fu Y, Zhu Y, Liao J-F, Shu L, Ren B, Ding Y, Han B, He Z, Kuang D-B, Wang K, Zeng H, Fan Z (2023) High-efficiency, flexible and large-area red/green/blue all-inorganic metal halide perovskite quantum wires-based light-emitting diodes. Nat Commun 14(1):4611
4. Liu W, Lustig WP, Li J (2019) Luminescent inorganic-organic hybrid semiconductor materials for energy-saving lighting applications. EnergyChem 1(2):100008
5. Sun Y, Rogers JA (2007) Inorganic semiconductors for flexible electronics. Adv Mater 19(15):1897–1916
6. Rai S, Mishra H, Singh VK, Gupta TK, Srivastava A (2023) Chapter 5—2D materials for flexible electronics. In: Das S (ed) 2D materials for electronics, sensors and devices. Elsevier, pp 169–206
7. Yu KJ, Yan Z, Han M, Rogers JA (2017) Inorganic semiconducting materials for flexible and stretchable electronics. NPJ Flex Elect 1(1):4

8. Han C, Li X, Liu Y, Tang Y, Liu M, Li X, Shao C, Ma J, Liu Y (2021) Flexible all-inorganic room-temperature chemiresistors based on fibrous ceramic substrate and visible-light-powered semiconductor sensing layer. Adv Sci 8(23):2102471
9. Ren Z, Yang J, Qi D, Sonar P, Liu L, Lou Z, Shen G, Wei Z (2021) Flexible sensors based on organic-inorganic hybrid materials. Adv Mat Technol 6(4):2000889
10. Peng X, Ai F, Yan L, Ha E, Hu X, He S, Hu J (2021) Synthesis strategies and biomedical applications for doped inorganic semiconductor nanocrystals. Cell Rep Phys Sci 2(5):100436
11. Stylianakis MM, Maksudov T, Panagiotopoulos A, Kakavelakis G, Petridis K (2019) Inorganic and hybrid perovskite based laser devices: a review. Materials 12(6):859
12. Zhu D, Humphreys CJ (2016) Solid-state lighting based on light emitting diode technology. In: Al-Amri MD, El-Gomati M, Zubairy MS (eds) Optics in our time. Springer International Publishing, Cham, pp 87–118
13. Jose A, Krishnapriya T, Biju PR (2023) Luminescence characteristics and energy transfer dynamics of rare-earth ion co-activated borosilicate glasses for solid-state lighting applications. In: Kumar V, Sharma V, Swart HC (eds) Advanced materials for solid state lighting. Springer Nature Singapore, Singapore, pp 375–402
14. Steigerwald ML, Brus LE (1990) Semiconductor crystallites: a class of large molecules. Acc Chem Res 23(6):183–188
15. Zhou J, Chen L, Wang Y, He Y, Pan X, Xie E (2016) An overview on emerging photoelectrochemical self-powered ultraviolet photodetectors. Nanoscale 8(1):50–73
16. Wan X, Xu Y, Guo H, Shehzad K, Ali A, Liu Y, Yang J, Dai D, Lin C-T, Liu L, Cheng H-C, Wang F, Wang X, Lu H, Hu W, Pi X, Dan Y, Luo J, Hasan T, Duan X, Li X, Xu J, Yang D, Ren T, Yu B (2017) A self-powered high-performance graphene/silicon ultraviolet photodetector with ultra-shallow junction: breaking the limit of silicon? NPJ 2D Mat Appl 1(1):4
17. Alaie Z, Nejad SM, Yousefi MH (2015) Recent advances in ultraviolet photodetectors. Mat Sci Semiconductor Process 29:16–55
18. Zheng Z, Gan L, Li H, Ma Y, Bando Y, Golberg D, Zhai T (2015) A fully transparent and flexible ultraviolet-visible photodetector based on controlled electrospun ZnO-CdO heterojunction nanofiber arrays. Adv Func Mater 25(37):5885–5894
19. Bai Z, Zhang Y (2016) Self-powered UV–visible photodetectors based on ZnO/Cu_2O nanowire/electrolyte heterojunctions. J Alloy Compd 675:325–330
20. Ning Y, Zhang Z, Teng F, Fang X (2018) Novel transparent and self-powered UV photodetector based on crossed ZnO nanofiber array homojunction. Small 14(13):1703754
21. Gupta TK, Singh BP, Dhakate SR, Singh VN, Mathur RB (2013) Improved nanoindentation and microwave shielding properties of modified MWCNT reinforced polyurethane composites. J Mat Chem A 1(32):9138–9149
22. Gupta TK, Singh BP, Teotia S, Katyal V, Dhakate SR, Mathur RB (2013) Designing of multi-walled carbon nanotubes reinforced polyurethane composites as electromagnetic interference shielding materials. J Polym Res 20(6):169
23. Gupta TK, Singh BP, Mathur RB, Dhakate SR (2014) Multi-walled carbon nanotube–graphene–polyaniline multiphase nanocomposite with superior electromagnetic shielding effectiveness. Nanoscale 6(2):842–851
24. Zhu J, Wang H, Ma L, Zou G (2021) Observation of ambipolar photoresponse from 2D MoS_2/MXene heterostructure. Nano Res 14(10):3416–3422
25. Singh A, Sikarwar S, Verma A, Yadav BC (2021) The recent development of metal oxide heterostructures based gas sensor, their future opportunities and challenges: a review. Sens Actuators, A 332:113127
26. Kannichankandy D, Pataniya PM, Zankat CK, Tannarana M, Pathak VM, Solanki GK, Patel KD (2020) Paper based organic–inorganic hybrid photodetector for visible light detection. Appl Surf Sci 524:146589
27. Wu G, Sahoo AK (2020) Influence of oxygen flow rate on channel width dependent electrical properties of indium gallium zinc oxide thin-film transistors. Nanomaterials 10(12):2357
28. Marimuthu G, Saravanakumar K, Jeyadheepan K, Mahalakshmi K (2022) Chemical growth and study of low intensity sensing ability of nanobranch and nanorod structured SnO_2 UV detector. J Environ Chem Eng 10(1):106981

29. Tripathi RK, Panwar OS, Rawal I, Dixit CK, Verma A, Chaudhary P, Srivastava AK, Yadav BC (2021) Study of variable range hopping conduction mechanism in nanocrystalline carbon thin films deposited by modified anodic jet carbon arc technique: application to light-dependent resistors. J Mater Sci: Mater Electron 32(2):2535–2546
30. Verma A, Chaudhary P, Singh A, Tripathi RK, Yadav BC (2022) ZnS nanosheets in a polyaniline matrix as metallopolymer nanohybrids for flexible and biofriendly photodetectors. ACS Appl Nano Mat
31. Mathew M, Rout CS (2021) Schottky diodes based on 2D materials for environmental gas monitoring: a review on emerging trends, recent developments and future perspectives. J Mat Chem C 9(2):395–416
32. Gautam C, Verma A, Chaudhary P, Yadav BC (2022) Development of 2D based ZnO–MoS_2 nanocomposite for photodetector with light-induced current study. Opt Mater 123:111860

N-Fluorobenzenesulfonimide (NFSI): A Multipurpose Reagent

Arushi Philips, Sheetal Gulia, Reena Singh, and Priyanka Chaudhary

1 Introduction

The past few decades, the synthetic chemistry has witnessed a resurgence of interest for the alternative and versatile reagents. Notably, N-fluorobenzenesulfonimide (NFSI) was found to have potential reagent for the formation of diverse C–N and C–F bonds, resulting in a broad range of double functionalized compounds. In the search for pharmaceuticals and agricultural chemicals, fluorine substitutes are of great significance. This evolution has taken place because of the extraordinary properties of fluorine. Target molecules physiochemical properties and, thus, biological activity, may be drastically altered. Xia B. et al. developed a protocol that permits the mild insertion of fluorine into target structures which had been found to have tedious work to achieve in past. Afterward they conventionally allow the subrogation of a carbon-hydrogen bond for a Carbon-fluorine covalent bond in a one-pot procedure, typically via a carbanionic intermediate, electrophilic fluorinating chemicals are particularly intriguing in this context [1]. Organo- and transition-metal catalysis has been principally responsible for utmost fundamental conceptual developments

A. Philips · S. Gulia · P. Chaudhary (✉)
AIAS, Amity University, Noida, Uttar Pradesh 201303, India
e-mail: priyankaciitbhu@gmail.com; chaudhary_priyanka@lkouniv.ac.in

A. Philips
e-mail: arushi.phillips@s.amity.edu

S. Gulia
e-mail: sheetal.gulia@s.amity.edu

R. Singh
Sunbeam College for Women, Bhagwanpur, Varanasi, Uttar Pradesh 221005, India
e-mail: reenas.rs.chy16@itbhu.ac.in

Present Address:
P. Chaudhary
Department of Chemistry, University of Lucknow, Lucknow, Uttar Pradesh 226007, India

S. Rattan et al. (eds.), *Emerging Trends in Synthesis and Catalysis in Chemistry*, Springer Proceedings in Materials 51, https://doi.org/10.1007/978-981-97-5169-3_15

in the field of fluorination. The optimum challenging transition is still the formation of the parent C–F bond. This is mostly due to the high hydration energy of fluoride, the durability of the metal-fluorine bonds, and its strongly polarized bindings.

Emsley described that the most fluorination reactions still lack predictability and economic viability [2a]. The fact that molecules containing nitrogen perform a range of biological roles makes the establishment of the C–N bond even more important. Liu and his co-workers described that there are opportunities to overcome obstacles in the pharmaceutical sector which has led to significant improvements in mild and efficient amidation reactions, with an emphasis on methods for creating intermediates with a nitrogen core, related processes, and novel synthetic chemical approaches. Among the chemicals previously used, compounds containing a nitrogen-fluorine bond have proven to be very useful [3]. As we recently demonstrated, *N*-Fluoro Sultams are easily accessible, stable, and crystalline compounds that allow for the selective fluorination of carbanions (Table 1). However, they are not reactive enough to fluorine less reactive nucleophiles like aromatics and enol ethers, which is a drawback. N-F nucleophiles do not react with these nucleophiles; however, they typically encounter access problems, lengthy synthesis periods, or carbanion-nucleophile side effects. Since it is simple to access a variety of nucleophiles, including carbanions and aromatics, an electrophilic fluorinating agent is particularly desired. NFSI is the only tool in the chemical industry that meets every criterion.

N-fluorodi(benzene sulfonyl)amine, commonly known as *N*-Fluorobenzenesulfonimide (NFSI), is a chemical compound that projects white, crystallized granules which is a reliable and uncomplicated. When it was 1991, Differding [2b] and Ofner prepared NFSI from benzenesulfonimide by fluorination using a one-tenth (v/v) nitrogen fluorides in ethanenitrile at negative 40 °C. During 1994, Wagner [2c] and colleagues engineered a high-volume synthesis that involved reacting sodium benzene sulphonyl fluoride containing 10% vol/vol nitrogen fluoride combination (Fig. 1).

NFSI has attracted the interest of synthetic chemists for exploitation due to its dual behavior, where it is presently utilized as a source for electrophilic fluorination and amidation (Fig. 2). We believe that by using NFSI, which is easily accessible on the market, we can examine current advancements, complications, and possible outputs in the discipline of chemical interactions [4].

Table 1 Importance of N-fluorobenzenesulfonamide (NFSI)

S. no	Importance of N-fluorobenzenesulfonamide (NFSI)
1	It is a mild, stable, and highly soluble crystalline solid
2	These compounds were more stable, safer, milder, easier to handle, cheaper, and selective sources of electrophilic fluorine
3	The amination reagent in various transformations of aliphatic and aromatic compounds
4	It is also used as external oxidant for the reactions

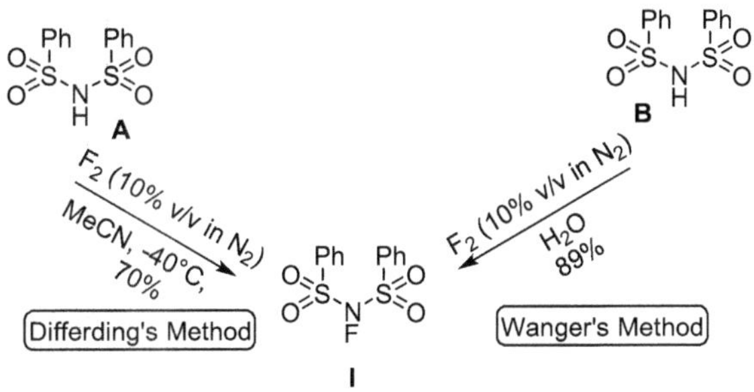

Fig. 1 Preparation of *N*-Fluorobenzenesulfonimide (NFSI)

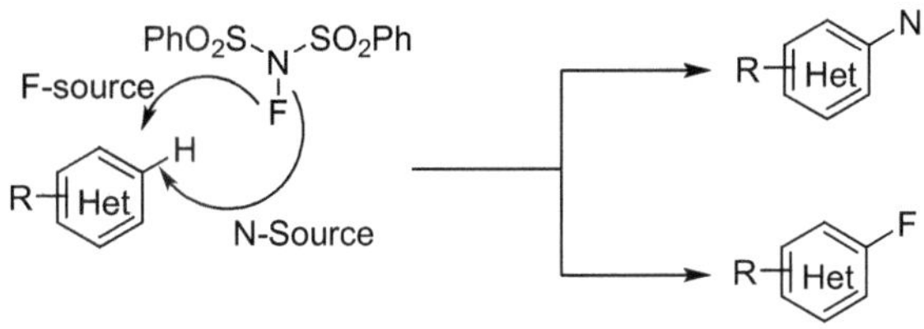

Fig. 2 Formation of Hetero-aromatic compound using NFSI

2 *N*-Fluorobenzenesulfonimide as a Nitrogen Compound

The construction of C–N bond could be explored with NFSI-mediated reactions which has shown promising results with the employment of effective nitrogen utilization. In both industry and pharmacology, nitrogen-containing compounds are frequently utilized as biologicals. These compounds are crucial intermediates in the synthesis of organic chemicals. Barker and colleagues mentioned, due to this, the creation of carbon-nitrogen bond is necessary for the process of synthesizing organic compounds, and research into new strategies for doing so is still appealing [5]. The stepwise, cost-effective synthesis of C–N bonds has made extensive use of direct coupling of amines and carbonyls using transition metals. Due to their novel catalytic processes, transition metals have recently gained a lot of interest. The amido functionalization of alkenes, initiated by the sensitive oxidant NFSI, which undergoes a chemical reaction with metals, resulting in radical formation with the element (N_2) at its center, was described by Agarwal et al. [6]. C–H amination utilizing various ammonia derivatives and metal complex catalysts is widely described in literature surveys. Wencel and kirk have described the reagent NFSI which serves as the reagent in the reaction of amidation to create functional molecules with a focus on biology and materials [7, 8].

2.1 Reactions with the Transition Metal Using NFSI

It has been discovered that amidation of hetero-cycles with N-fluorobenzenesulfonimide (NFSI) proceeds most likely by C–H bond activation. Copper and palladium-catalyst was used as the catalyst, a range of amidated heterocycle derivatives was achieved in satisfactory to outstanding efficiency. By means of direct carbon-nitrogen bond-coupling mechanism, this chemistry provided a practical way to synthesize valuable amidated heterocycles.

Beletskaya with his group reported that the compounds with aromatic amine and heteroaromatic amine blocks are often utilized namely biology, pharmacy, a process catalyzed by transition metals beside production of aryl Csp^2-Nitrogen bonds have been thoroughly analyzed across the last decade [9, 10]. Substantial carbon-nitrogen bond forming strategies have been identified thus far. One of them is the conventional metal-catalyzed arylation of amines, such as the copper-catalyzed production of carbon-nitrogen bonds of the Goldberg [11]. However, Louillat, Shang and their team said that this process requires high temperatures and arenes that have been pre-functionalized in addition to producing stoichiometric levels of byproducts. Metal-catalyzed direct C–H amination of arenes while exposed to oxidants is another method [12–14].

This atom-economical coupling necessitates specialized stoichiometric oxidants and groups-directing agents, which limits the scope of this method. The direct metal C–N coupling of heteroarenes with preactivated amino derivates, catalyzed by metals, was put forward as the third tactic by Kawano et al. [15]. However, this quick and effective transformation has been documented. Ritter and colleagues revealed the nucleophilic fluorination induced direct amidation of heteroarenes) as a nitrogen reagent in 2013 and Che and colleagues documented ruthenium-catalyzed direct amidation of heterocycles via $C(sp^2)$-H bond activation as a nitrogen reagent in 2004 [16].

2.1.1 Palladium-Assisted Reactions

Zhang and coworkers reported the C–N bond formation using NFSI as an aminating agent through the direct C–H amination of aromatic (Scheme 1) during the year 2011 [17]. Using NFSI and employing 10 mol% $Pd(OAc)_2$ catalyst, Sun K. and the team successfully carried out a variety of C–H amination reactions on 2-OR-substituted anilides [18]. The reaction was selectively directed at the para position and supplemented with 2.0 equiv. of $NaHCO_3$ in DCE, leading to the formation of nitrogen-substituted alkenes carrying conjugates in reasonable yields [19]. Effect of the substitute on the aromatic system's electronic and steric properties regulated the electrophilic palladation of $FPdN(SO_2Ph)_2$ to construct a dearomatized spirocyclic palladium complex that underwent ordered nucleophilic amination and dehydrogenation to give rise to directed ortho/para products.

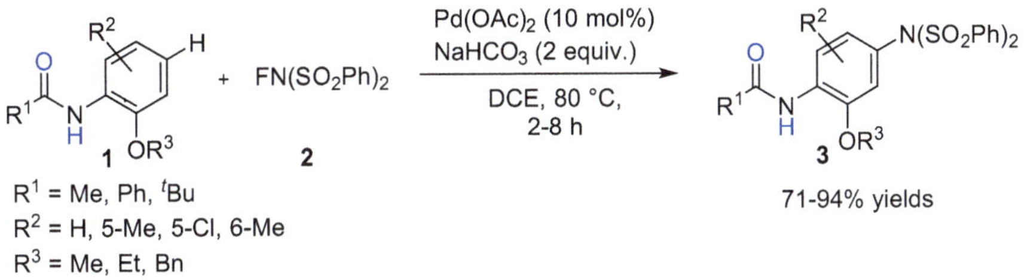

Scheme 1 Anilide C–H bond amination using palladium catalyst and NFSI

Ortho-sulfonamidoanilides of various types were produced in 44–90% yield by substrates with para–substituents, whereas the equivalent para–amination was produced in 71–94% yield by substrates with ortho–alkoxy–substituents. This transition served as precursor illustration of a targeted, palladium-mediated, extremely specific para-aromatic carbon-hydrogen amination. The existence of radical ions was not previously acknowledged in the process, according to mechanistic investigations. The authors suggest a plausible transformation pathway for this carbon-nitrogen bond formation reaction that implies a palladium (II)/palladium (IV) reaction cycle and a key iminoquinonemethide intermediate that undergoes nucleophilic amination by benzenesulfonimide to give 51–94% yield of corresponding products. The description of the chemical reaction resulting in the formation of a result product, presumably a compound, was provided by Zhou and the team [20].

Overall, the reaction involves the functionalization of the nitrogen-fluorine bond of NFSI, moreover the construction of a new carbon-nitrogen linkage between transition compound Y and $-N(SO_2Ph)_2$. The steps involve several key reactions in organometallic chemistry, including oxidative addition, coordination, electrophilic palladation, and hydrogen elimination.

2.1.2 Gold-Assisted Reactions

Gold salt is among the latest metal-based catalysts to participate in the electrophilic amination of arenes with NFSI story. Murakami and Itami demonstrated in 2018 (Scheme 2) that the inexpensive, commercially accessible AuCl efficiently catalyzed the aromatic C–H amination PAHs by utilizing NFSI as the amination reagent and C–H bond cleavage [21]. Bidentate ligand (BQ) has been used as a ligand in various metal-catalyzed reactions in the presence of DCE during the reactions, resulted in moderate yields of the mono-animated compounds, which are difficult to synthesize, and ranged to 55% from 13%. Notably, the reaction only proceeds if an *N*,*N*-bidentate ligand is present. In the lack of a bidentate ligand, nothing was discovered. A variety of Au-catalysts for the direct imidation of aromatic C–H bond was developed. Due to the present increase in the demand for aminated PAHs in organic LEDs, incorporating amino functional groups into PAHs is particularly crucial [22]. Additionally, it is unusual for Au-catalyzed reactions to pair Au(I) complex to be prepared with N,

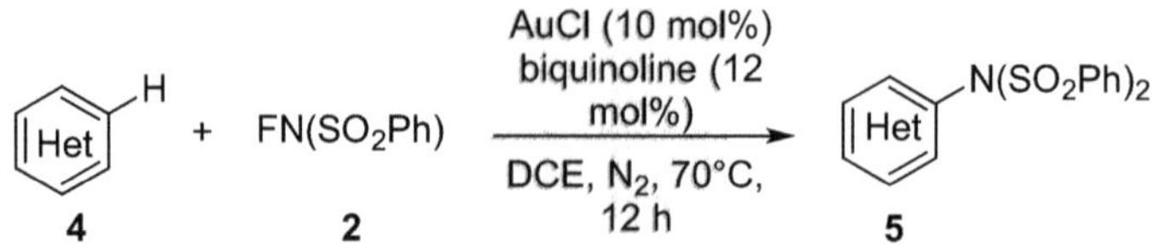

Scheme 2 Au-catalyzed direct amination of polycyclic aromatic hydrocarbons via direct C–H activation

N-bidentate ligands like bipyridine; thus, we anticipate that this system may have important potential outside of this C–H imidation chemistry.

The direct C–H imidation of PAHs using NFSI, catalyzed by AU(i), was described by Hirva P. and the team, where PAHs were functionalized with PAHs [23]. An imidyl radical might be created when the creation of an Au-ligand complex reacts with NFSI, which would then add radicals to the PAH substrates.

The resultant radical precursor undergoes a sole electron transfer to yield a positively charged intermediate, which subsequently converts to an aromatic compound to yield the product with incorporated amino group. The calculations, performed by Kawakami T. and the team, agree with the observed selectivity for a specific regioisomer in the reactions, indicating the involvement of an imidyl radical with arenes through radical addition [24].

2.1.3 Copper-Assisted Reactions

Over the past ten years, *N*-Fluorobenzenesulfonimide served as a nitrogen precursor, has drawn more vigilance. The use of NFSI for coupling of arenes by aryl Csp^2-N/Csp^3-N in the existence of Cu has been outlined by Zhang, Q along with many others [25]. There are few instances of direct amidation of heterocycles, even though the conjoining processes between heteroarenes and *N*-Fluorobenzenesulfonimide are well documented. Trenner J. and the team describe how cupric iodide served as an activator and *N*-Fluorobenzenesulfonimide served as a N_2 supply in the direct amidation of heteroaromatic compounds such as thiophene, furan, and pyrrole [26]. In this process, a variety of heterocycle substrates were quickly amidated in high yields with exceptional functional group flexibility using NFSI as the amine precursor. Thiophene and NFSI were coupled utilizing cupric iodide as the facilitator as an initial step in our experiment. The carbon-nitrogen coupling product obtained a yield of 52% by treating the thiophene with 1.2 equivalent of NFSI and 5 mol% of CuI in DCE for 8 h at room temperature. It is interesting that the product was obtained with an 88% yield when the temperature reached 60 °C. CuI was not superior to a few other copper facilitators, including copper chloride, copper bromide, copper cyanide and $CuCl_2$. As a result, we determined that the optimal conditions for producing with an 88% yield were copper iodide (5 mol%)/DCE/60 °C (Table 2).

Table 2 Thiophene amination through Cu-catalyzed reaction with NFSI

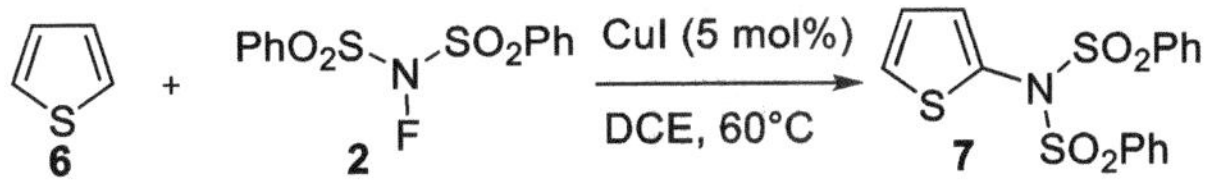

S. no.	Catalyst	Solvent	Temp (°C)	Yield (%)
1	CuI	DCE	25	52
2	CuI	DCE	40	76
3	**CuI**	**DCE**	**60**	**88**
4	CuCl	DCE	60	76
5	CuBr	DCE	60	80
6	CuCN	DCE	60	N.R
7	$CuCl_2$	DCE	60	56

2.2 *Metal-Free Chemical Syntheses with NFSI*

2.2.1 K_2CO_3

The first instance of aromatic hydrocarbon bonds being aminated without using the transition metal was stated by L. Yang and colleagues during 2013 [27]. Under oxidant-free environmental circumstances, the economically available *N*-fluorobenzenesulfonimide was used as the amino precursor in a regioselective and metal-free C-3 amidation of indoles. This amidation has great reactive functional unit resilience and regioselectivity and only needed the base catalytic quantity. It was advised that a free radical method may realize a C-3 regioselectivity. The reaction solution was heated at 60 °C for almost 24 h to produce optimum results (95% yield). Since more than 1.5 equiv. of NFSI were needed to catalyze the complete ingestion of 1-acetylindole, the yield was significantly reduced. Increasing the reaction temperature had no effect, while decreasing it resulted in a substantially lower yield with a greater percentage of the starting material remaining unreacted. The impact of solvents on this reaction was then examined, and low polarity DCE produced the highest yield.

Other typical solvents including THF, toluene, and ethanol were also useful but had lower yields. DMF and DMSO with higher polarities were almost useless. It is important to note that under all these circumstances, neither in situ gas chromatography-mass spectrometry nor one hydrogen nuclear magnetic resonance found any C-2 amidation products. The precursor range of the innovative metal-free C-3 amidation of indoles using NFSI via regioselective was investigated under optimal conditions. The first research impact of the *N*-protection of indole.

According to Lim [28], 3,3-difluoro-2-oxindole was obtained when the indole was substituted with *N*-protecting groups such as methyl, benzyl, and phenyl, resulting in boosting indole electron density. The direct C-3 amidation process went easily up to 95% in yields when the indole was *N*-substituted electron-withdrawing groups such as acetyl, benzoyl, pyrimidyl, and tosyl.

This suggests that procedure is distinct from that of the earlier described electrophilic fluorinating reaction [29] of alkyl substituted indoles by *N*-fluorobenzenesulfonimide. Higher yields were usually obtained in the transition metal-free C-3 amidation when there were more electron-poor indole substrates present. The 1-(acetylamino)-3methylindole and *N*-fluorobenzenesulfonimide reaction was undertaken to obtain insight into the reaction pathway of the transition metal-free C-3 amidation of indoles. Unusually, neither the C-2 amidation product nor an electrophilic fluorinated compound separation of the compound was achievable, and primary yield was identified as the benzylic amide product, which may indicate that the chemical synthesis is carried out through a free-radical pathway.

2.2.2 2,2,6,6-Tetramethyl-1-Piperidinyloxy (TEMPO)

It is stated that in a TEMPO-catalyzed reaction, *N*-Fluorobenzenesulfonimide served as the amination reagent for the direct amination of arene via carbon-hydrogen bond cleavage with high to exceptional yields and a comprehensive arene scope in the lack of any metal, ligand, or catalyst. This unique chemical synthesis at normal thermal state is the primary instance of carbon-hydrogen amination alongside *N*-Fluorobenzenesulfonimide that is realized by organocatalytic, as opposed to earlier metal-mediated amidation reactions, which typically need significant reaction temperatures.

Recent work by Li and colleagues uncovered the creation of C-5 aminated azoles at ordinary temperature through the direct amination of azoles via C–H bond activation using *N*-Fluorobenzenesulfonimide [30]. Under the conditions of TEMPO as the nitroxyl radical reactant and EtOAc as the carrier, the optimal reaction conditions were established. Many thiazole and oxazole compounds achieved effective C–N coupling reaction formation, the corresponding C5-aminated azoles were produced in acceptable to good yields (Scheme 3). Outcomes suggested this amination is likely an electrophilic substitution since substrates with more electrons provided better yields than those with less electrons. The ease of the reaction was remarkably unaffected by steric hindrance.

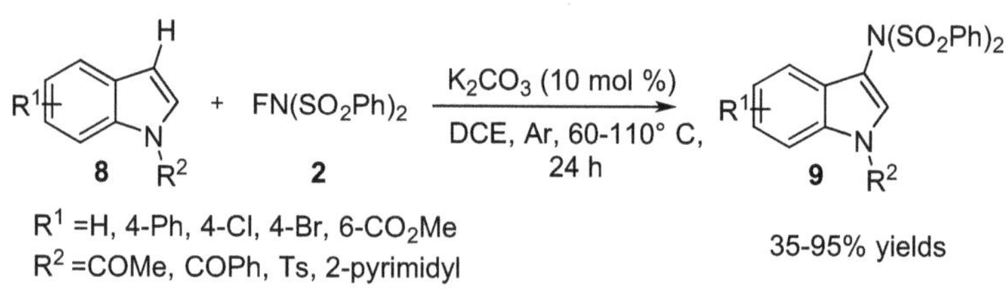

Scheme 3 Transition metal-free C3-amination of indoles

The authors also showed that their method could be used to directly aminated C(sp^2)-H bonds to thiopyran, furazan, pyrazole, Indene, benzofuran, benzothiazole, chromanone, and mesitylene. The amination of 2-dibenzofuran was studied using an intermolecular deuterium isotope fractionation effect experiment, and research discovered the rate-defining phase of reaction involves the breaking of the C–H bond.

3 NFSI as a Fluorine Source

The first electrophilic *N*-fluorinating reagents were presented by Umemoto and his team like *N*-fluoropyridinium salts. These substances provided sources of electrophilic fluorine that were more stable, secure, mild, manageable, affordable, and selective than the earlier reagents. Differding stated that there are already over 100 members of this class of fluorinating reagents, with many of them having found a market [31, 32].

Among the top well-known economically available nitrogen-fluorine reagents is *N*-Fluorobenzenesulfonimide (NFSI), was initially developed through Differding along with his colleagues. This is a moderate, reliable, and freely soluble crystalline dense served as an amination as well as a fluorination reagent in different transformation.

3.1 Reactions with the Transition Metal Using NFSI

3.1.1 Nickel-Assisted Reactions

With the use of N-Fluorobenzenesulfonimide as the "F" precursor and NiSO4 as the reactant, the first successful C–H selective fluorination of 8-aminoquinoline derivatives at the C-5 position has been accomplished using nickel catalyst. Under benign conditions, this technique displayed wide substrate tolerance, and the concept of a radical process included a credible SET (single electron transfer) mechanism.

By the usage of NFSI as the oxidant and fluorine precursor, Ding, Zhang, and Li illustrated the highly regioselective fluoridation of 8-amidoquinolines with Ni (Scheme 4) in 2017[33]. The only ingredient used in the procedure was $NiSO_4$, and neither an addition nor a ligand was necessary for the transformation. Several *N*-substituted 9-quinolines comprising aromatical, heterocyclic aromatic, acyclic, carbon chain, and bulky substituent-containing amides were used, to generate equivalent 5-fluorinated quinolines with yields varying from 32 to 46%. Later, Zeng and Li et al. observed that the transition relied on the presence of the amide's free N–H and the atom in the quinoline structure, as indicated by the outcomes [34, 35]. The co-authors demonstrated the amido groups can be effectively hydrolyzed under alkaline circumstances to yield the free amino (-NH_2) group without compromising

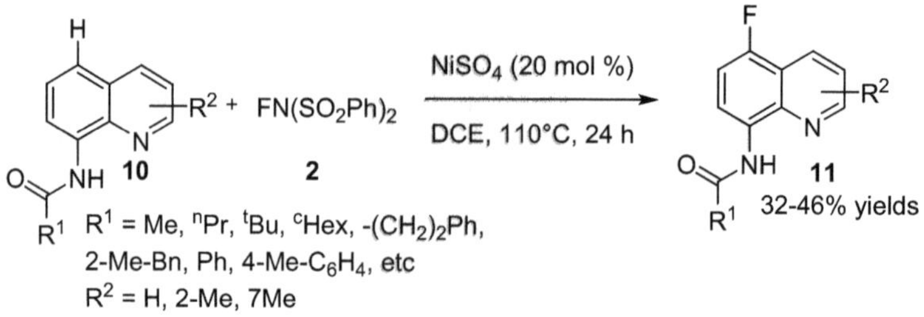

Scheme 4 Fluorinating 9-substituted quinolines via Ni-catalyzed reaction with NFSI

the C–F link. The control studies investigated whether a free radical cross-coupling route was involved in this fluorination reaction. To our understanding and literature survey, this was the initial and only instance where radical fluorination of C(aryl)-H chemical bonds was carried out utilizing NFSI using a nickel catalyst.

3.1.2 Zirconium-Assisted Reactions

Yamamoto and colleagues showed an effective approach for zirconium-catalyzed C–H fluorination of (hetero)arenes (Scheme 5) using directed method during 2005 [36]. With respect to comparable mild reaction factors, fluorination reactions also take place when *N*-fluorobenzene-sulfonimide (NFSI) is used as a source of fluorine. *N*-Fluorobenzenesulfonimide (NFSI), a moderate fluorinating compound, was chosen as the fluorine source. The fluorine source used was NFSI and it has been determined that Zirconium tetrachloride is the most efficient catalyst activated by Lewis's acid. Without using any extra chemicals, the reactions took place in DCM at room temperature and produced the matching mono-substituted with fluorine compounds in decent to admirable yields. To illustrate the fluorination process, there are only three examples provided. When NCS, NBS, or NIS were utilized, the procedure was also coherent for chloro-substitution, bromine addition, and iodine incorporation reactional process. The use of a significant amount of catalyst boosted the yield of fluorinated product, but it also raised the amount of some unidentified by-products.

3.2 *Solvent-Mediated Reactions and Metal-Free Reactions with NFSI*

Andreev et al. provided the initial broad document on the t fluorination of aromatic C–H bonds without the use transitions metal free utilizing NFSI as a fluorinating reagent in 2009 [37]. They demonstrated this type, in the lack of any inducer or substance, the reaction of a series of aromatics with electron-donating groups, such as methoxybenzenes, aromatic alcohols, toluols, and naphthalene derivatives, using

Scheme 5 Fluorination of arenes with NFSI using zirconium

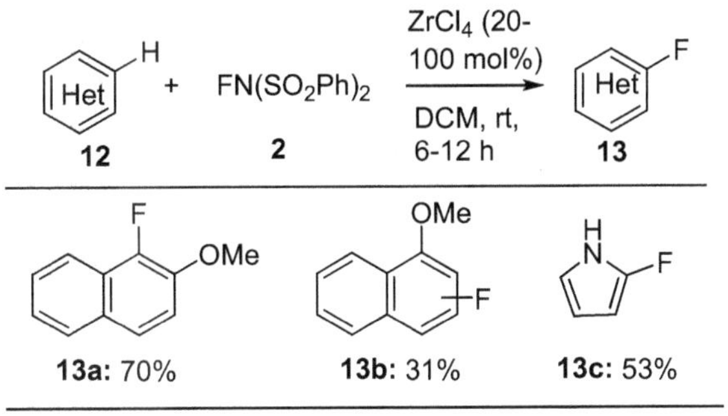

NFSI and under solvent-free factors yielded monofluorinated products in minimal to moderate forms.

3.2.1 Acetonitrile (CH_3CN)

Kobayashi et al.'s 1982 demonstration of mono-fluorination using gaseous fluorine was a vital milestone in the advancement of fluorination reactions. However, the utilization of gaseous fluorine can be challenging due to safety concerns, and the specialized equipment required to handle it [38].

Subsequently, a series of electrophilic fluorinating reagents was developed by R. E Banks and K. Adachi, along with their team, which are easier to handle and less toxic compared to gaseous fluorine [39–41]. These reagents include the commonly used Selectfluor and *N*-fluorobenzenesulfonimide (NFSI), among others. These reagents have made it easier and safer to perform fluorination reactions, expanding the range of synthetic possibilities in the field of organic chemistry.

N-protected pyridone derivatives are directly monofluorinated with fluorinating reagents such as *N*-fluorobenzenesulfonimide. F. Sakural, and the group highlighted the remarkable side selectivity observed for fluorine substitution in the pyridine substrate, specifically targeting the position opposite the carbonyl group, was found to be highly intriguing [42].

The report on the direct mono-substitute with fluorine of *N*-protected pyridone derivatives at the C5-position using NFSI under catalyst-free factors by Sakurai's research group is interesting (Scheme 6). The fact that the reactions were conducted using MeCN at 60 °C and withstood various *N*-substituted with alkyl group, –alpha-toluyl, and aryl-substituted mono and bicyclic 2-pyridinois while yielding the targeted C5-fluorinated products in decent yields demonstrates the method's potential applicability in synthesis of fluorinated pyridone derivatives with diverse substituent patterns.

Scheme 6 Catalyst-free-mono-fluorination of *N*-protected pyridones using NFSI with regioselectivity

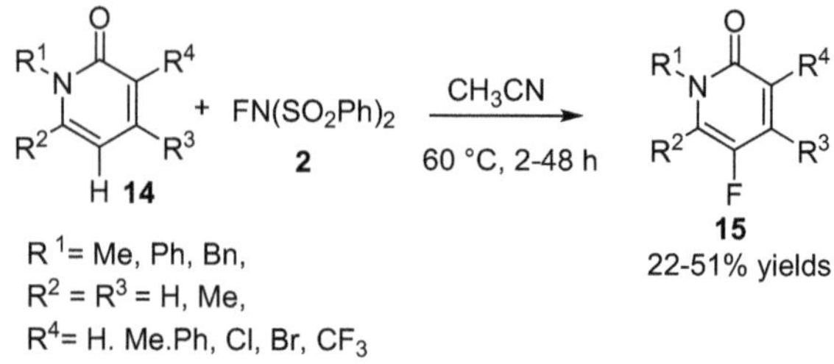

3.2.2 Bromo Benzene Solvent-Mediated Reaction

The reaction factors for the direct fluoridation of 2,5-diarylthiazoles were optimized with *N*-fluorobenzenesulfonimide as the fluorinating source, acetonitrile as the solvent, and a tiny amount of water in the reaction solution. Spur on the monofluorinated product development over tri-fluorinated product, water was added to the reaction solution. The improvement of the reaction was monitored using Thin Layer Chromatography and it was performed at 80 °C for 24 h as described by Hatfield [43].

X-ray crystallography was used to establish the crystal structure of one of the unusual 4,5-trifluoro-1,2-thiazole derivatives. According to the crystal lattice, the trifluoro thiazole product has a transoid shape regarding the aryl substituents. This transoid conformation is thought to be responsible for the 2,5-diarylthiazoles' particular reactivity to trifluorination.

The yield of mono and tri-fluorinated thiazole products remains restricted, as reported by S. M. Riyadh and team and the optimized reaction conditions and crystal structure studies provide important insights into the reactivity of 2,5-diarylthiazoles to direct fluorination (Scheme 7) employing N–F reagents [44].

The difficulties in separating the mono- and tri-fluorinated products by chromatography may have contributed to the low yields of mono-fluorine substituted products throughout the reaction of 2,5-diphenylthiphenes using NFSI in refluxing

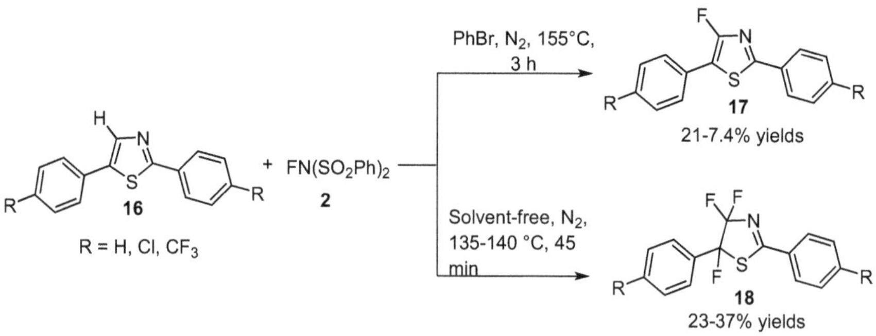

Scheme 7 NFSI-mediated mono and trifluorination of thiazoles

bromobenzene. However, group of Stephens revealed C–H bond fluorination of thiazoles without catalyst is possible, albeit with low yields [45].

3.2.3 With Selenium

It is worth noting that using selenium as a catalyst for amination processes has various advantages over other transition metals, including lower toxicity and cost, as well as the capacity to function at gentler reaction conditions. The production of a selenenyl amide intermediate, which then experiences a rearrangement to generate the targeted product, is a selenium-catalyzed amination reactions mechanism (Scheme 8).

Other reagents, such as *N*-fluoropyridinium salts and 2,2,6,6-tetramethylpiperidine-*N*-oxyl, can be employed for selenium-catalyzed amination processes in addition to NFSI as a terminal oxidant and nitrogen supply. J. Trenner with his team reported that these methods provide a variety of possibilities for efficiently and selectively nitrogenating alkenes, allowing the synthesis of a varied spectrum of nitrogen-containing molecules [46].

In our research toward the improvement of new protocols for the catalytic and chemoselective nitrogenation of non-reactive olefins, we were captivated by the utilization of organodiselanes as potent catalytic agent with redox properties, as described by Sharpless et al.48 and Tunge et al. [47] and reported that reactions of simple alkenes with *N*-halosuccinimides (halogen = Cl, Br) in the occurrence of selenium catalysts yield the related allylic and vinylic halides. To the best of our knowledge, no homologous intermolecular catalyzed by selenium reactions leading to allylic or vinylic imide derivatives was documented thus far, as stated by A. Toshimitsu and the team [48].

Intrinsic propensity for halogen transmission over nitrogen integration is a significant issue in designing such a substitution reaction employing *N*-haloimides as the terminal oxidizing agent. The possible solution for the challenge, E. Vessally describing a selenium catalysis technique for the efficient construction of a wide range of 2- and 4-substituted allylic and vinylic imides, employing *N*-fluorobenzenesulfonimide [49] as the terminal oxidizing agent and nitrogen source, respectively.

In the tests, alkene substrate 1 was underwent with 1 equivalent of N-fluorobenzenesulfonimide in THF (0.2 M) with catalyst PhSeBr (5 mol%). However, no product production was seen under these conditions. According to the literature, it has been reported that PhSeBr can be added to electron-neutral alkenes, leading to the transient formation of aryl alkylselanes, as described by Denis [50] These

Scheme 8 Selenium-catalyzed allylic imidation

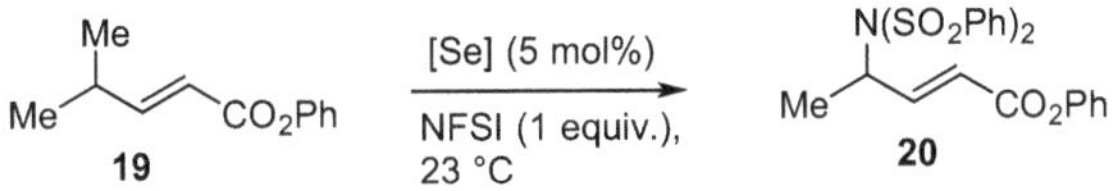

findings show NFSI is non-reactive enough to instigate dehydroselenylation, the key step leading to the formation of allylic imide.

3.2.4 Copper

Understanding the reaction pathway of the NFSI-spurred Cu-catalyzed aromatic C–H imide synthesis process is critical for optimizing reaction conditions and creating novel catalysts. It was proposed which the reaction takes place via radical mechanism, during which CuBr and the bipyridine ligand forms Cu(I) complex, that is activated by NFSI to produce a Cu (III)-imido intermediate. This intermediate is then combined with the aromatic substrate in sole electron transfer (SET) reaction to produce a radical cation intermediate, which is then nitrogen inserted to create the targeted imidation output. A radical cation intermediate can also renew the Cu(I) complex, completing the catalytic cycle.

Computational investigations have revealed details about the reaction mechanism and identified the major intermediates and transition states involved. According to density functional theory (DFT) calculations, the species that are in the reaction is the Cu (III)imido transition state, and the rate-controlling step is the SET process between the Cu (III)-imido and the aromatic substrate. The bipyridine ligand's nature has also been demonstrated to influence the catalyst's reactivity, with the 6,6'-Me-2-bpy ligands encouraging the production of the Cu (III)-imido intermediate via its strong-donating capacity and ability to stabilize the Cu(I) complex.

Itami and coworkers reported, in general, it is critical to optimize and improve the copper-catalyzed aromatic C–H imidizing reacting with *N*-fluorobenzenesulfonimide. The one-electron versus two-electron processes of NFSI in this reaction can be influenced by unique reaction conditions as well as the substrate utilized as emphasized by E. Haveli and Diau and their team [51–53].

The discovery of any intermediates or reaction pathways involved in the process can offer valuable into the selectivity and efficiency of the reaction, potentially enabling the creation of related catalytic reactions (Scheme 9), as emphasized by Surry and Shin [54, 55]. Furthermore, a greater knowledge of the reaction process can aid in the creation of new and improved ligands for the Cu catalyst, thereby widening the scope of the reaction even further.

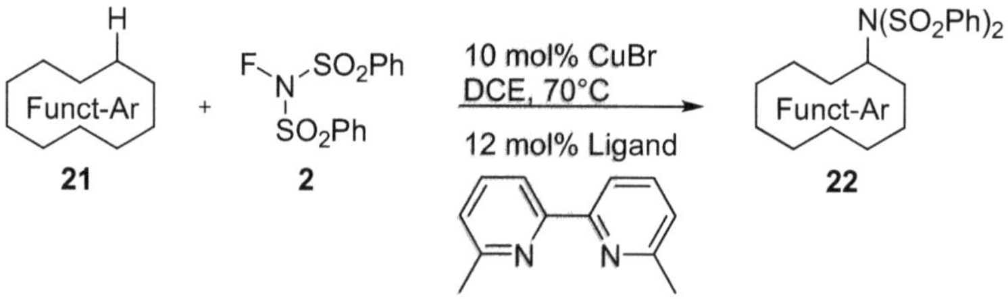

Scheme 9 CuBr-aromatization catalyst C–H imidization using NFSI

Segawa described, the suggested pathway inadequacy the reaction-specific information required to deliberately engineer more potent catalysts for highly selective C–H iminization and forecast the site-selective product formation [56]. A thorough knowledge of the pathway and controlling parameters of copper-induced aromatic C–H imidization by *N*-fluorobenzenesulfonimide was pursued through a collaborative theoretical and empirical effort, led by Zhang and their team [57].

4 Conclusion

N-fluorobenzenesulfonimide is a chemical reagent that is utilized in organic synthesis as a nitrogen and fluorine source. The multifaced NFSI is used to add fluorine or nitrogen atoms at specified places in C–H bonds in (hetero)aromatic compounds. It has been demonstrated that it possesses strong site-selectivity and broad functional group compatibility in a numerous reaction. Because of its capacity to emit fluorine gas upon reduction, NFSI can also be utilized as an oxidizing agent in specific reactions. NFSI is a commercially available and reasonably priced reagent that can be used in both academic and industrial contexts. To facilitate the direct of C–H bond functionalization via fluorination and amination in (hetero)aromatic compounds, *N*-fluorobenzenesulfonimide (NFSI) can be employed in combination with transition metal catalysts. Within these reactions, this transition catalyst facilitates the activation of the carbon-hydrogen bond, while the NFSI acts as a source of fluorine or nitrogen, depending on the reaction conditions. Transition metal catalysts increase the reaction efficiency and selectivity, allowing for softer reaction conditions and the synthesis of more complex compounds. Palladium, rhodium, and copper catalysts are some of the most often utilized transition metal catalysts in these processes.

N-fluorobenzenesulfonimide (NFSI) was employed within metal-free transition of C–H functionalization processes, including C–H bond fluorination and amination of (hetero)aromatic compounds through direct functionalization. This mechanism allows a moderate synthesis of a variety of fluorination and amination of (hetero)arenes and (hetero)aromatic compounds without requiring transition metals catalysts. *N*-fluorobenzenesulfonimide has a future potential as a flexible reagent for C–H functionalization processes which have been already explored. While NFSI has shown great promise in the C–H bond fluorination and amination of (hetero)aromatic compounds through direct functionalization, more efficient as well as milder catalytic systems that can broaden the scope of reactions and improve effectiveness, particularly in the preparation intricate compounds, are even so needed.

Furthermore, evolution of novel techniques that allow the use of NFSI in other types of transformations, such as selective oxidation or reduction processes, could be an intriguing field of investigation. The improvement of novel techniques that allow the use of NFSI in other types of transformations, such as selective oxidation or reduction processes, could be an intriguing field of investigation. The investigation of the mechanism of NFSI-mediated reactions and discovering new reactivity modes may aid in the advancement of highly efficient and distinct synthetic techniques.

The C–N bond formation has increased substantial developments from last few years that consider its enormous impact on the progress of modern approaches using certainly accessible reagents. The methods designated in the review disclosed that NFSI show dual behavior in the reaction. This review gave a notable prospect for the medicinal pharmacists to strategy new drug molecule in short reaction sequence and synthetic chemist will explore the structure of diversely products to increase the utility of this NFSI reagent.

Acknowledgements A.P., S.G., R.S., and P.C. would like to acknowledge AIAS, Amity University, Noida, Uttar Pradesh for the work done.

Conflict of Interest On behalf of all authors, the corresponding author states that there is no conflict of interest.

References

1. (a) Xia J-B, Zhu C, Chen C (2013) Visible light-promoted metal-free C–H activation: diarylketone-catalyzed selective benzylic mono- and difluorination. J Am Chem Soc 135:17494–17500. (b) Gua Q, Vessally E (2020) *N*-Fluorobenzenesulfonimide: a useful and versatile reagent for the direct fluorination and amination of (hetero)aromatic C–H bonds. RSC Adv 10:16756–16768. (c) Sushmita, Aggarwal T, Kumara S, Verma AK (2020) Exploring the behavior of the NFSI reagent as a nitrogen source. Org Biomol Chem 18:7056–707
2. (a) Emsley (1980) Very strong hydrogen bonding. J Chem Soc 9:91–124. (b) Differding E, Ofner H (1991) *N*-Fluorobenzenesulfonimide: a practical reagent for electrophilic fluorinations. Synlett 3:187–189 (c) Umemoto T, Yang Y, Hammond GB (2021) Development of N-F fluorinating agents and their fluorinations: historical perspective. Beilstein J Org Chem 17:1752–1813
3. (a) Garai S, Kulkarni PM, Schaffer PC, Leo LM, Brandt AL, Zagzoog A, Black T, Lin X, Hurst DP, Janero DR, Abood ME, Zimmowitch A, Straiker A, Pertwee RG, Kelly M, Szczesniak A-M, Denovan-Wright EM, Mackie K, Hohmann AG, Reggio PH, Laprairie RB, Thakur GA (2020) Application of fluorine- and nitrogen-walk approaches: defining the structural and functional diversity of 2-phenylindole class of cannabinoid 1 receptor positive allosteric modulators. J Med Chem 63(2):542–568. (b) Hagmann WK (2020) The many roles for fluorine in medicinal chemistry. J Med Chem 63(2):542–568. (c) Kerru N, Gummidi L, Maddila S, Gangu KK, Jonnalagadda SB (2020) A review on recent advances in nitrogen-containing molecules and their biological applications. Molecules 25(8):1909–1951
4. Liang T, Neumann CN, Ritter T (2013) Introduction of fluorine and fluorine-containing functional groups. Angew Chem Int Ed 52:8214–8264
5. Barker TJ, Jarvo ER, Barker TJ, Jarvo ER (2011) Synthesis 24:3954–3964
6. Yadav S, Aggarwal T, Kumar S, Verma AK (2020) Nitrogen containing heterocycles as anticancer agents: a medicinal chemistry perspective. Org Biomol Chem Pharmaceuticals 16(2):299
7. Wencel-Delord J, Drage T, Liu F, Glorius F (2011) Towards mild metal-catalyzed C-H bond activation. Chem Soc Rev 40:4740–4761
8. Kirk KL (2008) Fluorination in medicinal chemistry: methods, strategies, and recent developments. Org Process Res Dev 12(2):305–321
9. (a) Beletskaya IP, Cheprakov AV (2004) Copper in cross-coupling reactions: the post-Ullmann chemistry. Coordination Chem Rev 248:2337–2364. (b) Ley SV, Thomas AW (2003) Modern

synthetic methods for copper-mediated C(aryl)-O, C(aryl)-N, and C(aryl)-S bond formation. Angew Chem Int Ed Engl 115:5558 (c) Muci AR, Buchwald SL (2002) Topics in current chemistry 219:131–209. (d) Hartwig JF (1998) Carbon−heteroatom bond-forming reductive eliminations of amines, ethers, and sulfides. Acc Chem Res 31(12):852–860. (e) Hartwig JF (1998) Transition metal catalyzed synthesis of arylamines and aryl ethers from aryl halides and triflates: scope and mechanism. Angew Chem Int Ed 37:2046–2067

10. (a) Hili R, Yudin AK (2006) Making carbon-nitrogen bonds in biological and chemical synthesis. Nat Chem Biol 2:284–287. (b) Ricci A (2008) Amino group chemistry: from synthesis to the life science. Wiley-VCH, Weinheim, Germany. (c) Taylor EC, Jones RA (1990) The chemistry of heterocyclic compounds: pyrroles. Wiley, New York. (d) Sundberg RJ (1996) The chemistry of indoles. Academic, New York. (e) Smith MB, March J (2001) March's advanced organic chemistry: reaction, mechanisms, and structure, 5th edn. Wiley, New York
11. For selected reviews on copper-mediated Ullmann-type amination/amidation, see: (a) Qiao JX, Lam PYS (2011) Copper-promoted carbon-heteroatom bond cross-coupling with boronic acids and derivatives. Synthesis, 829–856. (b) Surry DS, Buchwald SL (2010) Diamine ligands in copper-catalyzed reactions. Chem Sci 1:13. (c) Monnier F, Taillefer M (2009) Catalytic C-C, C-N, and C-O Ullmann-type coupling reactions. Angew Chem Int Ed 48:6954. (d) Ma D-W, Cai Q (2008) Copper/amino acid catalyzed cross-couplings of aryl and vinyl halides with nucleophiles. Accounts Chem Res 41(11):1450–1460. (e) Evano G, Blanchard N, Toumi M (2008) Copper-mediated coupling reactions and their applications in natural products and designed biomolecules synthesis. Chem Rev 108(8):3054–131
12. For selected reviews on oxidative direct C−H amination of arenes with amines, see: (a) Louillat ML, Patureau FW (2014) Oxidative C–H amination reactions. Chem Soc Rev 43:901–910. (b) Jeffrey JL, Sarpong R (2013) Intramolecular C(sp^3)–Hamination. Chem Sci 4:4092–4106. (c) Wencel-Delord J, Dröge T, Glorius F (2011) Towards mild metal-catalyzed C–H bond activation. Chem Soc Rev 40:4740. (d) Cho SH, Kim JY, Kwak J, Chang S (2011) Recent advances in the transition metal-catalyzed twofold oxidative C–H bond activation strategy for C–C and C–N bond formation. Chem Soc Rev 40:5068–5083
13. For selected recent examples, see: (a) Shang M, Sun S-Z, Yu J-Q (2014) Cu(II)-mediated C–H amidation and amination of arenes: exceptional compatibility with heterocycles. J Am Chem Soc 136(9):3354–3357. (b) Tran BL, Li B-J, Driess M, Hartwig JF (2014) Copper-catalyzed intermolecular amidation and imidation of unactivated alkanes. J Am Chem Soc 136(6):2555–2563. (d) Mei T-S, Leow D, Xiao H, Yu J-Q (2013) Synthesis of Indolines via Pd(II)-catalyzed amination of C–H bonds using $PhI(OAc)_2$ as the bystanding oxidant. Org Lett 15:3058–3061. (e) Kantak AA, Potavathri S, Barham RA, Ramano KM, DeBoef B (2011) Metal-free intermolecular oxidative C–N bond formation via tandem C–H and N–H bond functionalization. J Am Chem Soc 133:19960–19965. (f) Cho SH, Yoon J, Chang S (2011) Intramolecular oxidative C−N bond formation for the synthesis of carbazoles: comparison of reactivity between the copper-catalyzed and metal-free conditions. J Am Chem Soc 133:5996. (g) Mei T-S, Wang X-S, Yu J-Q (2009) Pd(II)-catalyzed amination of C-H bonds using single-electron or two-electron oxidants. J Am Chem Soc 131:10806. (h) Nadres ET, Daugulis O (2012) Heterocycle synthesis via direct C H/N H coupling. J Am Chem Soc 134:7–10
14. (a) Monguchi D, Fujiwara T, Furukawa H, Mori A (2009) Direct amination of azoles via catalytic C−H, N−H coupling. Organic Lett 11(7):1607–1610. (b) Cho SH, Kim JY, Lee SY, Chang S (2009) Silver-mediated direct amination of benzoxazoles: tuning the amino group source from formamides to parent amines. Angew Chem Int Ed 48:9291–9294. (c) Kim JY, Cho SH, Joseph J, Chang S (2010) Cobalt- and manganese-catalyzed direct amination of azoles under mild reaction conditions and the mechanistic details. Angew Chem Int Ed 49:9899. (d) Wang Q, Schreiber SL (2009) Copper-mediated amidation of heterocyclic and aromatic C−H bonds. Org Lett 11:5178. (e) Chen X, Hao X-S, Goodhue CE, Yu J-Q (2006) Pd-catalyzed ortho C–H hydroxylation of benzaldehydes using a transient directing group. J Am Chem Soc 128:6790
15. (a) Kawano T, Hirano K, Satoh T, Miura M (2010) A new entry of amination reagents for heteroaromatic C−H bonds: copper-catalyzed direct amination of azoles with chloroamines at

room temperature. J Am Chem Soc 132:6900–6901. (b) Ng K-H, Chan ASC, Yu W-Y (2010) Pd-catalyzed intermolecular C–H amination with alkylamines. J Am Chem Soc 132:12862. (c) Yoo EJ, Ma S, Mei TS, Chan KSL, Yu J-Q (2011) Pd-catalyzed intermolecular C-H amination with alkylamines. J Am Chem Soc 133:7652. (d) Grohmann C, Wang H, Glorius F (2012) Rh[III]-catalyzed direct C–H amination using N-chloroamines at room temperature. Org Lett 14:656. (e) Allen LJ, Cabrera PJ, Lee M, Sanford MS (2014) N-acyloxyphthalimides as nitrogen radical precursors in the visible light photocatalyzed room temperature C–H amination of arenes and heteroarenes. J Am Chem Soc 136:5607

16. He L, Chan PWH, Tsui W-M, Che C-M (2004) Ruthenium(II) porphyrin-catalyzed amidation of aromatic heterocycles. Org Lett 6(14):2405–2408
17. Xiong T, Li Y, Lv Y, Zhang Q (2010) Remote amide-directed palladium-catalyzed benzylic C-H amination with N-fluorobenzenesulfonimide. Chem Commun 46:6831
18. Tan Y, Hartwig JF (2010) Palladium-catalyzed amination of aromatic C−H bonds with oxime esters. J Am Chem Soc 132(11):3676–3677
19. Sun K, Li Y, Xiong T, Zhang J, Zhang Q (2011) Palladium-catalyzed C−H aminations of anilides with N-fluorobenzenesulfonimide. J Am Chem Soc 133:1694
20. Qui S, Xu T, Zhou J, Guo Y, Liu G (2010) Palladium-catalyzed intermolecular aminofluorination of styrenes. J Am Chem Soc 132:2856
21. Yip SJ, Kawakami T, Murakami K, Itami K (2018) Gold-catalyzed C−H imidation of polycyclic aromatic hydrocarbons. Asian J Org Chem 7:1372–1375
22. Examples on Au–bipyridine complexes: (a) Chernyshev AN, Chernyshev MV, Hirva P, Kukushki VY, Haukka M (2015) Weak aurophilic interactions in a series of Au(III) double salts, Dalton (Trans) 44:14523–14531. (b) Harper MJ; Arthur CJ, Crosby J, Emmett EJ, Falconer RL, Fensham-Smith AJ, Gates PJ, Leman T, McGrady JE, Bower JF, Russell CA (2018) Oxidative addition, transmetalation, and reductive elimination at a 2,2′-Bipyridyl-Ligated gold center. J Am Chem Soc 140:4440–4445
23. Amine-substituted PAHs are widely utilized in organic ELs: (a) Shirota Y, Kageyama H (2007) Charge carrier transporting molecular materials and their applications in devices. Chem Rev 107(4):953–1010. (b) Tao Y, Yang C, Qin (2011) Organic host materials for phosphorescent organic light-emitting diodes. J Chem Soc Rev 40:2943. (c) Figueira-Duarte TM, Müllen K (2011) Pyrene-based materials for organic electronics. Chem Rev 111:7260–7314
24. Haines BE, Kawakami T, Kuwata K, Murakami K, Itami K, Musaev DG (2017) Cu-catalyzed aromatic C-H imidation with N-fluorobenzenesulfonimide: mechanistic details and predictive models. Chem Sci 8:988–1001
25. For related Cu-catalyzed reactions using NFSI as a nitrogen source, see: (a) Ni Z-K, Zhang Q, Xiong T, Zheng Y-Y, Zhang J-P, Liu Q (2012) Highly regioselective copper-catalyzed benzylic C-H amination by N-fluorobenzenesulfonimide. Angew Chem Int Ed 51:1244. (b) Zhang B, Studer A (2014) Copper-catalyzed intermolecular aminoazidation of alkenes. Organic Lett 16(6):1790–1793
26. Trenner J, Depken C, Weber T, Breder A (2013) Direct oxidative allylic and vinylic amination of alkenes through selenium catalysis. Angew Chem Int Ed 52:8952–8956
27. Liu HH, Wang Y, Deng G, Yang L (2013) Transition metal-free regioselective C-3 amidation of indoles with N-fluorobenzenesulfonimide. Adv Synth Catal 355:3369–3374
28. Sibbald PA, Rosewall CF, Swartz RD, Michael FE (2009) Mechanism of N-fluorobenzenesulfonimide promoted diamination and carboamination reactions: divergent reactivity of a Pd(IV) species. J Am Chem Soc 131(43):15945–15951
29. Lim YH, Ong Q, Duong HA, Nguyen TM, Johannes CW (2012) Direct conversion of indoles to 3,3-difluoro-2-oxindoles via electrophilic fluorination. Org Lett 14(22):5676–5679
30. Miao Q, Shao Z, Shi C, Ma L, Wang F, Fu R, Gao H, Li Z (2019) Metal-free C-H amination of arene with *N*-fluorobenzenesulfonimide catalysed by nitroxyl radicals at room temperature. Chem Commun 55:7331–7334
31. Differding E, Lang RW (1988) New fluorinating reagents—I. The first enantioselective fluorination reaction. Tetrahedron Lett 29:6087–6090

32. Differding E, Lang RW (1989) New fluorinating reagents. Part II†. Preparation and synthetic application of a Saccharin derived N-Fluorosultam. Chem Acta 72:1248–1252
33. Ding J, Zhang Y (2017) Li, Nickel-catalyzed selective C-5 fluorination of 8-aminoquinolines with NFSI. J Org Chem Front 4:1528–1532
34. Cong XF, Zeng XM (2014) Iron-catalyzed, chelation-induced remote C-H allylation of quinolines via 8-amido assistance. Org Lett 16:3716–3719
35. Chen H, Li P, Wang M, Wang L (2016) Nickel-catalyzed site-selective C-H bond difluoroalkylation of 8-aminoquinolines on the C5-position. Org Lett 18(19):4794–4797
36. Zhang Y, Shibatomi K, Yamamoto H (2005) Lewis acid catalyzed highly selective halogenation of aromatic compounds. Synlett 18:2837–2842
37. Andreev RV, Borodkin GI, Shubin VG (2009) Fluorination of aromatic compounds with N-fluorobenzenesulfonimide under solvent-free conditions. Russ J Org Chem 45:1468–1473
38. Kobayashi Y, Kumadaki I, Yamashita T (1982) Experimental and theoretical studies on rhodium-catalyzed coupling of benzamides with 2,2-difluorovinyl tosylate: diverse synthesis of fluorinated heterocycles. Heteroc 17:429430
39. Banks RE, Mohialdin-Khaffaf SN, Lal GS, Sharif I, Syvret R (1992) G. 1-Alkyl-4-fluoro-1,4-diazoniabicyclo[2.2.2]octane salts: a novel family of electrophilic fluorinating agents. J Chem Soc Chem Commun, 595–596
40. Umemoto T, Nagayoshi M, Adachi K, Tomizawa G (1998) Synthesis, properties, and reactivity of N, N'-difluorobipyridinium and related salts and their applications as reactive and easy-to-handle electrophilic fluorinating agents with high effective fluorine content. J Org Chem 63(10):3379–3385
41. Differding E, Ofner H (1991) *N*-Fluorobenzenesulfonimide: a practical reagent for electrophilic fluorinations. Synlett 1991(3):187–189
42. Sakurai F, Yukawa T, Taniguchi T (2019) Direct and regioselective monofluorination of N-protected pyridone derivatives using N-fluorobenzenesulfonimide (NFSI). Org Lett 21(18):7254–7257
43. Hatfield JM, Stephens CE (2013) Mono- and trifluorination of the thiazole ring of 2,5-diarylthiazoles using N-fluorobenzenesulfonimide (NFSI). Tetrahedron Lett 54(8):1025–1028
44. Riyadh SM, Fuchigami T (2002) Trolytic partial fluorination of organic compounds. 64.1 anodic mono- and difluorination of thiazolyl sulfides. J Org Chem 67:9379–9383
45. Hatfield JM, Eidell CK, Stephens CE (2013) Mono- and trifluorination of the thiazole ring of 2,5-diarylthiazoles using N-fluorobenzenesulfonimide (NFSI). Tetrahedron Lett 54(8):1025–1028
46. Trenner J, Depken C, Weber T, Breder A (2013) Direct oxidative allylic and vinylic amination of alkenes through selenium catalysis. Angew Chem Int Ed 52(34):8952–8956
47. Tunge JA, Mellegaard SR (2004) Selective selenocatalytic allylic chlorination. Org Lett 6(8):1205–1207
48. For representative stoichiometric procedures, see (a) Toshimitsu A, Aoai T, Owada H, Uemura S, Okano M (1981) Oxy- and aminoselenation of alkenes utilizing an isolable selenenyl iodide. J Org Chem 46:4727. (b) Tiecco M, Testaferri L, Santi C, Tomassini C, Marini F, Bagnoli L, Temperini A (2000) Asymmetric amidoselenenylation of alkenes promoted by camphorselenenyl sulfate: a useful synthetic route to enantiopure oxazolines. Eur J Org Chem, 3451
49. Gu Q, Vessally E (2020) *N*-Fluorobenzenesulfonimide: a useful and versatile reagent for the direct fluorination and amination of (hetero)aromatic C-H bonds. RSC Adv 10:16756–16768
50. Denis JN, Vicens J, Kief A (1979) New synthetic routes to β-hydroxyselenides and β-azidoselenides. Tetrahedron Lett 20:2697–2700
51. (a) Halevi EA (2015) The role of electrostatic induction in secondary isotope effects on acidity. New J Chem 39:1522–1524. (b) Perrin CL (2015) The role of electrostatic induction in secondary isotope effects on acidity. New J Chem 39:1517–1521
52. (a) Hartwig JF (2008) Ion of a fourth generation catalyst for the amination and thioetherification of aryl halides. Acc Chem Res 41(11):1534–1544. (b) Ley SV, Thomas AW (2003) Modern synthetic methods for copper-mediated C(aryl)-O, C(aryl)-N, and C(aryl)-S bond formation. Angew Chem Int Ed 42:5400–5449

53. Li LL, Diau EWG (2013) Porphyrin-sensitized solar cells. Chem Soc Rev 42:291–304
54. Surry DS, Buchwald SL (2008) Biaryl phosphane ligands in palladium-catalyzed amination. Angew Chem Int Ed 47(34):6338–6361
55. Shin K, Kim H, Chang S (2015) Transition-metal-catalyzed C-N bond forming reactions using organic azides as the nitrogen source: a journey for the mild and versatile C–H amination. Acc Chem Res 48(4):1040–1052
56. Segawa Y, Maekawa T, Itami K (2015) Synthesis of extended π-systems through C-H activation. Angew Chem Int Ed 54:66–81
57. Zhang SL, Bie WF, Huang L (2014) Theoretical insights into mechanisms for copper(I)-catalyzed C-P coupling of diarylphosphines with aryl halides: a combined solvent and ancillary ligand effect on the identity of active catalyst and reaction mechanism. Organometallics 33:5263–5271

A Facile Method for Removal of Methylene Blue Dye from Wastewater Using Biochar-Modified Chitosan Beads

Shivani Vaishnav, Mousumi Sen, and Christine Jeyaseelan

1 Introduction

Water is a very crucial natural resource for the continuity of life on earth. Water shortage is a global problem due to contamination and drought that is intensifying, and the world is worried about [1–3]. The demand for water is rising daily as a result of the population growth rate, as well as industrialization and urbanization are speeding up quickly, leading to a significant rise in contaminated effluent waste. Pollutants and pollution have reached dangerous levels which have an immediate negative impact on human health while also harming the environment. Accordingly, it is most important to obtain efficient methods to remove heavy metals, dyes, and toxic chemicals from wastewater. Some methods have been used for this purpose such as physical adsorption, chemical adsorption, electrocoagulation, membrane filtration, and so on. Different methods are utilized according to the need or demand. Among all these methods, the surface phenomena or adsorption process is a low-priced method as well as gives successful productivity. This is an effective process due to the use of inexpensive, nontoxic, easy adsorbability and recyclable materials. Major water pollutants are dyes, metals, toxic chemicals, plastic, pharmaceutical, nitrates, phosphate, etc. [4]. These pollutants affect both surface and ground water. Among all the pollutants listed, dyes are one of the most important classes of pollutants. These pollutants affect human internal organs and cause health problems. Using biochar adsorbent was found to be a facile and effective method to adsorb dyes from aqueous solutions due to its excellent adsorption properties, small size, high degree of micro porosity. Volume of pores is very small in size that enhances the surface area of the material, which is easily available for chemical reactions and any adsorption process [5, 6].

S. Vaishnav · M. Sen · C. Jeyaseelan (✉)
Department of Chemistry, Amity Institute of Applied Sciences, Amity University, Noida, India
e-mail: cjeyaseelan@amity.edu

S. Rattan et al. (eds.), *Emerging Trends in Synthesis and Catalysis in Chemistry*,
Springer Proceedings in Materials 51, https://doi.org/10.1007/978-981-97-5169-3_16

Biochar is utilized in various fields such as purification of gas, gold processing, metal extraction, water purification, drug delivery, medicine tissue or cell engineering, sewage or wastewater treatment, in gas mask filter of air and respirators, compressed air filtration, and many other applications. Formation of biochar can be produced from various rich materials used as carbon sources that include nutshells, coconut husks, peat, wood, coir, coal, rice husk, corn cob, petroleum pitch, etc. [5]. Typically, heated gases such as steam, carbon dioxide, nitrogen, etc., are applied to the carbon material to create biochar. Hot gases are supplied using muffle furnaces. Carbonization is the method of using a sample with a high carbon content and pyrolyzing it at temperatures between 500 and 1000 °C [5–7]. Thus, peanut shell biochar beads are a great choice as a low cost-effective method and capable of removal of pollutants from wastewater.

Adsorption is a method that involves mass transfer between the two phases, such as when liquids and gases come into contact on the surface, or between solid gas and solid liquid. Adsorbents serve to remove a particular pollutant (adsorbate) from contaminated water due to the intermolecular force of attraction that exists between the two substances [7]. This process is easy, fast, and cheap. They are of two types: if the connection exhibits a weak physical behavior, such as Vander Waals forces and a reversible process, the process is referred to as physisorption. The interaction of chemical bonds between a solid surface and an adsorbate surface is a component of chemisorption. Typically, it only exists as a monolayer, and because of the strong attraction forces present, adsorbates are virtually never eliminated [7, 8]. There are numerous parameters that influence the adsorption process, such as molecule surface area, starting adsorbate material concentration, and solution pH, (acidic, basic, or neutral), temperature rate, interfering substances, dose of adsorbent, etc. Materials made from inexpensive and easily accessible solid agricultural wastes such rice husk, straw, coconut husk, and sawdust fruit peels have been tested for the removal of pollutants with the help of adsorption technology [8, 9]. In essence, wastewater can be cleaned by thick bacterial biomass being produced from dissolved organic materials [9, 10].

Chitosan is the natural and second most abundant biopolymer on earth which is available in the form of powder or flakes and low-cost alternative biopolymer. It possesses properties such as nontoxic, linear, heterogeneous (mixture of two components), cationic, easily biodegraded, and high molecular weight polysaccharide. In essence, it is a linear polysaccharide composed of a randomly distributed mixture of N-acetyl-D-glucosamine and -(14)-linked D-glucosamine (deacetylated units) [11]. Shrimp and other crustacean chitin shells are treated with a base to make chitosan. This chitin is mainly extracted from the exoskeleton of sea creatures like crabs, prawns, crayfish, etc., and from the fungal species [11].

Chitosan possesses one big property which is to function as the best adsorbent for the elimination of impurity from the water bodies, because of the presence of various functional groups on it (amino groups and hydroxyl group). Basically, chitosan is produced by the deacetylation of chitin [11]. Higher the degree of deacetylation means higher the adsorption capacity. Since the amount of deacetylation significantly influences the ability of chitosan to adsorb substances that means protonation of high amounts of amino groups increases the pollutant adsorption capacity [11, 12].

Biochar is a solid black residue, lightweight, carbonaceous rich material that is produced through the conversion of organic mass through thermochemical decomposition. Various methods for thermochemical decomposition include pyrolysis, hydrothermal carbonization, microwave heating, etc. [13]. The basic example is charcoal, derived from wood which contains lignin, cellulose, and hemicellulose. There are various types of biochar depending on the original components from which they are derived. Each type of highly rich carbon substance results in a very specific and different kind of biochar production, depending on the physical and chemical features of the parent material. Mostly, biochar preparation is carried out by heating the biomass or waste material through a process called pyrolysis. It includes thermal and chemical decomposition of biomass. Different types of biomass have been generated for the manufacturing of biochar, consisting waste materials from forestry and agriculture (such as rice hulls, cotton seed hulls, and wood pellets, nut shells, orange peel, straw, and oak), industrial by-products (like bagasse, paper sludge, and pulp), animal wastes (like dairy waste, cow dung, and swine manure), sewage sludge, etc. [13, 14]. Biochar is a cost effective, good sorbent for organic and inorganic pollutants removal from the environment, good physical and chemical properties, large specific area, and high porosity. To develop a potentially useful composite biosorbent for eliminating the hazardous, harmful dye methylene blue from an aquatic environment. The aim of this work is to create a biochar-modified chitosan beads composite. The most important adsorption parameters that were modified included adsorbent dose, solution pH, dye concentration, contact duration on methylene blue adsorption, and temperature influence. The isotherm and adsorption kinetics were also established. The thermodynamic quantities of Gibbs free energy ($G°$), enthalpy ($H°$), and entropy ($S°$) were investigated.

2 Materials and Methods

2.1 Preparation of Composite

For preparation the biochar beads, 2 g of chitosan powder was taken along with1 g of biochar powder in the beaker, both the powders were mixed into the 1% of acetic acid solution (CH_3COOH). This solution was left on a magnetic stirrer for stirring for about 2–3 h, for the complete homogeneous mixing of the solution [15]. After the mixing of the solution to prepare the biochar beads out of the chitosan-biochar solution, the solution was filled in the 2 or 5 ml syringe and dropped into the 2 M NaOH, this step was done very carefully and slowly in order for them to assume their right spherical shape. Then the beads were left overnight into the sodium hydroxide solution. NaOH solution was important for the beads to neutralize the acid content. After that beads were filtered using filtered paper and then washed the beads with distilled water several times (5–6) to remove the excess NaOH solution. At a temperature of 60 °C, the beads were dried in an oven (4–5 h) for the removal of moisture content from the beads.

2.2 Preparation of Methylene Blue Dye Solution

Stock solution of 10 ppm methylene blue was prepared. For this, 0.001 g of methylene blue was dissolved in 100 ml of distilled water, for each experiment 25 ml of it was taken out from the stock solution of 10 ppm and poured into the flask.

2.3 Techniques Used

The SEM images were taken through ZEISS model after the gold coating of biochar beads were done under the vacuum condition by electrodeposition [16]. SEM images were analyzed or investigated at 1.00 kx magnifications and observed, to have an idea about the surface characteristics of the biochar beads. It was found that the biochar beads after the methylene blue dye was adsorbed on it has a surface which was smooth and continuous [16]. Similarly, the Fourier transform infrared spectrophotometer was used and the infrared spectra of both samples were analyzed using FTIR (Perkin Elmer Spectrum Two) and KBr pellet. Using a twin beam UV–VIS spectrophotometer (UV–VIS. Shimazu 1800), the concentration of methylene blue solutions was determined. The aqueous pH was measured using a pH meter (Systronics).

2.4 Batch Adsorption Studies

Batch adsorption studies were used to remove methylene blue from its aqueous solutions using chitosan beads that had been biochar-modified. For each batch, a 250 mL beaker was filled with 100 mL of methylene blue solution in varied concentrations, pH levels, and adsorbent amounts. A magnetic stirrer was used to stir the liquid for an additional 120 rpm at room temperature. The adsorbents were separated by glass wool filtering at regular intervals. The filtrate's absorbance was quantified using a UV–VIS spectrophotometer with a max 497 nm wavelength to determine the residual methylene blue concentration. The rate of removal in percentage terms was computed using

$$\text{Removal Efficiency}(\%) = \frac{C_i - C_f}{C_i} * 100,$$

where, C_i represents the starting concentration (in mg/L) and C_f represents the final concentration (in mg/L), At time t, the sorption capacity was acquired by the equation

$$q_t = \frac{(C_i - C_f)}{M} * V,$$

where C_i was taken into account as the methylene blue initial liquid phase concentration (mg/L). At time 't', C_f was regarded as the liquid phase concentration in mg/L. V represents the volume of the solution in liters, and M as the adsorbent mass absorbed (g).

The equation for calculating the equilibrium adsorption q_e is

$$q_e = \frac{(C_i - C_e)}{M} * V,$$

C_e is the concentration at equilibrium in this scenario.

2.5 Adsorption Isotherms

2.5.1 Langmuir

The Langmuir adsorption isotherm demonstrates a relationship between high adsorption and monomolecular adsorbate layer on the adsorbent molecule's surface, with constant adsorption energy. With increasing distance from the adsorption surface, the intermolecular interactions are negligible or minor and all adsorption sites are energetically equivalent or identical [17, 18].

The equation below represents the Langmuir isotherm model:

$$q_t = \frac{q_m K_L C_e}{1} + K_L C_e,$$

where q_m (mg g^{-1}) is the highest adsorption capability. K_L is the Langmuir constant (L/mg). The Langmuir model has the following linear form:

$$\frac{C_e}{q_e} = \frac{C_e}{q_m} + \frac{1}{q_m * K_L},$$

q_m = Maximum capacity for adsorption, K_L = Langmuir adsorption constant,
C_e = Adsorbate equilibrium concentration,
q_e = Capacity for adsorption that is adsorbed at equilibrium.

The ability of the Langmuir isotherm to convey the dimensionless constant known as the balance parameter, which is defined by the following equation, is one of its distinctive features.

$$R_L = 1(1 + (K_L \times C_i)),$$

where C_i represents the initial concentration (mg/L) and R_L displays the equilibrium parameter (dimensionless). The R_L values show the type of adsorption that can be expected. Adsorption exhibits favorable condition if $R_L < 1$, shows unfavorable conditions if $R_L > 1$, linear if $R_L = 1$, and indicates irreversible if $R_L = 0$ [17, 18].

2.5.2 Freundlich

The Freundlich model is portrayed (non-empirically) by blaming the surface's heterogeneity and variations in heat absorption for the change in the equilibrium constant of the binding method [17, 18].

The Freundlich equation are given below:

$$Q_e = K_F + C_e^n.$$

Freundlich's constant, denoted by K_F in mg/g, in order to demonstrate how the Freundlich exponent relates to the degree of adsorption. It has no dimension. Q_e: the equilibrium amount of adsorbed (mg/g).

The equation of this model can be linearized are represented as given below:

$$\log Q_e = \frac{1}{n} \log C_e + \log K_F.$$

The slope and intercept of the values of the constants K_F and n are calculated using the graphing lines on the ordinate ln Q_e and the abscissa ln C_e.

The nonlinearity between the concentration of the solution and the adsorption is demonstrated by the $1/n$ interval, which has values among 0 and 1. The adsorption is linear if the value of $1/n$ is equal to 1 [18, 19].

2.5.3 Temkin

The Temkin isotherm model has a factor that specifically takes into account the relationship between the species and the adsorbate in terms of adsorption [18, 19]. The Temkin model is described by the following equation:

$$q_e = \left(\frac{\mathrm{RT}}{b}\right) \ln(AC)_e, \quad q_e = \left(\frac{\mathrm{RT}}{b}\right) \ln A + \left(\frac{\mathrm{RT}}{b}\right) \ln C_e.$$

The equation gives this model's linear form:

$$q_e = A + b \ln C_e,$$

where q_e is the amount of dye adsorbent per specific amount (in mg/g), (mg/L), and C_e is the concentration at equilibrium. Temkin constants A and b relate to the heat of adsorption (J/mol) and the equilibrium binding constant (Lg^{-1}), respectively. The constants b and RT are determined by the slope and point at which the lines drawn on the X-axis and q_e plotted on the Y-axis connect [19].

2.5.4 D–R

Dubinin–Radushkevich isotherm (D–R) does not give an idea about a homogeneous surface or a constant adsorption potential [18–20]. This model is explained by the following expression:

$$q_e = q_s - K_{\mathrm{DE}\varepsilon^2},$$
$$q_e = q_s - K_{\mathrm{DR}\varepsilon^2}.$$

q_s is the adsorption capacity (mol g^{-1}), where K_{DR} is the D–R constant (mol^2/KJ^2), and ε is linked by the following equation:

$$\varepsilon = \mathrm{RT}\ln(1 + C_e).$$

ε is the potential of Polanyi, C_e is the equilibrium concentration (mol L^{-1}), R is the universal gas constant (8.314 103 kJ mol^{-1} K^{-1}), and q_e is the equilibrium concentration in the adsorbent (mol g^{-1}). T is the absolute temperature in (°K) [19, 20].

The following expression defines the D–R adsorption isotherm in linear form

$$\ln q_e = K_{\mathrm{DR}\varepsilon^2} + \ln q_s,$$

q_s is the isotherm saturation capacity in (mg g^{-1}). The slope and intercept for the plot of ln q_e and ε^2 on the Y-axis and the X-axis, respectively, yield the adsorption isotherm constant q_s and K_{DR}.

2.5.5 Elovich

The Elovich adsorption isotherm [19, 20] is a model that explains multilayer adsorption that is explained by following equation

$$\ln\frac{q_e}{C_e} = \ln(k_e q_m) - \frac{q_e}{q_m},$$

where the Elovich constant (L/mg) and adsorption capacity, respectively, are K_e and q_m is given in (mg/g). This model believes that adsorption sites increase exponentially with the adsorption and cannot be used to get the equilibrium adsorption since it is unlike lower-order equations which assume monolayer adsorption.

2.6 Adsorption Kinetics

It explains the resident time of the adsorbate and the rate at which the solute is adsorbed on the solid liquid surface at a given parameter.

2.6.1 Pseudo-First-Order Kinetics

It has been extensively used to study the adsorption kinetics of dye. The pseudo-first-order kinetics are expressed as follows:

$$\log(q_e - q_t) = \log q_e - \left(\frac{k_{1t}}{2.303}\right).$$

According to the plot of $\log(q_e - q_t)$ versus time, the rate constant can be determined where q_t and q_e represent the dye's equilibrium concentration at time t.

2.6.2 Pseudo-Second-Order Kinetics

Pseudo-second-order kinetics has been greatly used and is represented by the following expression:

$$\frac{T}{q_t} = \frac{1}{k_2} qe^2 + \frac{t}{q_e},$$

where K_2 (g/mg/min) is the pseudo-second-order kinetics constant, and plotting of graph gives the value of K_2.

2.6.3 Thermodynamic Parameters

It involves change in free energy, entropy, and enthalpy linked to adsorption method and were determined by the following expression.

$$\Delta G = -\text{RT} \ln K,$$

where T is temperature, the equilibrium constant is k, while R is the universal gas constant.

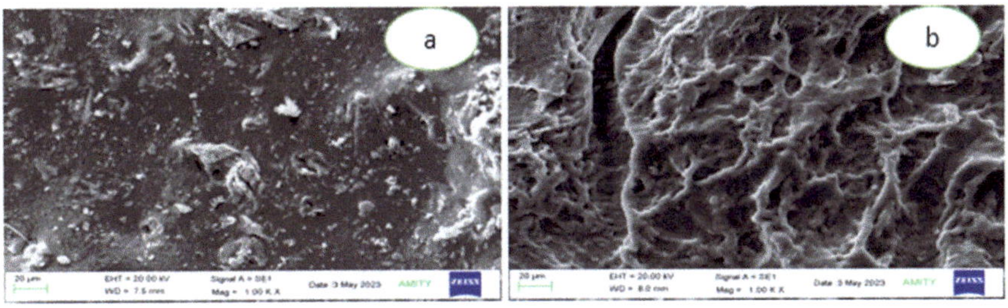

Fig. 1 **a** SEM image of biochar beads, **b** Beads with methylene blue dye

$$\ln k = \frac{\Delta S^0}{R} - \frac{\Delta H^0}{\mathrm{RT}},$$

S° and H° can be calculated from the slope and intercept.

3 Results and Discussion

3.1 Scanning Electron Microscope (SEM)

SEM analysis was done to learn more about the morphological and surface characteristics of the chitosan that had been impregnated with biochar as well as what happened when the dye methylene blue was adsorbed on it. Before the dye adsorption, it was seen that the chitosan beads impregnated with biochar (Fig. 1a) had a porous, extremely rough surface, and the surface was smooth and continuous when the dye methylene blue was deposited onto it (Fig. 1b).

3.2 Fourier Transmission Infrared Red (FTIR)

The surface groups that aid in the process of dye adsorption are identified by FTIR investigations, this also includes the existence of distinctive peaks and active areas. According to Fig. 2, Fig. 2a displays the spectra of biochar beads, while Fig. 2b displays the spectra of compost with methylene blue dye deposited on it. Biochar beads have a prominent peak in the FTIR at 3418 cm^{-1} that is associated with the (hydroxyl group, H bound OH stretch). Then the next peak was absorbed at 2919 and 2862 cm^{-1}which means (H–C–H Asymmetric and symmetric stretch). The bands at 1728 cm^{-1} show that C=O stretching which means aldehydes and ketonic functional group is present. The band at 1572 and 1435 cm^{-1} indicates that (N=O stretching is present), which means nitro or amides functional group are attached. Then peak was observed at 1062 cm^{-1} corresponding to the C–O stretch that shows ester and

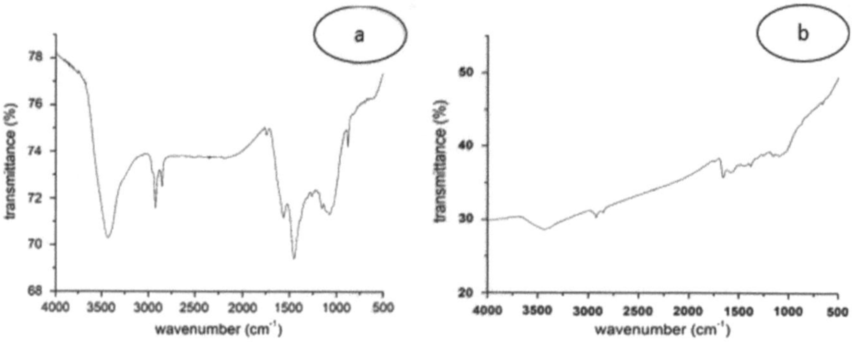

Fig. 2 **a** FT-IR spectra of biochar beads, **b** biochar beads with methylene blue dye

ether functional group is attached. At 874 cm^{-1}, a small band for OH bending is also visible.

In the FTIR spectra of biochar beads with methylene blue, firstly other characteristics peaks of methylene blue can be observed. The characteristic peak at 3416 cm^{-1} corresponds to the N–H Stretch (similar to amines) that can be observed. The bands at 2913 and 2848 cm^{-1} show that (H–C–H asymmetric and symmetric stretch vibration) is present. A small peak is observed at 1649 cm^{-1} that corresponds to (C=O stretch). Then next band at 1234 and 1032 cm^{-1} correspond to the (C–O stretch vibrations, respectively).

3.3 Effect of Initial Dye Concentration

Depending on the dye concentration (2–12 mg/L); (0.1 g) of biochar-chitosan beads were added and 30 min were spent shaking at ambient temperature in an orbital shaker. It measured the absorbance. Because there were fewer adsorption sites available, the percentage of dye that was removed reduced as dye concentration increased. The highest removal was observed at a concentration of 2 mg/L (Fig. 3a).

3.4 Effect of pH

The adsorption of the dye was significantly influenced by the pH of the solution. The solution's pH was raised from 3 to 9 to conduct the experiment. The beads' capacity to adsorb pigment first increased as the pH increased before declining. At pH 6, the largest amount of color was removed. This is brought on by the protonated adsorbent and anionic dye's increased electrostatic interaction at lower pH. (Fig, 3b). An electrostatic interaction caused by the presence of positively charged amine groups is seen at pH 6, and it was discovered that this interaction gets weaker as pH increases.

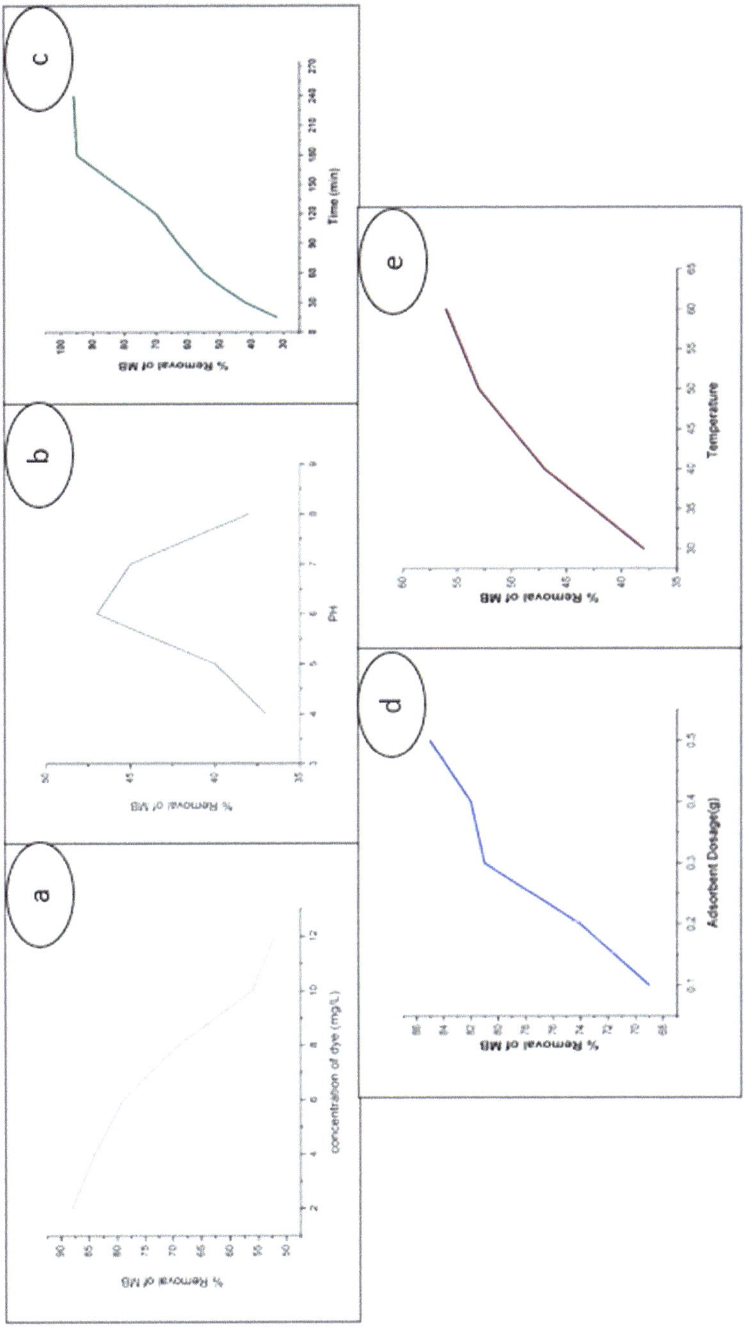

Fig. 3 **a** Effect of dye concentration, **b** effect of pH, **c** effect of contact time, **d** effect of adsorbent dosage, **e** effect of temperature

The decrease could be brought on by deprotonation of the positively charged groups on the adsorbent and electrostatic attraction between the negatively charged sites. At high pH values, the OH-ions may compete with the anionic dye species.

3.5 Effect of Contact Time

The dye mixture (25 ml), 10 mg/L, was maintained with 0.01 g of biochar-chitosan beads at various contact durations (15–240 min). The absorbance was then calculated. As the contact duration increased, the dye solution's absorbance declined and the beads' ability to absorb additional dye increased. No discernible increase in adsorption was seen after 200 min, which may be because there were fewer active sites available after a given amount of time (Fig. 3c).

3.6 Effect of Adsorbent Dosage

The dye solution (25 ml) was shaken for 30 min at room temperature in an orbital shaker with varying dosages of adsorbent (0.1–0.5 g), and then the absorbance was measured. The number of adsorption sites rose with increasing adsorbent dosage, allowing for more dye to be adsorbed and lowering absorbance (Fig. 3d).

3.7 Effect of Temperature

After shaking 0.1 g of biochar-chitosan beads in a 10 mg/L solution of methylene blue (25 ml) for 30 min at a temperature range of (30, 40, 50, and 60 °C), the dye's absorbance was determined both before and after the various temperatures, absorbance was noticed. As the temperature rose, the beads were able to absorb more color, and the solution's absorbance dropped (Fig. 3e).

3.8 Adsorption Isotherms

Different models have been developed, and the collected data is fitted in to those models to compare and determine the efficacy of the adsorbent material. Using the equations presented in the previous part, the Langmuir, Freundlich, and three other isotherm models were applied in this investigation. For chitosan + biochar + methylene blue dye, Langmuir, Freundlich, and Temkin adsorption isotherms have the best fit experimental data. When R^2 values are seen, Langmuir provides a superior model for the adsorption system, indicating that the adsorbent is homogeneous and condensed. Temkin isotherm suggests that the decline in sorption heat is linear rather than logarithmic (Fig. 4a–e).

The isotherms were analyzed by using regression and least square fit that are shown in the Table 1. Langmuir adsorption isotherm for biochar beads and methylene blue dye have best fitted experimental data.

3.9 Adsorption Kinetics

Figure 5a and b, respectively, depicts the kinetics for pseudo-first-order and pseudo-second-order, respectively. Adsorption for the combination of chitosan, biochar, and methylene blue dye was investigated for contact times between 15 and 180 min. The data collected matched pseudo-first-order dynamics very well.

This model helps in interpreting the impact of contact time for methylene blue dye removal with the help of biochar beads. Hence to interpret the data, the controlling rate and mechanism of kinetics include for the adsorption process.

From the above Table 2, it is observed that the value of R^2 obtained from the pseudo-first-order comes closer to 1, so pseudo-first-order kinetics best suited kinetic model.

3.10 Thermodynamic Studies

Temperature effect on chitosan at (313,323,333) K gives the thermodynamic aspect of the adsorption process (Fig. 6). Entropy and enthalpy are calculated by plotting the graph between ln K versus $1/T$, where T is absolute temperature and K has been evaluated by the ratio of q_e/C_e [where, C_e is the remaining equilibrium dye concentration (mg/L); q_e is the concentration of dye adsorbed in the adsorbent (mg/g)], provides the value of slope and intercept giving the value of $\Delta S^\circ = -0.1315$ kJ mol^{-1}, $\Delta H^\circ = -40.252$ kJ mol^{-1}.

The positive value of free energy represents the reaction feasibility, and the positive value of enthalpy and entropy signifies that randomness of the molecules increases as the methylene blue dye moves from solution to the adsorbed condition and the adsorption process shown is endothermic. The maximum adsorption capacity of biochar composite for methylene blue dye removal was found to be 2.36 mg/g in comparison with 50.07 [21] and 35.21 [22] mg/g previously reported adsorbents.

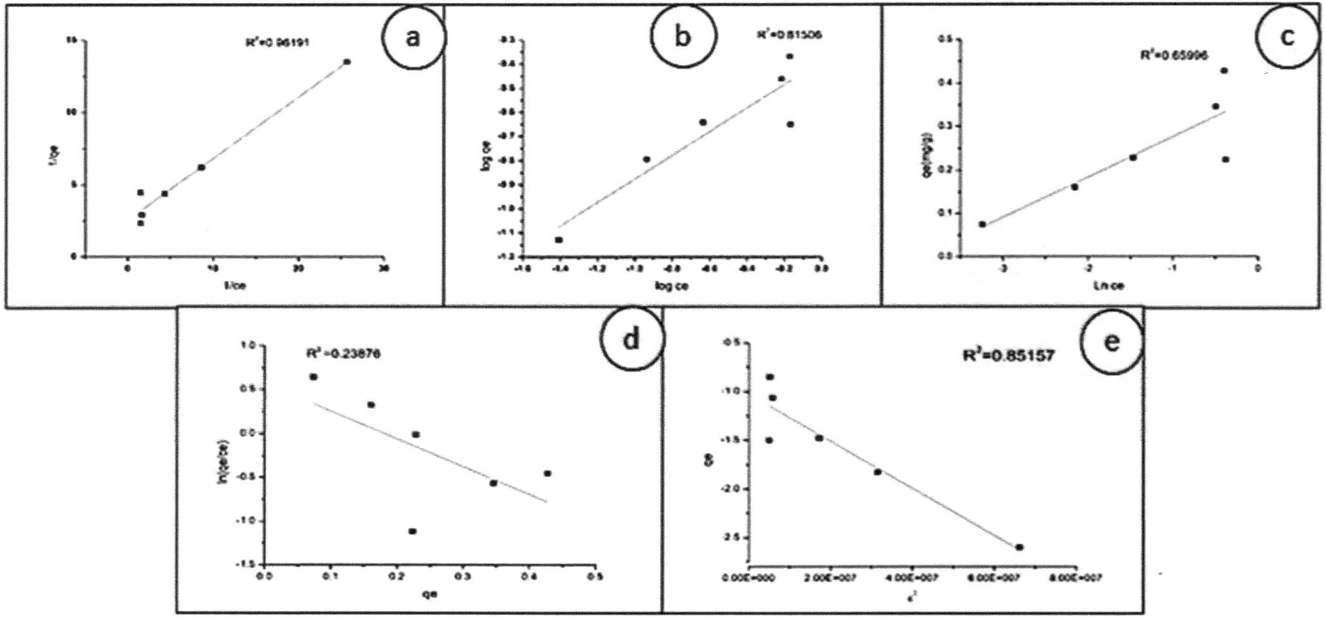

Fig. 4 **a** Langmuir adsorption isotherm, **b** freundlich adsorption isotherm, **c** Temkin adsorption isotherm, **d** Elovich adsorption isotherm, **e** (D–R) adsorption isotherm

Table 1 Isotherm constants for biochar composite and methylene blue

Isotherm	R^2	K_L	a_L	K_F	n	B (J/mol)	A_T (L/g)	q_m (mg/g)	K_e (Lmg^{-1})	β ($mol^2\ kJ^{-2}$)	E ($kJ\ mol^{-1}$)
Langmuir	0.9695	6.0682	0.3290					2.36			
Freundlich	0.8150			0.41096	2.044						
Temkin	0.72796					0.09258	5.3732				
Elovich	0.23876							0.3154	−0.1824		
D–R	0.85157							0.3580		4.8219	0.4553

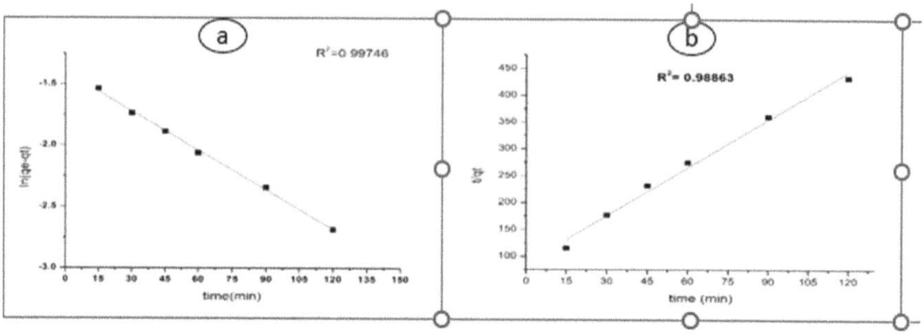

Fig. 5 **a** Pseudo-first-order kinetics, **b** Pseudo-second-order kinetics

Table 2 Kinetics of biochar composite and methylene blue dye

Pseudo-first-order kinetics	R^2 0.9974	K_1 −0.0003
Pseudo-second-order kinetics	R^2 0.9886	K_2 0.1029

Fig. 6 Van't Hoff plot of Ink versus 1/*T*

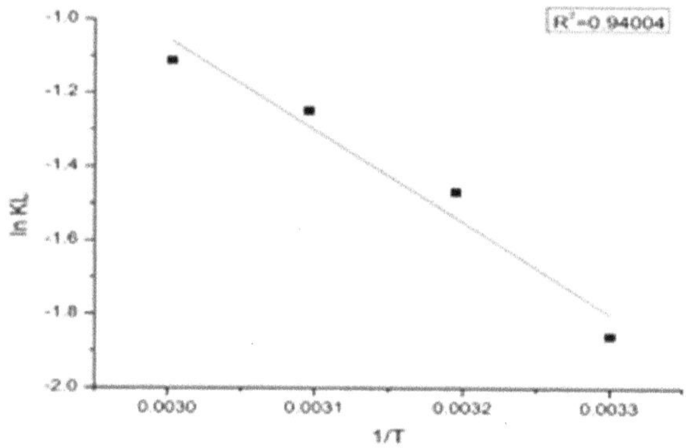

4 Conclusion

The present study shows a good potential for the removal of chemicals and dyes from wastewater. The high-cost adsorbents can be replaced by the inexpensive adsorbent (peanut shell waste) that is not just resourceful, cheaper, more effective, nontoxic, and biodegradable, but also one of the most effective and environmentally friendly techniques to remove dangerous dyes from wastewater. 75–85% of the maximal elimination was achieved. The studies followed Langmuir adsorption isotherm and pseudo-first-order kinetics. The removal efficiencies of the harmful dye methylene blue are moderately influenced by processing conditions.

References

1. Abeer N, Khan SA, Muhammad S, Rasool A, Ahmad I (2020) Health risk assessment and provenance of arsenic and heavy metal in drinking water in Islamabad. Pakistan. Environ Technol Innov 20:101171
2. Genchi G, Carocci A, Lauria G, Sinicropi MS, Catalano A (2020) Nickel: human health and environmental toxicology. Int J Environ Res Public Health 17:1124
3. Senevirathna STMLD, Goncher AM, Hollier A (2019) Assessment of drinking water quality in regional New South Wales, Australia. J Water Supply Res Technol 68:708–717
4. Li S, Han K, Li J, Li M, Lu C (2017) Preparation and characterization of super activated carbon produced from gulfweed by KOH activation. Microp Mesop Mater 243:291–300
5. Bagheri AR, Ghaedi M, Asfaram A, Bazrafshan AA, Jannesar R (2017) Comparative study on ultrasonic assisted adsorption of dyes from single system onto Fe_3O_4 magnetite nanoparticles loaded on activated carbon: experimental design methodology. Ultrason Sonochem 34:294–304
6. Chen B, Chen Z (2009) Sorption of naphthalene and 1-naphthol by biochars of orange peels with different pyrolytic temperatures. Chemosphere 76(1):127–133
7. Lehmann J, Rillig MC, Thies J et al (2011) Biochar effects on soil biota: a review. Soil Biol Biochem 43(9):1812–1836
8. Chan KY, Van Zwieten L, Meszaros I et al (2008) Agronomic values of greenwaste biochar as a soil amendment. Soil Res 45(8):629–634
9. Woolf D, Amonette JE, Street-Perrott FA, Lehmann J, Joseph S (2010) Sustainable biochar to mitigate global climate change. Nat Commun 1(3):118–124
10. Abdullah H, Wu H (2009) Biochar as a fuel: properties and grindability of biochars produced from the pyrolysis of Mallee wood under slow-heating conditions. Energy Fuels 23(8):4174–4181
11. Marco M-P, de Alda ML (2004) Biosensors for environmental monitoring of endocrine disruptors: a review article. Anal Bioanal Chem 378:588–598
12. Gerald S (2000) Dyes, general survey. In: Ullmann's encyclopedia of industrial chemistry. Wiley, Amsterdam
13. Rodriguez-Amaya DB (2016) Natural food pigments and colorants. Curr Opin Food Sci 7:20–26
14. Duarte FJ, Hillman LW (eds) (1990) Dye laser principles. Academic, New York
15. Lehmann J, Joseph S (2009) Biochar for environmental management: science and technology. Earthscan Ltd., London
16. Blanchard G, Maunaye M, Martin G (1984) Removal of heavy metals from waters by means of natural zeolites. Water Res 18:1501–1507
17. Rafatullah M, Sulaiman O, Hashim R, Ahmad A (2010) Adsorption of methylene blue on low-cost adsorbents: a review. J Hazard Mater 177(1–3):70–80
18. Bello OS, Adegoke KA, Olaniyan AA, Abdulazeez H (2015) Dye sorption using biomass wastes and natural adsorbents: overview and future prospects. Desalin Water Treat 53:1292–1315
19. Hameed BH, Din ATM, Ahmad AL (2007) Adsorption of Methylene blue onto bamboo-based activated carbon: kinetics and equilibrium studies. J Hazard Mater 141:819–825
20. Yang Q, Li Z, Lu X, Duan Q, Huang L, Bi J (2018) A review of soil heavy metal pollution from industrial and agricultural regions in China: Pollut Risk Assess. Sci Total Environ 642:690–700
21. Dindorkar SS, Patel RV, Yadav A (2022) Adsorptive removal of methylene blue dye from aqueous streams using photocatalytic CuBTC/ZnO chitosan composites. Water Sci Technol 85(9):2748–2760. https://doi.org/10.2166/wst.2022.142
22. Utomo HD, Phoon RYN, Shen Z, Ng LH, Lim ZB (2015) Removal of methylene blue using chemically modified sugarcane bagasse. Nat Resour 6(4):2158–7086

Smart Sustainable Solutions to Mitigate Water Pollution in Yamuna River

Gazala Ruhi, Panchami Muraleedharan, Sneha Kohli, Nyasha Malik, Aleena Shakreen, Riti, Anjali, Mahima, Pinkey B. Gandhi, and Haritma Chopra

Abbreviations

CPCB	Central pollution control board
MLD	Minimal liquid discharge
YAP	Yamuna action plan
DO	Dissolved oxygen
BOD	Biochemical oxygen demand
MoEF	Ministry of environment and forests
TDS	Total dissolved solids
TSS	Total suspended solids
STL	Standard limit
BIS	Bureau of Indian Standards
WHO	World Health Organization
ppm	Parts per million
WWTPs	Wastewater treatment plants
STP	Sewage treatment plant
UK	United Kingdom
CNT	Carbon nanotube
NF	Nanofiltration
MWCO	Molecular weight cut-off
CWs	Constructed wetlands

G. Ruhi (✉) · P. Muraleedharan · S. Kohli · N. Malik · A. Shakreen · Riti · Anjali · Mahima · P. B. Gandhi · H. Chopra
Department of Chemistry, Maitreyi College, University of Delhi, Delhi 110021, India
e-mail: gruhi@maitreyi.du.ac.in

S. Rattan et al. (eds.), *Emerging Trends in Synthesis and Catalysis in Chemistry*, Springer Proceedings in Materials 51, https://doi.org/10.1007/978-981-97-5169-3_17

1 Introduction

Yamuna is an immensely significant and venerated river in Hindu mythology. The divine river, originating from Champasar Glacier in Uttarakhand is treated as a goddess and is believed to wash away the misdeeds and arrogance of all beings through her pious water. This river has nurtured numerous civilizations and allowed their cultures to flourish without any biases. However, the ever-growing population, extensive industrialization, and reckless agricultural practices have severely compromised the water quality of Yamuna. As a result, slowly but inevitably the river of elixir has transformed into a large water body of filth and toxins.

The Yamuna River originates from the Yamunotri glacier in the high Himalayas at a height of 4421 m. The 1376-km-long Yamuna flows across three states: Uttarakhand, Uttar Pradesh, and Haryana. It enters Delhi at Palla village, where it is pretty much in good condition, and leaves from Okhla in a much worse state (total stretch 48 km). Delhi, being one of the most populous cities depends heavily on the Yamuna River for its daily water needs. More than 70% of Delhi's water consumption comes from the Yamuna River. But sadly, the 22 km urban stretch of the river in Delhi (between Wazirabad Barrage and Okhla Barrage) (Fig. 1) reflects 80% of the pollution load [1]. The residential, as well as industrial areas along the river, dump a significant amount of wastewater into the river. In this stretch, the river receives sewage from almost 22 drains.

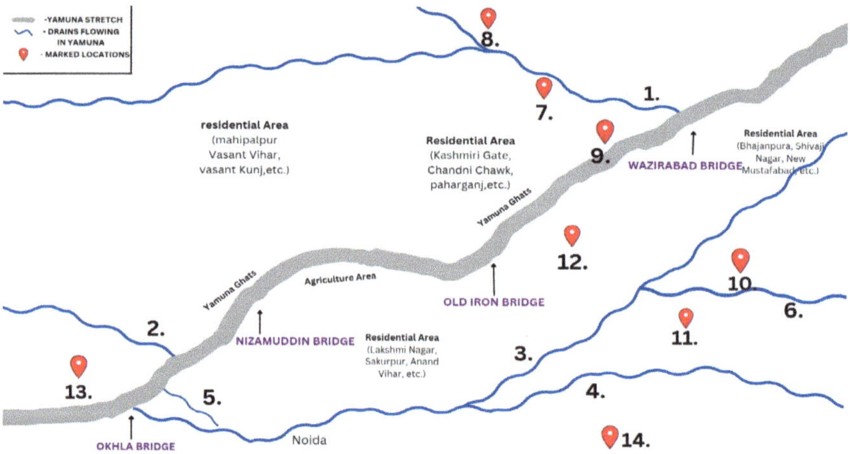

Fig. 1 Schematic of the stretch of Yamuna in Delhi. The marked locations are: (1) Najafgarh drain, (2) Barapullah drain, (3) Shahdara drain, (4) Hindon drain, (5) Hindon cut canal, (6) Ghazipur drain, (7) Small-scale industries cluster (Shalimar Bagh, Patel Nagar, Shastri Nagar, Nehru Nagar), (8) Dye industries (Ranhola, Khayala, Bindapur, Matiala, Nilothi), (9) Majnu ka Tila colony, (10) Jhilmil inustrial area, (11) Ghazipur market dumping ground and dye industrial area, (12) Dye industries (Shiv Vihar, Tukhmirpur, Seelampur), (13) Dye industries (Badarpur, Meethapur), (14) Industrial area near Hindon (Map not to scale)

Yamuna is the primary source of drinking water for the majority of the population of Delhi. However, the discharge of by-products and waste materials produced from industries, domestic settlements, and agricultural activities into Yamuna results in water contamination. The quality of water of rivers in India has been classified into five classes, namely; class A: river water is fit for drinking after disinfection with chlorine or bleaching powder, class B: river water is fit only for bathing, class C: river water is fit for drinking only after proper treatment (screening), class D: river water is fit only for fish and wildlife, and class E: river water is satisfactory for industrial cooling. Surprisingly, the water quality of the Yamuna River falls under class E, revealing that its untreated water is extremely unhealthy for human consumption or usage. The drains at Najafgarh and Shahdara contribute the most to the pollution of Yamuna River. The Najafgarh drain is connected to various sub-drains that carry sewage into it. The heavily populated residential areas add around 1340 MLD wastewater, followed by 150 MLD from private enterprises [2]. On the other hand, waste from the Shahdara industrial area (Jhilmil) and Hindon drain, which merges with it near the New Ashok Nagar city forest is carried by the Shahdara drain.

2 Sources of Pollution

The sources of pollution in the Yamuna River can be broadly classified into two categories; active sources and passive sources. Active sources include industrial effluents, domestic and agricultural waste, whereas, passive sources include some religious practices like the immersion of idols, holy bathing, dumping of dead bodies, open defecation, etc. The literature reports that a total of 3663 MLD of wastewater is generated from domestic waste alone, out of which only 61% is treated [3]. In addition to this, around 13,468 MLD of wastewater is produced by industries, out of which only 60% is being treated [3]. The untreated sewage discharged into Yamuna causes extremely adverse effects on the flora and fauna in its vicinity. It leads to a decrease in the level of dissolved oxygen (DO), an increase in biological oxygen demand (BOD) levels, total coliform count, and an increase in the concentration of total suspended particles and heavy metals [3]. According to the report of the Central Pollution Control Board (CPCB) of the year 2021, the BOD and DO levels near the Okhla region were found to be 9–10 and 3.9–4.5 mg/l, respectively (the standard values set by the Ministry of Environment, Forest and Climate Change are, BOD < 3 mg/l and DO > 5 mg/l). A wide variety of organic, and inorganic pollutants like heavy metal ions, pesticides, petroleum, detergents, human/animal faeces, etc., contaminate the water of Yamuna. Among these pollutants, the presence of heavy metals deteriorates the quality of water many folds because of its toxicity, bioaccumulation, and non-biodegradable nature [4]. Activities like mining, agricultural practices, industrial/traffic emissions, and sewage disposal allow the entry of these metals into the river system. It is suggested to closely monitor the amount of arsenic, cadmium, chromium, lead, etc., in the river ecosystem. Malik et al. [5] have discussed that the order of average heavy metal concentration in river water is Fe > Cr > Mn > Zn > Pb > Cu >

Table 1 Important parameters with their standard limits and their present concentrations in two different locations

Parameter/heavy metal and ions	STL (BIS/WHO/CPCB)/ppm	Present concentration near Wazirabad Barrage/ppm	Present concentration near Okhla Barrage/ppm
Fe	0.3	4.1	6.3
Cr	0.05	0.1	0.53
Mn	0.1	0.39	0.37
Zn	5.0	0.28	0.43
Pb	0.01	0.40	0.51
Cu	0.05	0.17	0.22
Ni	0.02	0.20	0.47
Cd	0.003	0.031	0.056
$(SO_4)^{2-}$	200	61.9	79.9
NO_3^-	45	2.4	3.2
Cl^-	250	115.0	236.2
F^-	1.0	0.48	0.62
$(PO_4)^{3-}$	0.5	0.2	0.8
TDS	500.57	519.59	712.81
pH	6.5–8.5	7.5	7.0
EC	2000	788	1265

Ni > Hg > As > Cd and are present much above the permissible limit. Table 1 exhibits the comparative data on the standard concentration of various metals and ions and their actual concentrations near Wazirabad Barrage and Okhla Barrage [5]. The data reveals that the present concentration of metals and ions are many folds higher than the standard concentration set by CPCB/WHO/BIS.

3 Government Initiatives to Mitigate Pollution in Yamuna

Over the past several decades, there have been several attempts made to clean Yamuna River. The initial evaluation by the Central Pollution Control Board (CPCB) in 1977–1978 showed that the pollution status of river Yamuna over its 1200 km course is mainly due to the discharge of domestic wastewater. Other important polluting causes were noted to be industrial activities and agricultural practices. The outcome of this study led to the launch of one of the largest river restoration projects, the Yamuna Action Plan (YAP) under the mega project of Ganga Action Plan Phase-II. After YAP entered Phase III in 2012, the government proposed a six-point action plan with sewage treatment as a primary focus to reinforce the existing plan and to tackle the burden on sewage discharge in the Yamuna due to the fast-growing population, industrial and agricultural growth. Table 2 presents the details of the various government initiatives taken to mitigate the pollution level in Yamuna River.

4 Conventional Sewage Treatment Process

The wastewater coming out from urban and industrial sewage must be treated before being discharged into the river. The Sewage Treatment Plants (STPs) are established to serve this purpose. Conventional sewage treatment system includes steps like sedimentation, flocculation, filtration, adsorption, and biological/chemical treatment. Various STPs have been installed at different locations in Delhi to treat the wastewater before discharging it to the Yamuna River. Figure 2 shows the working of conventional STPs. Screening through iron bars is the first process that removes the large floating solid materials like branches of the tree, leaves, plastic bags, etc. After this step, the wastewater is collected in a collection tank from where it is sent to the primary clarifier for removal of organic waste which settles down as sludge. The wastewater is then subjected to secondary treatment in the aeration tank and secondary clarifier. The process of aeration involves pumping oxygen into the tank which promotes microbial growth. The bacteria decompose the organic matter and form flocs which can easily settle down in the secondary clarifier in the form of activated sludge. This activated sludge gets back to the aeration tank to fasten the rate of decomposition. The water from the secondary classifier is sent to a chlorine contact tank where potential pathogens are removed. After the process of disinfection, water undergoes dual media filtration and is discharged back into the environment for later use.

The dependency on conventional physical and chemical methods for the treatment of wastewater is huge. But, these methods are time-consuming and require a lot of space and energy. It is interesting to note that the conventional water treatment processes focus on the removal of suspended solids, colloidal impurities, and microbes from the wastewater. However, this process is not ideal for the treatment of wastewater containing dissolved organic pollutants and heavy metal ions. In view of this, it can be stated that conventional wastewater treatment methods alone are not sufficient to maintain water quality standards. Hence, smart sustainable methods/ technologies for the treatment of wastewater must be developed.

5 Nanotechnological Interventions for Wastewater Treatment

Nanotechnology aims to design particles, composites/hybrid materials, and devices with superior optical, magnetic, electronic, mechanical, and electrochemical properties by altering the morphology/size/surface properties at the molecular or atomic

Table 2 Government initiatives taken so far to control pollution levels in the Yamuna River

S. No	Initiatives	Timeline	Activities undertaken
1	Yamuna Action Plan		
	Phase I	1993–2003	(i) Construction of 29 STPs, 58 pumping stations, and 179 km of sewers (ii) Installation of 1282 public toilet complexes, 96 crematoria, river front development, public awareness [6]
	Phase II	2004–2008	(i) Abatement of severe pollution of Yamuna River by raising STP capacity [7] (ii) Constructing new and expanding capacity of old STPs and laying and rehabilitating sewers to increase treatment capacity, particularly in Delhi and Agra [7]
	Phase III	2012–onwards	(i) 9 packages of rehabilitation and upgradation of WWTPs in Rithala, Kondli, and Okhla catchments as well as rehabilitation of trunk and outer sewer lines and rising mains in these catchments at critical stretches [6] (ii) Construction of 124 MGD WWTPs at Okhla, rehabilitation and upgrading of Rithala WWTP and Kondli WWTP [6]
2	Mukhyamantri Muft Sewer Connection Yojna	2019	(i) Around 2,34,000 houses in 787 colonies are set to benefit from the scheme
3	Construction of 14 mini STPs under Najafgarh drain zone	2019	Treated water from these plants to be used to restore 56 waterbodies in the area
4	Yamuna Cleaning Cell	2021	Six-Point Action Plan for Yamuna Cleaning (i) Sewage treatment plants, (ii) Drains, (iii) Industrial waste, (iv) Jan Suvidha complexes, (v) Individual sewer connections, (vi) Desilting and rehabilitation of entire sewerage work
5	Treated Yamuna water to Rejuvenate lakes	2021	Polluted water from the neighbouring states flowing into the Yamuna is to be treated in Narela STP. The water is then to be released into the lake in Tikri Khurd. The aim is to achieve the recycling of 15 MGD of polluted water
6	Sealing of polluting industries	2022	Committee was formed to inspect and seal the unauthorized industries like, small-scale dyeing industries, pickling industries, etc.
7	Completion of the Interceptor Sewage Project (ISP)	2022	Treatment of 238 MGD of sewage out of the targeted 248 MGD from the minor drains was accomplished

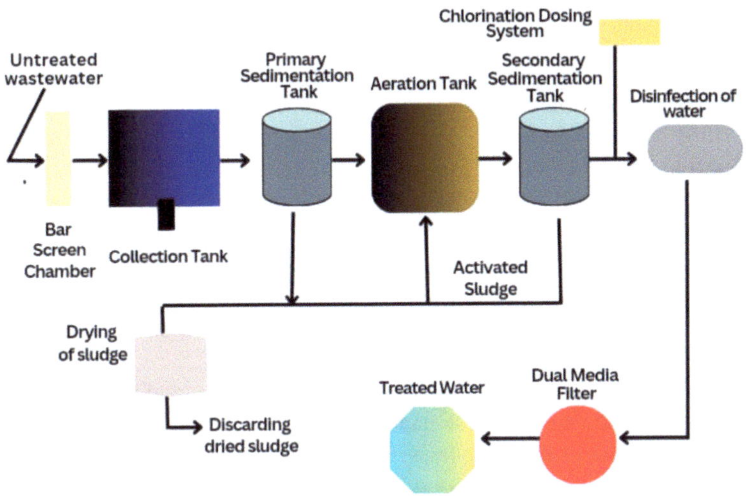

Fig. 2 Schematic representation of the working of the Sewage Treatment Plant

level. The wide array of applications of nanotechnology has resulted in its prevalent use in the past few decades in various fields, including the mitigation of water pollution. The nanotechnological interventions have paved the way to synthesize materials of nanoscale having high surface area, ease of functionalization, remarkably high adsorption properties, high reactivity, etc. These properties can be harnessed to either reinforce the pre-existing technologies of wastewater treatment or to adopt a totally innovative approach with much more sustainable than the earlier approaches. The present section presents a detailed discussion of the potential role of nanotechnological interventions in combatting water pollution through efficient wastewater treatment.

5.1 Nanoparticles: A Potent Alternative to Treat Wastewater

Designing of nanostructured materials in the form of particles, rods, sheets, etc., is one of the important aspects of nanotechnological interventions to address wastewater treatment problem. The high surface area to volume ratio of these nanomaterials has rendered its high applicability as a potentially advanced alternative for the removal of organic/inorganic/aesthetic pollutants, decolourization of dyes, killing of microorganisms, removal of heavy metal, etc. The application of nanoparticles as nanophotocatalysts, nanofiltration membranes, and nanoadsorbents has gained momentum in the past few decades.

Metal/metal Oxides Nanoparticles

Among various nanoparticles, silver nanoparticles (Ag-Nps) are the most studied nanoparticles because of their antimicrobial properties [8–11]. Ancient practices include silver as a prevalently used antimicrobial agent. Reducing the size of silver particles to the nanoscale further improves its effectiveness to destroy/deactivate the microorganisms by easily attaching and permeating through the cell membranes of the pathogen, resulting in the blocking of metabolic activities and damaging the cellular DNA [9]. Ag-Nps-impregnated filtration membranes, concrete pebbles loaded with Ag-Nps, and nanocomposites of cement and silver are cost-effective and efficient alternatives for wastewater treatment. In addition to Ag-Nps, other metal/ metal oxide nanoparticles like Fe [12], ZnO [13], MgO [14], Fe_3O_4 [13], TiO_2 [15], etc., are found to be efficient in removing heavy metals, dyes, and ions from wastewater. The large surface area of these nanoparticles offers effective adsorption of the heavy metal ions due to the complexation between the negatively charged oxygen of metal oxides and the cationic metal ions present in wastewater [16]. Apart from this, the high reactivity of nanoparticles (zero valent zinc and iron) leads to the faster degradation of the pollutants like CCl_4, octa-chloro-dibenzo-p-dioxine, etc., into its lower chlorinated constituents. Recent research has shown the utilization of nanocomposites and quantum dots for the removal of a wide variety of pollutants from wastewater. TiO_2-based nanocomposites have been found to exhibit excellent photocatalytic properties with a high degradation rate of ammonia nitrogen under visible and ultraviolet light [17, 18]. ZnO-biochar-based nanocomposites have synergistically combined properties of ZnO (adsorption and catalysis) and biochar (natural adsorbent with a high surface area) for the removal of pharmaceutical pollutants [19, 20]. Hu et al. [20] discussed that the mechanism of adsorption of ZnO/biochar is based on cation exchange, electrostatic interaction, and $\pi - \pi$ stacking interaction.

Literature reports there are certain nanoparticles, like, TiO_2, ZnO, ZrO_2, ZnS, CdS, etc., that act as photocatalysts [21, 22]. A photocatalyst absorbs photons from sunlight and initiates a chemical reaction without being consumed during the reaction. These photocatalysts are employed to carry out advanced oxidation processes (AOPs), which facilitate the in-situ generation of strong oxidizing agents to oxidize the organic pollutants like dyes, estrogens, crude oil, pesticides, organic acids, etc., present in the wastewater [23]. Ag doped-TiO_2 nano photocatalyst designed by Mahmoudian-Boroujerd et al. [21] was found to be effective in the degradation of recombinant DNA (rDNA) which was present in the wastewater discharged from Hepatitis B surface antigen production plant. In another work, Karimi and coworkers carried out a comparative study of the photodegradation ability of Fe_3O_4 Nps with ZnS quantum dots and N, S-doped graphene quantum dots on Victoria blue R (an organic dye pollutant) [24]. In addition to this, the filtration or precipitation units impregnated with these photocatalytic nanoparticles are found to be effective in the removal of metals like mercury [25].

Carbon Nanoparticles

Recent research has showcased carbon nanoparticles (carbon nanotubes and graphene) as potent alternatives to the metal/metal oxide nanoparticles for the treatment of industrial wastewater laced with heavy metals, dyes and organic/inorganic compounds. Carbon nanotubes (CNTs) and graphene exhibit high surface area, making them ideal for the adsorption of a variety of pollutants. The types of adsorption, namely, physisorption and chemisorption involve binding with the solute particles (pollutants) via van der Waals forces, electrostatic interaction, and chemical bonding. The carbon nanoparticles can be functionalized to induce positive or negative charges on the molecule to facilitate electrostatic interaction with the pollutants. The functionalization process increases their adsorption efficiency many folds. Mubarak et al. [26] reported the enhanced copper removal efficiency of carbon nanotubes, functionalized with nitric and sulphuric acid at pH 5.5. He discussed that the functionalization of CNT allows the attachment of carboxyl groups which attracts the copper ions. Carbon nanotubes are unique nanomaterials having the morphology of tubes of length in nanometers to centimeters and radius in the nanometers range. The application of functionalized CNTs as an efficient adsorbent is basically due to the presence of negative charge on the surface, enabling its extensive interaction with the positively charged heavy metal ions like Pb^{2+}, Ni^{2+}, Cu^{2+}, Cd^{2+}, and Zn^{2+} [27, 28]. Apart from this, CNTs are also observed to be efficient in adsorbing textile dyes between the pH range of 2–10. Another carbon-based nanomaterial; graphene exhibits a two-dimentional honeycomb lattice structure with superior adsorption efficiency of pollutants. The functionalized graphene nanosheets are observed to have an excellent adsorption tendency towards lead and cadmium in wastewater [29]. In addition to these two well-studied carbon nanoparticles, the nano-porous activated carbons have also exhibited their efficiency in removing aesthetic pollutants from wastewater. The nano-porous activated carbon can be easily synthesized from agricultural wastes by functionalizing it with acid or alkali. The small size of the pores (1–2 nm) increases the surface area drastically, facilitating the adsorption of even stubborn and reactive dyes. Ahmed and Hameed [30] synthesized granular activated carbon from bamboo waste for the effective removal of reactive black dye (RB5). The adsorption capability of activated carbon was found to be much more than neutral zeolite.

5.2 *Biochar: A Sustainable Adsorbent Material*

Biochar is a highly porous carbonaceous material obtained from different types of biomass feedstock via thermochemical decomposition (chemical treatment followed by pyrolysis/carbonization/gasification) [31, 32]. It is a sustainable material with almost no residual impact on the environment and is a low-cost technological solution for wastewater treatment. It is an excellent adsorbent due to its high surface area and the presence of surface functional groups [31, 33]. Ambaye et al. [34] discussed the mechanism of biochar adsorption which includes partitioning, electrostatic interaction between biochar surface and pollutant, pore filling, interaction between electron donating and accepting moieties, and hydrophobic interaction for removal of organic pollutants. The adsorption efficiency of biochar is dependent on various parameters, like, the type of feedstock, the temperature at which thermochemical decomposition is carried out, pretreatment with acid/base/plasma/microwave/ultrasound, etc. The findings of Zhao et al. [35] highlight that the acid or base pretreatment functionalizes the surface of biochar resulting in enhanced pore volume and surface area for better adsorption ability. The composites of biochar with chitosan are found to be efficient in the removal of heavy metals like lead, cadmium, and copper from wastewater.

5.3 Nanofiltration Membranes for Wastewater Treatment

The filtration process reduces the concentration of suspended particles, chemical pollutants, microorganisms, etc., making it fit for drinking and for other usages. Membrane technology has become popular in the last few decades, because of its excellent filtration efficiency and durability. The filtration membranes are classified on the basis of their ability to filter pollutants present in wastewater. Microfiltration membranes (pore size 0.1–5 μm), ultrafiltration membranes (20 nm–0.1 μm), nanofiltration membranes (0.5–20 nm), and reverse osmosis membranes (allows only water to pass through them) are the different categories commercially available nowadays. Microfiltration membranes cannot filter out viruses and very fine pollutants but can be used with the other three fine filters to protect them from biofouling. The ultrafiltration membranes (made of polyesters and polystyrenes) are prevalently used by pharma and textile industries to treat their wastewater because the affluents of these industries lie in the ultrafilter range. Further, advancement in this field led to the designing of nanofiltration (NF) membranes. These membranes have a molecular weight cut-off (MWCO) for nanometer-range charged particles. These membranes of pore size 0.5–20.0 nm allow the nanofiltration process; a pressure-driven membrane process (pressure between 5 and 20 bars) with the ability to reject the divalent and higher valency ions [36, 37]. The availability of a wide variety of membranes and unique filtration mechanism has made it one of the widely employed technological interventions for wastewater treatment. It can filter out pollutants ranging from organic to inorganic contaminants efficiently at a low cost [38]. Ceramic membranes and polymeric membranes are mainly two categories of NF membranes. The ceramic membranes offer high thermal, mechanical and chemical durability [39, 40]. In an interesting work, Parham et al. designed the CNT-ceramic composite NF for filtration of yeast [41]. The designed filter showed 98% filtration efficiency for yeast and nearly 100% removal of heavy metal ions. The graphene-based NF membranes of thickness 22–53 nm, designed by Han et al. [42] were found to be effectively impermeable to organic dye and ionic substances. Polymeric membranes are also a low-cost, fine mesh filtration alternative. Polysulfone-based filters work efficiently in the ultrafiltration range; however, it is difficult to design membranes in the nanometer range. Polyamide-based membranes have helped to design NF membranes of 1 nm pore size. Two major reasons that NF membranes are preferred over reverse osmosis membranes are their low cost and comparatively less energy consumption. However, nanofiltration membranes cannot filter monovalent ions. The present membrane-based filtration technologies are a revolution in the field of wastewater treatment.

6 Construction of Wetlands as a Sustainable Approach to Treat Wastewater

Constructed wetlands (CWs) or artificial treatment systems are engineered in such a manner to design an ecosystem utilizing natural resources, like, wetland vegetation, soil, pebbles, and the associated microbial population to help treat wastewater. In other words, the CW system replicates the functions of natural wetlands in a more controlled fashion. This decentralized natural methodology of treatment of wastewater can be easily adopted by the communities for the betterment of the flora and fauna of that region. The wastewater generated from domestic usage like bathing, washing, cooking, etc., is classified as grey water. It can be treated by proper mechanism and can be reused for irrigation purposes, domestic use for toilet flushing, and recreational activities [43]. Additionally, treating grey water reduces the number of pathogens and inorganic micropollutants in the wastewater. The CW systems are of two types, namely surface flow wetlands (wastewater flows horizontally over the wetland sediment) and infiltration wetlands (wastewater flows vertically through the sediment layer and is collected in the drains). The ecosystem designed in wetlands has special characteristics which have the efficiency to treat wastewater [44]. Firstly, the treatment systems consist of shallow (usually less than 1 m deep) beds or channels planted with aquatic plants and they rely upon natural, biological (decomposition), physicochemical (adsorption), and chemical (oxidation) processes between plants and substrates such as soil, sand, etc., to treat wastewater. Secondly, the vegetation facilitates the breakdown of nutrient pollutants such as nitrogen and phosphorus. It aids in the removal of solid contaminants by trapping them in the roots of plants. The CW ecosystem supports tall emergent vegetation (helophytes) that has the capacity to take up large amounts of nutrients and are able to aerate soil rhizosphere through their special type of aerenchyma roots. These constructed wetlands exhibit high efficiency in the removal of phosphorous, nitrogen, COD, BOD, and microbial population. The introduction of absorbing material to the sediments of the wetland reinforces the pollutant removal efficiency. The schematic of the surface flow of the constructed wetland is shown in Fig. 3.

Neela Hauz lake revival has set a significant example for constructed wetlands. The system includes in-situ remediation of wastewater through two oxidation and stabilization ponds, a filtration zone, and aquatic plants with zero energy input. The wetland constructed in Neela Hauz has two oxidation ponds, a gradient channel with

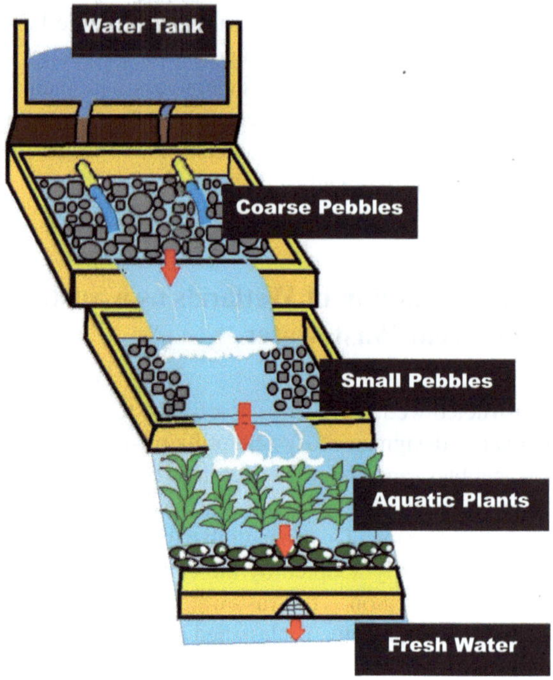

Fig. 3 Schematic representation of surface flow constructed wetland

a sieve and boulders which act as filters, three physical treatment tanks (the first treatment tank with stone filters, the second one with larger river bed pebbles, and the third one with smaller river bed pebbles), ridges of gravel (20 mm) and furrows having 25 aquatic plants, and other microbial communities. No energy consumption takes places in these processes. The constructed wetland unit of Neela Hauz carries out treatment of a million litres of sewage water every day using sustainable natural processes. Many countries across the globe have adopted this sustainable CW approach to treating wastewater. The Netherlands has 1.3 ha constructed wetland located near Lauwersoog as a water treatment zone for recreational housing projects. In Germany, numerous CW were designed during 1980s and 1990s to treat dairy wastewater, feedstock wastewater, and phenol wastewater. In North America, almost 1000 constructed wastewater systems are operational to treat not only municipal wastewater but almost all types of wastewater.

7 Lessons Learned from Other Countries

Han River located in Seoul, South Korea was once a heavily polluted river due to the discharge of untreated chemical wastes, industrial wastewater, and domestic trash. As a result of this, lives of almost 12 million people relying on Han for water were severely affected. The collaborative effort of the government and native dwellers has not only improved the river's conditions but has also revived the count of flora and fauna in the river. Besides the Han River Renaissance Project, the government has adopted local initiatives like, building forests around the river, introducing wetlands, and dividing areas around the river into restoration areas, conservation areas, and user areas. Another example is of Thames River, which was declared biologically dead in 1957. Human waste and toxic runoffs from tanneries were dumped into the river during the great industrial revolution. An event called the Great Stink in 1858 was a result of human sewage flowing into the Thames. This forced the British parliament to design and construct an effective wastewater disposal system. The environmental agency has established the Thames Estuary 2100 plan to mitigate, monitor, and respond to potential flood risks. Thames River is now one of the cleanest rivers in England. In comparison with the Han River and Thames River, the Yamuna has a bigger stretch with many small-scale and large-scale industries dumping their untreated waste directly into it. Apart from this, the population density along the Yamuna River is almost five times more than Han. Therefore, the implementation of wastewater treatment initiatives has additional challenges.

8 Conclusions

The stretch of Yamuna River in Delhi-NCR has become a dumping water body for treated/untreated wastewater discharged by industries, human settlements, and agricultural practices resulting in severe pollution. The existing wastewater treatment technologies are not sufficient under present conditions. This paves the way for advanced sustainable technological interventions to address the wastewater treatment problem of the city. Utilizing nanotechnological solutions like the use of nanoparticles/nanocomposites, activated nano carbonaceous materials, nano biochar particles with advanced wastewater treatment efficiency could be done to mitigate the effect of growing pollution conditions. In addition to this, sustainable approaches, like employing natural processes with zero energy requirement are the demand of the moment. However, selection of any one technology or the combination of various technologies will depend on the set of conditions of the pollution treatment.

Acknowledgements The authors express their sincere thanks to the Centre for Research, Maitreyi College for the research funding of the project under the Annual Research Programme.

References

1. Parween M, Ramanathan AL, Khillare PS, Raju NJ (2014) Persistence, variance and toxic levels of organochlorine pesticides in fluvial sediments and the role of black carbon in their retention. Environ Sci Pollut Res 21:6525–6546

2. Paul A, Jain P, Handa K (2014) Najafgarh drain-sampling and analysis. Int J Adv Res Eng Appl Sci 3(10):1–11
3. Christopher D, Kaur S, Singh R (2012) Water quality status of river Yamuna in Delhi with reference to presence of heavy metals: a review. Int J Pharm Med Bio Sci 1(2):266–271
4. Parween M, Ramanathan AL, Raju NJ (2017) Waste water management and water quality of river Yamuna in the megacity of Delhi. Int J Environ Sci Technol 14:2109–2124
5. Malik DS, Thakur J, Singh RK, Kaur A, Nijhwan S (2014) Heavy metal pollution of the Yamuna River: an introspection. Int J Curr Microbiol Appl Sci 3(10):856–863
6. Sharma D, Kansal A (2011) The status and effects of the Yamuna Action Plan (YAP). In: Proceedings of the TERI–Yale workshop on the Yamuna River "The Yamuna river: a confluence of waters, a crisis of need
7. Yamuna Action Plan (2020) Journal of India. Homepage. https://journalsofindia.com/yamuna-action-plan/
8. Slawson RM, Van Dyke MI, Lee H, Trevors JT (1992) Germanium and silver resistance, accumulation, and toxicity in microorganisms. Plasmid 27(1):72–79
9. Borrego B, Lorenzo G, Mota-Morales JD, Almanza-Reyes H, Mateos F, López-Gil E, Bogdanchikova N (2016) Potential application of silver nanoparticles to control the infectivity of Rift Valley fever virus in vitro and in vivo. Nanomed Nanotechnol Biol Med 12(5):1185–1192
10. Gera M, Kumar R, Jain VK (2014) Fabrication of a pocket friendly, reusable water purifier using silver nano embedded porous concrete pebbles based on green technology. J Bionanosci 8(1):10–15
11. Suman N, Kumar R, Jain VK (2013) Development of electroless, potable water sterilisation device based on nanophase modified concrete pebbles. Mater Res Innov 17(4):244–249
12. Wu Q, Yang H, Ke L (2017) Application of nanotechnology in wastewater treatment. Nanosci Nanotechnol 1(1):19–25
13. Feng L, Cao M, Ma X, Zhu Y, Hu C (2012) Superparamagnetic high-surface-area Fe_3O_4 nanoparticles as adsorbents for arsenic removal. J Hazard Mater 217:439–446
14. Gao C, Zhang W, Li H, Lang L, Xu Z (2008) Controllable fabrication of mesoporous MgO with various morphologies and their absorption performance for toxic pollutants in water. Cryst Growth Des 8(10):3785–3790
15. Luo T, Cui J, Hu S, Huang Y, Jing C (2010) Arsenic removal and recovery from copper smelting wastewater using TiO_2. Environ Sci Technol 44(23):9094–9098
16. El-Sayed ME (2020) Nanoadsorbents for water and wastewater remediation. Sci Total Environ 739(15):139903
17. Luo M, Lin H, He Y, Li B, Dong Y, Wang L (2019) Efficient simultaneous removal of cadmium and arsenic in aqueous solution by titanium-modified ultrasonic biochar. Bioresour Technol 284:333–339
18. Yousaf B, Liu G, Abbas Q, Wang R, Ullah H, Mian MM, Rashid A (2018) Enhanced removal of hexavalent chromium from aqueous media using a highly stable and magnetically separable rosin-biochar-coated TiO_2@C nanocomposite. RSC Adv 8(46):25983–25996
19. Hu H, Sun L, Gao Y, Wang T, Huang Y, Lv C, Zhang YF, Huang Q, Chen X, Wu H (2020) Synthesis of ZnO nanoparticle-anchored biochar composites for the selective removal of perrhenate, a surrogate for pertechnetate, from radioactive effluents. J Hazard Mater 387:121670
20. Hu Y, Zhu Y, Zhang Y, Lin T, Zeng G, Zhang S, Wang Y, He W, Zhang M, Long H (2019) An efficient adsorbent: simultaneous activated and magnetic ZnO doped biochar derived from camphor leaves for ciprofloxacin adsorption. Bioresour Technol 288:121511
21. Mahmoudian-Boroujerd L, Karimi-Jashni A, Hosseini SN, Paryan M (2019) Optimization of rDNA degradation in recombinant Hepatitis B vaccine production plant wastewater using visible light excited Ag-doped TiO_2 nanophotocatalyst. Process Saf Environ Protect 122:328–338
22. Malakootian M, Nasiri A, Asadipour A, Faraji M, Kargar E (2019) A facile and green method for synthesis of $ZnFe_2O_4$@ CMC as a new magnetic nanophotocatalyst for ciprofloxacin removal from aqueous media. Methods 6:1575–1580

23. Mehrjouei M, Müller S, Möller D (2015) A review on photocatalytic ozonation used for the treatment of water and wastewater. Chem Eng J 263:209–219
24. Karimi H, Rajabi HR, Kavoshi L (2020) Application of decorated magnetic nanophotocatalysts for efficient photodegradation of organic dye: a comparison study on photocatalytic activity of magnetic zinc sulfide and graphene quantum dots. J Photochem Photobiol A Chem 397:112534
25. McLain AA (2019) Photocatalytic properties of zinc oxide and graphene nanocomposites. In: Proceedings of the Wisconsin space conference
26. Mubaraka N, Daniela S, Khalid M, Tana J (2012) Comparative study of functionalized and non-functionalized carbon nanotube for removal of copper from polluted water. Int J Chem Environ Eng 3(5):1154
27. Jeon I-Y, Chang DW, Kumar NA, Baek J-B (2011) Functionalization of carbon nanotubes. Carbon Nanotubes Polym Nanocomp 6:91–110
28. Rao GP, Lu C, Su F (2007) Sorption of divalent metal ions from aqueous solution by carbon nanotubes: a review. Separ Purif Technol 58(1):224–231
29. Deng X, Lü L, Li H, Luo F (2010) The adsorption properties of Pb(II) and Cd(II) on functionalized graphene prepared by electrolysis method. J Hazard Mater 183(1–3):923–930
30. Ahmad A, Hameed B (2010) Fixed-bed adsorption of reactive azo dye onto granular activated carbon prepared from waste. J Hazard Mater 175(1–3):298–303
31. Zhang Y, Chen P, Liu S, Peng P, Min M, Cheng Y, Anderson E, Zhou N, Fan L, Liu C, Chen G, Liu Y, Lei H, Li B, Ruan R (2017) Effects of feedstock characteristics on microwave-assisted pyrolysis: a review. Bioresour Technol 230:143–151
32. Park J-H, Ok YS, Kim S-H, Cho J-S, Heo J-S, Delaune RD, Seo D-C (2016) Competitive adsorption of heavy metals onto sesame straw biochar in aqueous solutions. Chemosphere 142:77–83
33. Wang B, Gao B, Fang J (2017) Recent advances in engineered biochar productions and applications. Crit Rev Environ Sci Technol 47(22):2158–2207
34. Ambaye T, Vaccari M, van Hullebusch ED, Amrane A, Rtimi S (2021) Mechanisms and adsorption capacities of biochar for the removal of organic and inorganic pollutants from industrial wastewater. Int J Environ Sci Technol 24:1–22
35. Zhao N, Yang X, Zhang J, Zhu L, Lv Y (2017) Adsorption mechanisms of dodecylbenzene sulfonic acid by corn straw and poplar leaf biochars. Materials 10(10):1119
36. Wu L, Wang H, Xu T-W, Xu Z-L (2017) Polymeric membranes. Membrane-based separations in metallurgy. Elsevier, Amsterdam, pp 297–334
37. Abdel-Fatah MA (2018) Nanofiltration systems and applications in wastewater treatment. Ain Shams Eng J 9(4):3077–3092
38. Shon HK, Phuntsho S, Chaudhary DS, Vigneswaran S, Cho J (2013) Nanofiltration for water and wastewater treatment: a mini review. Drink Water Eng Sci 6(1):47–53
39. Mostafavi ST, Mehrnia MR, Rashidi AM (2009) Preparation of nanofilter from carbon nanotubes for application in virus removal from water. Desalination 238(1–3):271–280
40. Asif MB, Zhang Z (2021) Ceramic membrane technology for water and wastewater treatment: a critical review of performance, full-scale applications, membrane fouling and prospects. 418:129481
41. Parham H, Bates S, Xia Y, Zhu Y (2013) A highly efficient and versatile carbon nanotube/ ceramic composite filter. Carbon 54:215–223
42. Han Y, Xu Z, Gao C (2013) Ultrathin graphene nanofiltration membrane for water purification. Adv Funct Mater 23(29):3693–3700
43. ElZein ZAKAA, Abdou A, Abd-ElGawad I (2016) Constructed wetlands as a sustainable wastewater treatment method in communities. Proced Environ Sci 34:605–617
44. Vymazal J (2011) Constructed wetlands for wastewater treatment: five decades of experience. Environ Sci Technol 45(1):61–69

Synthesis and Structural Characterization of Schiff Base-Based Transition Metal Complexes

Mukesh Kumar, Vagisha Mishra, and Ravindra Singh

1 Introduction

The prospective applications of the nitrogen and sulfur donor-chelating ligands in medicinal chemistry [1–3], anti-bacterial action [4], anti-fungal activity [5], and antiviral activity [6] have led to extensive research in recent years. The 4-amino-5-mercapto-1,2,4-triazoles have been investigated for their antimicrobial, anticancer, and antiviral activities and are used for variety of compounds because of their potential use in medicinal chemistry [7, 8]. Due to their antitubercular properties, thiosemicarbazide and thiosemicarbazones have also been the subject of substantial research [9, 10]. Several drugs have been found to have dramatically increased activity when taken with metal ions when they are used in the form of metal complexes. [11, 12]. Thus, as the research progresses in this area, several metal chelators have been found to inhibit tumor growth [13]. The metal chelator of thiosemicarbazone is also used in the treatment of cancer [14].

The coordination behavior of thiosemicarbazone with various metals has also been studied in the past few years due to various donor sites for the metal ion [15–18]. The advantage of using these ligands is that they contain sulfur and nitrogen donor in one moiety which chelates with metal ions and forms a five-membered ring. This five-membered ring provides extra stability to the metal complex. They also contain the thioamide group (–CS–NH–) which can perform the thione-thiol tautomerism so they provide several donor sites for the coordination of the metal ion [19].

Thiosemicarbazide and its derivatives' coordination behavior with Ni(II), Pd(II), and Pt(II) have been extensively researched [20]. Co(III) complexes with

M. Kumar (✉) · R. Singh
P. G. Department of Chemistry, Jai Prakash University, Saran Bihar, Chapra 841302, India
e-mail: mkajit101988@gmail.com

V. Mishra (✉)
Institute of Information Technology Pune (IIIT Pune), Pune 411041, India
e-mail: drvagisha@iiitp.ac.in

S. Rattan et al. (eds.), *Emerging Trends in Synthesis and Catalysis in Chemistry*, Springer Proceedings in Materials 51, https://doi.org/10.1007/978-981-97-5169-3_18

thiosemicarbazide have been obtained in both cis and trans forms [21]. Several manganese compounds with Mn(thiosemicarbazide) are described in another publication where X_2 ($X = Cl^-$, Br^-, SO_4^{2-}) [22]. Pt(IV) complexes containing 1-phenylthiosemicarbazide have also been documented [23]. The literature has looked into thiosemicarbazone, which is made from 2-acetyl thiophene and metal ions. So, in this article, we describe that two different metal ions (nickel and cobalt) have been found to chelate with thiosemicarbazone, which is produced from 2-acetyl thiophene. The structure of the complex has been established using several spectroscopic techniques, including IR, UV–Vis, and LC–MS spectroscopy. Additionally, the complex's magnetic and electronic spectrum behaviors have been used to establish its geometry.

2 Experimental Section

2.1 *Materials and Methods*

2-acetyl thiophene, thiosemicarbazide, $CoCl_2$, and $NiCl_2$.6H2O, which were used directly without additional purification was purchased from Sigma-Aldrich. Under anaerobic and anhydrous conditions, all experiments were conducted using conventional Schleck synthesis methods. Prior to usage, ethanol, acetonitrile, methanol, and ether were freshly distilled under a nitrogen environment and purified according to usual technique.

2.2 *Instrumentation and Characterization*

The Agilent 6540 instrument's mass-to-charge ratio (m/z) was scanned from 100 to 1700 to acquire mass spectra in the positive ion mode. Each peak's envelope matched the isotopic distribution that was predicted. Thermo Scientific Nicolet iS50 with iD5-ATR attachment was used to conduct Fourier transform infrared (FT-IR) in the 4000 to 400 cm^{-1} range with a 4 cm^{-1} resolution. A UV–Vis-NIR spectrophotometer (UV-3600 Shimadzu) was used to record the UV–Vis spectra in the spectral range of 220–800 nm.

3 Synthetic Procedure

3.1 Preparation of Ligand

2-acetyl thiophene thiosemicarbazone (L) was synthesized from 2-acetyl thiophene and thiosemicarbazide as starting materials [24]. The process for synthesizing is as follows. In a 250 mL round bottom flask with two necks, 40 mL of ethanol was used to dissolve 0.76 g (1 mmol) of 2-acetyl thiophene. Thiosemicarbazide was dissolved in 50 mL of ethanol at 0.66 g (1 mmol). The 2-acetyl thiophene solution received dropwise additions of the thiosemicarbazide solution. The resultant mixture was left for 5 h to reflux. Over time, the white precipitate was seen. After the reaction was finished, the solvent was expelled under reduced pressure, yielding the desired white crystalline solid. Yield: 0.76 g (63%). M.P.: 180 °C. Mass data (m/z):[M + H]$^+$ $C_7H_9N_3S_2$ = 199.57. Calc. for ($C_7H_9N_3S_2$): C, 41.33; H, 3.22; N, 20.22; and S, 32.43. Found for (C_6H_7N3S2): C, 41.34; H, 3.53; N, 22.14; and S, 31.15. IR analysis (cm^{-1}): ν(CH) (very weak): 2986, ν(NH) (broad): 3320,3222, ν(C=N) (medium):1682, ν(SH) (weak): 2512, ν(NH_2) (sharp): 3362, ν(N–N) (weak): 1050, ν(C=N) (very weak): 1585, ν(C=S): (sharp) 840. Absorption band (nm): 212, 278.

3.2 Preparation of $[Co(C_7H_9N_3S_2)_2Cl_2]Cl_2$ Complex

Cobalt(II) chloride (0.15 mmol) and 2-acetyl thiophene thiosemicarbazone (0.30 mmol) were added to 500 mL of acetonitrile in a two-neck round bottom flask. For two hours, the mixture was left to stir continuously. After 2 h, the reaction mixture was kept for refluxing for another 4 h. The solvent was decreased under vacuum to around 2 mL. Diethyl ether (2 mL) was used to precipitate the final product, and the resulting brown solid was then dried in vacuo. Yield: 0.174 g (72%). M.P.: 220 °C. Mass data (m/z): [M + H]$^+$ $[C_{14}H_{18}Cl_2CoN_6S_4]^{2+}$ = 281.50. Elemental data analysis: cal. C, 27.60; H, 2.83; Cl, 24.51; Co, 10.33; N, 14.55; and S, 21.49. Found C, 27.91; H, 2.95; Cl, 23.59; Co, 9.78; N, 13.91; and S, 21.35. IR bands (cm^{-1}): ν(NH) (broad): 3312,3226, ν(CH) (very weak): 2986, ν(SH) (weak): 2512, ν(C=N) (medium): 1694, ν(NH_2) (sharp): 3358, ν(N–N) (weak): 1034, ν(C=S): (sharp) 836, ν(Co–Cl) (very weak): 324, ν(Co–N) (very weak): 238, ν(Co–S) (very weak): 212 UV–Vis (nm): 215, 278.

3.3 Preparation of $[Ni(C_7H_9N_3S_2)_2Cl_2]Cl_2$ Complex

$[Ni(C_7H_9N_3S_2)_2Cl_2]Cl_2$ complex was synthesized following the similar of $[Co(C_7H_9N_3S_2)_2Cl_2]Cl_2$ complex mentioned above. However, instead of cobalt(II) chloride, nickel(II) chloride hexahydrate was used. After completion of the reaction, the final product was obtained using diethyl ether (5 mL) and the product was afforded an off a light green solid that was dried in a vacuum. Experimental yield: 0.204 g (78%). M.P.: 230 °C. Mass data (m/z): $[M + H]^+$ $[C_{14}H_{18}Cl_2NiN_6S_4]^{2+}$ = 275.42. Elm. analysis: cal. C, 24.923; H, 2.922; Cl, 25.232; Ni, 10.673; N, 14.813; S, 22.641. Found C, 25.17; H, 2.72; Cl, 24.73; Ni, 10.22; N, 14.62; and S, 22.49. IR bands (cm^{-1}): ν(NH) (broad): 3312,3226, ν(CH) (very weak): 2986, ν(C=N) (medium):1 690, ν(SH) (weak): 2512, ν(NH_2) (sharp): 3375, ν(NN) (weak): 1022, ν(C=S): (sharp) 831, ν(Ni–Cl) (very weak): 345, ν(Ni–N) (very weak): 241, and ν(Ni–S) (very weak): 217 UV–Vis (nm): 207, 271.

4 Results and Discussion

4.1 Preparation and Characterization of Ligand

The ligand, 2-acetyl thiophene thiosemicarbazone, required for this study was synthesized using 2-acetyl thiophene and thiosemicarbazide as starting ingredients using the reported method (Scheme 1). The detailed synthesis procedures are mentioned in the experimental section.

The ligand was thoroughly characterized using infrared spectroscopy. The assignment of the IR bands has been tabulated in Table 1. The band at 3220–3320 cm^{-1} can be attributed to ν_{NH} symmetric and antisymmetric modes of vibration [25]. The very weak band at 2984 cm^{-1} in the spectrum can be assigned to $\nu_{C\text{-}H}$ (aromatic) vibration mode [26, 27]. In addition, 2500 cm^{-1} is solely responsible for $\nu_{S\text{-}H}$ mode [28]. Similar way, a weak band at 990 cm^{-1}, and a strong band at 950 cm^{-1} are responsible for $\nu_{N\text{-}N}$, $\nu_{C\text{-}S}$ vibration mode [29, 30].

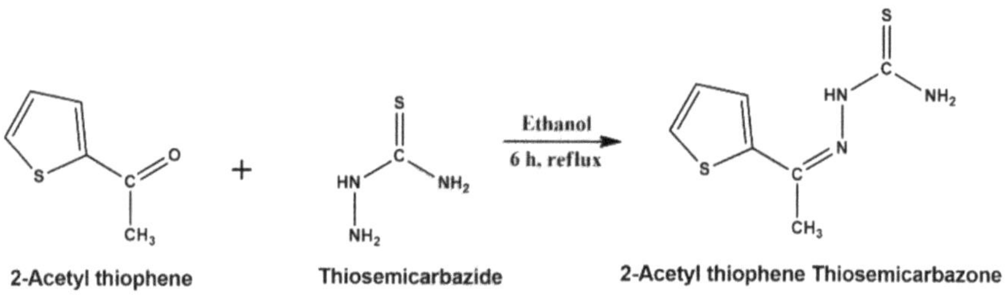

Scheme 1 Synthetic scheme for the preparation of 2-acetyl thiophene thiosemicarbazone

Table 1 Infrared spectral band of 2-acetyl thiophene thiosemicarbazone in cm^{-1} [25]

Band position [25]	Intensity	Assignment
3320, 3220	br br	ν_{NH} (antisym) ν_{NH} (sym)
2984	vw	$\nu_{C\text{–}H}$ (aromatic)
2500	w	$\nu_{S\text{–}H}$ (aromatic)
1680	m	$\nu_{C=N}$ (exothermic)
990	w	$\nu_{N\text{–}N}$
950	s	$\nu_{C\text{–}S}$

s—strong; m—medium; vw—very weak; w—weak, s—sharp; and br—broad

4.2 *Uv–Visible Spectrum of 2-Acetyl Thiophene Thiosemicarbazone:*

The UV–visible spectrum of 2-acetyl thiophene thiosemicarbazone was recorded in ethanol. The structure of 2-acetyl thiophene thiosemicarbazone contains a lone pair on S- and N-atom and the double bond (π bonds) in C=S, C=N. So, n–π^* and π–π^* transitions are possible. So as expected there are two absorption values at 210 and 275 nm were observed [31]. The strong band at 210 nm was assigned to π–π^* transition. It is well-known that higher energy absorption occurs at a lower wavelength. In addition, there is another absorption band found at 275 nm, which can be attributed to n–π^* transitions. Since π–π^* are always observed on the lower wavelength and n–π^* transition occurs at higher wavelength, the explanation given here is reasonable.

4.3 *Preparation and Characterization of the Complexes*

The cobalt complex of 2-acetyl thiophene thiosemicarbazone was prepared using cobalt chloride and 2-acetyl thiophene thiosemicarbazone as the starting material as shown in Scheme 2. The detailed experimental procedure is mentioned in the experimental section. To determine the structure of the complex various characterization was performed.

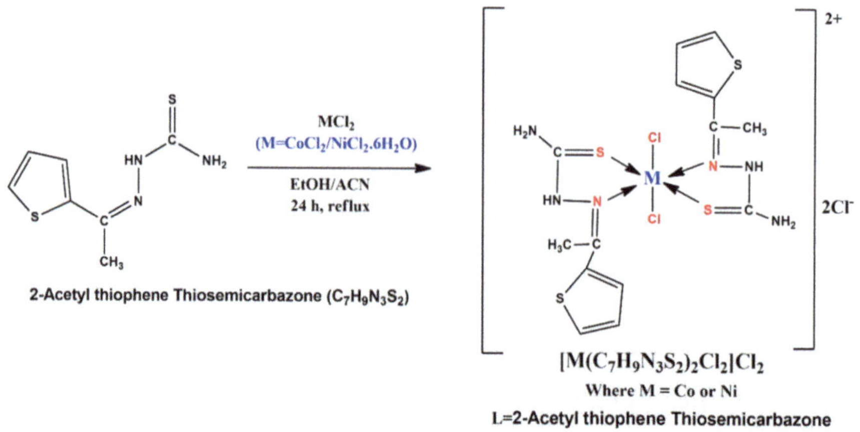

Scheme 2 Synthetic scheme for the preparation of cobalt complex of 2-acetyl thiophene thiosemicarbazone

5 Characterization of the Complex

5.1 *Magnetic Behavior of [Co($C_7H_9N_3S_2$)$_2$$Cl_2$]$Cl_2$ Complex*

Cobalt(II) comes under $3d^7$ system and the magnetic behavior of cobalt(II) complexes lies from 4.3 to 5.2 B.M [32]. In addition, some of the complexes were found in the 1.8–2.9 BM. Since Co^{2+} carries three unpaired electrons in the group state and spin only magnetic behavior can be calculated using the formula, $\mu s = \sqrt{4S(S+1)}$. The value comes to be 3.88 B.M. The excess value of magnetic behavior can be attributed to the various degrees of orbital contribution.

Co^{2+} forms various geometry of complexes such as a square planer, octahedral, and tetrahedral (inner and outer orbital types) complexes, and all are paramagnetic in nature as shown in Table 2. Table 2 shows the theoretical magnetic moment, μ_s (calculated from the formula), and experimental, μ_{obs} value of the various geometry of complexes which is being reported in the literature [34]. It is possible to distinguish the spin-paired and spin-free complexes from the magnetic moment. Nyholm et al. [35] have reported that the magnetic behavior of the complex can be altered due to the stereochemistry of the metal ion. In the tetrahedral complex of Co(II), the ground state of the Co(II) complex is found to be $^4A_2$, which is non-degenerate. Due to this, it does not allow any orbital contribution to the magnetic behavior and there is the quenching of orbital contribution. However, the higher level such as $^4T_2$, $^4T_1$, and $^4A_2$ gets mixed together, so due to this mixing at a higher level, there are some orbital contributions comes under the picture, and the value lies higher (4.3–4.7 B.M.) in comparison with theoretical μ_s value (3.88 B.M.) [34]. In the square planer complex of Co(II), the experimental value is found to be 2.1–2.8 B.M., which is again high in comparison with the theoretical value (1.73 B.M.). Although the ground state of the

Table 2 Calculated and observed magnetic behavior value of Co^{2+} [33]

Complex	Bonding	Unpaired electron	μ_s (B.M.)	μ_{obs} (B.M.)
Co^{2+}, free ion	–	3	3.88	–
Tetrahedral (sp^3)	Ionic	3	3.88	4.3–4.88
Octahedral (sp^3d^2)	Ionic	3	3.88	4.88–4.20
Sq. Planer (dsp^2)	Covalent	1	1.73	2.10–2.90
Octahedral (d^2sp^3)	Covalent	1	1.73	1.73–2.0

square planer complex is also degenerate, $^2A_{1g}$ the high value is due to the mixing of the first level $^4A_{2g}$ with the partly unquenched orbital contribution. In the case of spin-free octahedral complex and tetragonal distorted octahedral complex, the ground state is found to be $^4T_{1g}$. So, a large expected variation in the spin magnetic moment is expected in comparison with the theoretical value as shown in Table 2 [34–36]. To deduce the structure, the magnetic moment of the synthesized Co(II) complex was found to be 4.88 B.M. This observation suggests to us that cobalt is paramagnetic in nature and the geometry of the complex is octahedral.

5.2 *Magnetic Behavior of [$Ni(C_7H_9N_3S_2)_2Cl_2]Cl_2$ Complex*

The outer d-orbital configuration of Nickel(II) is $3d^8$ which means it is found in ^3F as a free ion in group state. Ni^{2+} is found in the octahedral, tetrahedral, and square plane configurations. In the literature, Reddy et al. [37] have also extensively published their pioneer work on Ni(II) and Cu(II) complexes and showed their diamagnetic. So, $[Ni(C_7H_9N_3S_2)_2Cl_2]Cl_2$ magnetic moment was found to be 2.93 B.M. indicating octahedral nature.

6 Conductance Behavior of Complex

To ascertain the complex's structure, the electrical conductance was also measured. DMF was employed to measure the electrical conductance because the complexes were found to be poorly soluble in organic solvents such as CCl_4, CH_3OH, C_2H_5OH, and THF. $Co(C_7H_9N_3S_2)_2Cl_2$ and $Ni(C_7H_9N_3S_2)_2Cl_2$ complexes were discovered to have electrical conductance values of 184 and 173 Ohm^{-1} cm^2 $mole^{-1}$, respectively. These results show that the coordination chemicals in both complexes are 1:2 electrolytic.

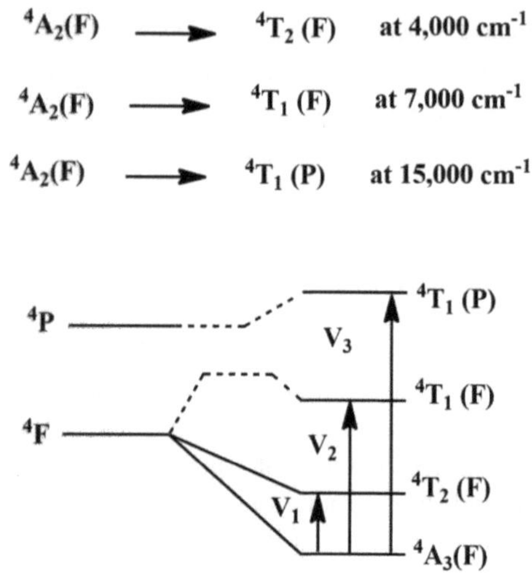

Fig. 1 Tetrahedral splitting of ^4F and ^4P and Orgel diagram of tetrahedral splitting

7 Spectral Behavior of the Complexes

7.1 Spectral Behavior of $[Co(C_7H_9N_3S_2)_2Cl_2]Cl_2$ Complex

The Co(II) ion in the free state has the 4_F term in the ground state and the 4_p term in the first excited state. The 4_F term gets split into $^4T_{1g}$ (F), $^4T_{2g}$ (F), $^4A_{2g}$ (F) state, and the 4_P transforms to $^4T_{1g}$ (P) state only under the influence of the cubic field. In the tetrahedral field, the Orgel diagram explains the three transitions of 4_F and 4_P as shown in Fig. 1. The $^4A_2(F) \rightarrow T1(P)$ transition occurs in the red part of the spectrum about 14,000–16,000 cm^{-1}. The other two absorptions are found to be beyond visible regions. Whereas the octahedral splitting of 4_F and 4_P takes place as shown in Fig. 2. It is reported that the Co(II) spin-free octahedral complex has three bands shown in the range of 8000–11,000, 17,000–22,000, and 11,000–17,000 cm^{-1}, respectively, as shown in Fig. 3 [38]. In addition, it was observed that the third absorption band occurs at a lower intensity when compared with the two other bands in the spin-free octahedral complex.

The Co(II) complex's electronic spectra fall in the region of 16,000–25,000 cm^{-1}. As shown in Fig. 4, the intensity of our cobalt complexes has three transitions and is well-agreed with the spin-free octahedral complex.

7.2 Spectral Behavior of $[Ni(C_7H_9N_3S_2)_2Cl_2]Cl_2$ Complex

Figure 5 shows the three d–d transitions of Ni(II) in a weak octahedral field. 3F and 3p are the free ground and first excited states of Ni(II). Similar way, the Ni(II) tetrahedral complex gives three d-d transitions as shown in Fig. 6.

Fig. 2 Octahedral splitting of ^4F and ^4P and Orgel diagram of octahedral splitting

$^4T_{1g}(F) \longrightarrow {}^4T_{2g}(F)$ at 8,000 cm^{-1}

$^4T_{1g}(F) \longrightarrow {}^4A_{2g}(F)$ at 20,000 cm^{-1}

$^4T_{1g}(F) \longrightarrow {}^4T_1(P)$ at 22,000 cm^{-1}

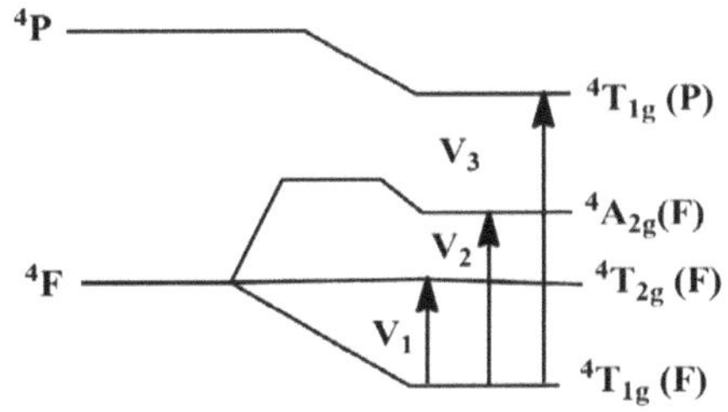

Fig. 3 Transition of an octahedral complex reported by Lever et al.

$^4T_{1g}(F) \longrightarrow {}^4T_{2g}(F)$, v_1=8,000-11,000 cm^{-1}

$^4T_{1g}(F) \longrightarrow {}^4T_{1g}(P)$, v_2=17,000-22,000 cm^{-1}

$^4T_{1g}(F) \longrightarrow {}^4A_{2g}(F)$, v_3=11,000-17,000 cm^{-1}

Fig. 4 Electronic transition of $[Co(L)_2]Cl_2$ complex

$^4T_{1g}(F) \longrightarrow {}^4T_{2g}(F)$, v_1=8,000-8650 cm^{-1}

$^4T_{1g}(F) \longrightarrow {}^4T_{1g}(P)$, v_2=20800-21580 cm^{-1}

$^4T_{1g}(F) \longrightarrow {}^4A_{2g}(F)$, v_3=16200-17500 cm^{-1}

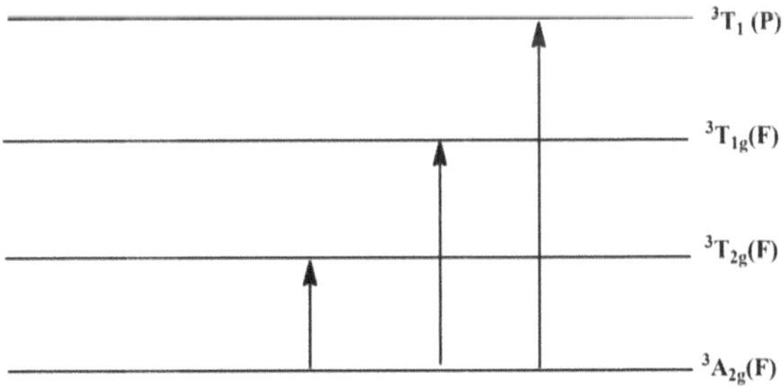

Fig. 5 d–d transition in the weak octahedral field in Ni(II)

So, electronic spectra of as-synthesized $[Ni(C_7H_9N_3S_2)_2Cl_2]Cl_2$ are shown in Fig. 7 and well-matched with the literature. So, from this observation, it is found that $[Ni(C_7H_9N_3S_2)_2Cl_2]Cl_2$ of 2-acetyl thiophene have an octahedral geometry.

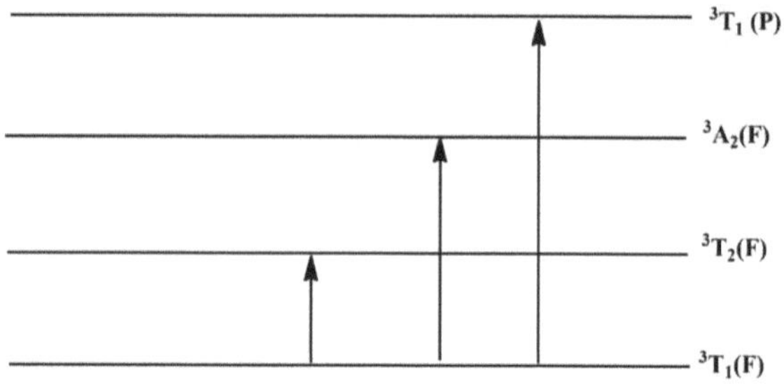

Fig. 6 d–d transition in the tetrahedral field in Ni(II)

Fig. 7 Electronic transition of $[Ni(L)_2]Cl_2$ complexes

$^4A_{2g}$ (F) ⟶ $^3T_{2g}$ (F), ν_1=9500-10,415 cm^{-1}

$^3A_{2g}$(F) ⟶ $^3T_{1g}$ (F), ν_2=14,200-14,940 cm^{-1}

$^3A_{2g}$(F) ⟶ $^3T_{1g}$ (P), ν_3=23,500-24,000 cm^{-1}

8 Infrared Spectrum of the Complex

The IR spectrum of the complex was also recorded to determine the bonding mode and coordination behavior of the ligand toward the metal ion. (Table 3) A very broad band of 3320–3220 cm^{-1} was found in both the complexes which indicates that the -NH_2 group is free in the complex and not involved in the bonding with a metal ion.

The $\nu_{C=N}$ (exothermic) is found to be at around 1695 cm^{-1} in both the complexes which is 15 cm^{-1} higher than the free ligand which indicates that the C=N group

Table 3 Infrared spectral band of $[M(C_7H_9N_3S_2)_2\ Cl_2]Cl_2$ complex [39]

Band position (cm^{-1}) for $[Co(C_7H_9N_3S_2)_2Cl_2]Cl_2$	Band Position (cm^{-1}) for $[Ni(C_7H_9N_3S_2)_2Cl_2]Cl_2$	Intensity	Assignment [39]
3312,3226	3312,3226	br, br	ν_{NH} (antisym) ν_{NH} (sym)
2984	2984	vw	ν_{C-H} (aromatic)
2500	2500	w	ν_{S-H} (aromatic)
1695	1695	m	$\nu_{C=N}$ (exothermic)
1035	1035	w	ν_{N-N}
837	837	s	ν_{C-S}
–	610	vw	ν_{Ni-N}
–	350	vw	ν_{Ni-Cl}
325	–	vw	ν_{Co-Cl}
240	240	vw	ν_{M-S}
210	210	vw	ν_{M-N}

s—strong; m—medium; vw—very weak; w—weak, s—sharp; and br—broad is coordinating with the metal ion with from the nitrogen side. This observation also has been reported in the literature previously [39]. In both compounds, the ν_{C-S} thioamide band is seen at 837 cm^{-1}. In light of this fact, it is obvious that the coordination of a metal ion also involves the sulfur atom of C–S. The novel bands at 240 and 210 cm^{-1}, which can be attributed to Co–N and Co–S, were discovered in the far infrared region of the $[M(L)_2]Cl_2$ complex and show that the C=S (via sulfur atom) and C=N (nitrogen atom) coordinate with a metal ion. In a similar vein, additional bands were also discovered and attributed to the M–Cl vibrational mode at 325 cm^{-1} (for $[Co(C_7H_9N_3S_2)_2Cl_2]Cl_2$) and 350 cm^{-1} (for $[Ni(C_7H_9N_3S_2)_2Cl_2]Cl_2$). This indicates that the coordinate bond between the metal ion and Cl- ion also involves Cl-ion.

9 Structure of the Complex

Thus, it can be said that the complex is octahedral in shape based on IR, UV–Vis absorption data, conductance, magnetic moment, and spectral investigations. Two equivalent of ligand was coordinated with cobalt metal. Each equivalent of the ligand is coordinated to the metal C=N (via two nitrogen atoms), and C=S (via two sulfur atoms). Cobalt is found in the +2-oxidation state. The counter ion is chlorine atoms. Overall, the structure of the complex is also shown in Scheme 2.

Conclusion: Metal chlorides and 2-acetyl thiophene thiosemicarbazone were used to synthesize the complexes of cobalt and nickel. The compound was subjected to several characterizations to identify its structure. Indicated by magnetic moment values of 4.92 B.M. for $[Co(C_7H_9N_3S_2)_2Cl_2]Cl_2$ and 2.93 B.M. for $[Ni(C_7H_9N_3S_2)_2Cl_2]Cl_2$, the complexes are paramagnetic by nature and have an octahedral geometry. Electrical conductance measurements for $[Co(C_7H_9N_3S_2)_2Cl_2]Cl_2$ and $[Ni(C_7H_9N_3S_2)_2Cl_2]Cl_2$ (184–173 Ohm^{-1} cm^2 $mole^{-1}$, respectively) revealed that the two equivalents of the ligand are linked to the metal ions. The $[Co(C_7H_9N_3S_2)_2Cl_2]Cl_2$ spin-free octahedral complex was suggested by the electronic spectrum values (8000–8650, 20,800–21,580, and 16,200–17,500 cm^{-1}). Similar to how they indicated that the $[Ni(C_7H_9N_3S_2)_2Cl_2]Cl_2$ is in octahedral geometry, the electronic spectrum values (9500–10,415, 14,200–14,940, and 23,500–24,000 cm^{-1}) were also significant. Finally, it was discovered using infrared spectroscopy that each equivalent of the ligand is linked to the metal by two C=N and two C=S nitrogen and sulfur atoms. We also received confirmation for the νM-Cl mode of vibrations from a new band at 325–350 cm^{-1} (325 for $[Co(C_7H_9N_3S_2)_2Cl_2]Cl_2$ and 350 cm^{-1} for $[Ni(C_7H_9N_3S_2)_2Cl_2]Cl_2$. This indicates that the coordinate binding with a metal ion also involves the Cl^- ion. Finally, the structure and synthesis of novel metal (Co or Ni) complexes based on 2-acetyl thiophene thiosemicarbazone ligands are confirmed by this study.

Conflict of Interest The authors state that there is no conflict of interest.

References

1. Siddiqui E, Jamal I, Azad AR, Khan T (2019) Khan, Thiosemicarbazone complexes as versatile medicinal chemistry agents: a review. J Drug Deliv Ther 9(3):689–703
2. Surrah A, Adnana S, Kettunen M (2006) Platinum group antitumor chemistry: design and development of new anticancer drugs complementary to cisplatin. Curr Med Chem 13(11):1337–1357
3. Bugarčić ZD, Bogojeski J, Petrović B, Hochreuther S, Eldik RV (2012) Mechanistic studies on the reactions of platinum(II) complexes with nitrogen-and sulfur-donor biomolecules. Dalton Trans 41(40):12329–12345
4. Varma RS, Nobles WL (1967) Synthesis and antiviral and antibacterial activity of certain Ndialkylaminomethylisatin β-thiosemicarbazones. J Med Chem 10(5):972–974
5. Bajaj K, Buchanan RM, Grapperhaus CA (2021) Antifungal activity of thiosemicarbazones, bis (thiosemicarbazones), and their metal complexes. J Inorg Biochem 225:111620
6. Mohan C, Sharma K, Chandra S (2017) A Zinc(II) PVC-membrane sensor based on synthesized thiophene-2-aldehyde semicarbazone (TAS) As an Ionophore. Jordan J Chem 12(1):39–50
7. Banerjee S, Ganguly S, Sen KK, Bardhamann BW (2013) A review on 1, 2, 4-triazoles. J Adv Pharm Educ Res 3(3):102–116
8. Mohan C, Sharma K, Chandra S (2017) Cd(II) ion-selective electrode based on 2-acetylthiophene semicarbazone in polymeric membrane. Anal Bioanal Electrochem 9(1):35–46
9. Bakır SE, Celik S, Şahin M, Ozel AE, Akyuz S, Ülküseven B (2021) Synthesis, molecular modelling, FT-IR, Raman and NMR characterization, molecular docking and ADMET study of new nickel(II) complex with an N4-tetradentate thiosemicarbazone. J Biomol Struct Dyn 39(12):4212–4224
10. Khanye SD, Baojie W, Franzblau SG, Gut J, Rosenthal PJ, Smith GS, Chibale K (2011) Synthesis and in vitro antimalarial and antitubercular activity of gold(III) complexes containing thiosemicarbazone ligands. J Organometal Chem 696(21):3392–3396
11. Rodrigo A, Quintana C, Biot C, Medina ME, Kremer SC, Kremer L, Klahn AH (2015) Palladium(II) and platinum(II) complexes containing organometallic thiosemicarbazone ligands: synthesis, characterization, X-ray structures and antitubercular evaluation. Inorg Chem Commun 55:139–142
12. Netalkar PP, Netalkar SP, Revankar VK (2014) Nickel(II) complexes of thiosemicarbazones: synthesis, characterization, X-ray crystallographic studies and in vitro antitubercular and antimicrobial studies. Trans Metal Chem 39(5):519–526
13. Axel S, Adam C, Sedgwick S, James T, Brewster C, Yan KC, Shang Y, Knoll DM, Zúñiga GIV, He XP, Tian H, Jonathan L (2020) Transition metal chelators, pro-chelators, and ionophores as small molecule cancer chemotherapeutic agents. Chem Soc Rev 49:3726–3747
14. Yu Y, Gutierrez E, Kovacevic Z, Saletta F, Obeidy P, Rahmanto YS, Richardson DR (2012) Iron chelators for the treatment of cancer. Curr Med Chem 19(17):2689–2702
15. Chauhan N, Sharma N, Buhari S, Das PP, Kumar S, Chalana A (2024) Facile synthesis of copper telluride nanoparticles as a robust nanoenzymatic catalyst for intrinsic peroxidase-like activity. ChemistrySelect 9(14):e202304361
16. Nyawade EA, Sibuyi NRS, Meyer M, Lalancette R, Onani MO (2021) Synthesis, characterization and anticancer activity of new 2-acetyl-5-methyl thiophene and cinnamaldehyde thiosemicarbazones and their palladium(II) complexes. Inorg Chim Acta 515:120036
17. Şen B, Kalhan HK, Demir V, Güler EE, Kayal HA, Subaş E (2019) Crystal structures, spectroscopic properties of new cobalt(II), nickel(II), zinc(II) and palladium(II) complexes derived from 2-acetyl-5-chloro thiophene thiosemicarbazone: anticancer evaluation. Mater Sci Eng C 98:550–559

18. Chalana A, Rai RK, Karri R, Jha KK, Kumar B, Roy G (2022) Interplay of the intermolecular and intramolecular interactions in stabilizing the thione-based copper(I) complexes and their significance in protecting the biomolecules against metal-mediated oxidative damage. Polyhedron 215:115647
19. Jakobsen HJ, Lawesson SO (1965) Thiophene chemistry—X: preparation, tautomeric structure and hydrogen bonding studies of thioisomaltol (2-acetyl-3-hydroxythiophene) and some related 3hydroxythiophenes. Tetrahedron 21(12):3331–3349
20. Chalana A, Karri R, Das R, Kumar B, Rai RK, Saxena H, Gupta A, Banerjee M, Jha KK, Roy G (2019) Copper-driven deselenization: a strategy for selective conversion of copper ion to nanozyme and its implication for copper-related disorders. ACS Appl Mater Interf 11(5):4766–4776
21. Sau DK, Butcher RJ, Chaudhuri S, Saha N, Chaudhuri S, Saha N (2004) Synthesis and spectroscopic characterization of new cobalt(III) complexes with 5-methyl-3-formyl pyrazole 3hexamethyleneiminyl thiosemicarbazone ($HMPz_3Hex$): X-ray crystallographic identification of $HMPz_3Hex$ and $[Co(MPz_3Hex)\ 2]\ Br{\cdot}H_2O$ with evidence for unusual rotation about the azomethine double bond of the ligand on complexation with cobalt(III). Polyhedron 23(1):5–14
22. Sharma NK, Agarwal M, Kohli S, Tiwari B, Gupta JN, Agarwal B (2010) Synthesis, structure and physico-chemical studies of Mn(II) complexes of salicylaldehyde derived semicarbazone and thiosemicarbazone. Orient J Chem 26:103–108
23. Bhowmik SR, Gangopadhyay S, Gangopadhyay PK (2005) Platinum coordination compounds of thiosemicarbazide derivatives: a new class of platinum blues. J Coord Chem 58:795–801
24. Ardakani MM, Niassari MS, Sadeghi A (2018) Highly selective thiocyanate PVC membrane electrode based on Ni(II) complex of 2-acetyl thiophene thiosemicarbazone schiff base (NATS) as ionophore. Eurasian J Anal Chem 13(6):44–56
25. Rao CN, Venkataraghavan R (1964) Contribution to the infrared spectra of five-membered N-and N, S-heterocyclic compounds. Can J Chem 42:43–49
26. Koji N, Philippa H (1977) Solomon. Infrared absorption spectroscopy. Holden-day
27. Alpert NL, Keiser WE, Szymanski HA (2012) IR: theory and practice of infrared spectroscopy. Springer, New York
28. Barbuceanu SF, Ilies DC, Saramet G, Uivarosi V, Draghici C, Radulescu V (2014) Synthesis and antioxidant activity evaluation of new compounds from hydrazinecarbothioamide and 1, 2, 4-triazole class containing diarylsulfone and 2, 4-difluorophenyl moieties. Int J Mol Sci 15(6):10908–10925
29. Vallette MMC, Rodríguez MJ, Chapanoff MA, Clavijo E, Jeria JSG, Aliaga AE, Jara GP, Celis F, Paipa C, Leyton P (2017) SERS spectrum of red dyes in the Mapuche belts from the beginning of the XXth century. J Raman Spectrosc 48(7):958–965
30. Xuan X, Wang Y, Wang N (2011) Density functional theory calculations of structure, FT-IR and Raman spectra of S-phenyl thioacetate. Spectrochim Acta A Mol Biomol Spectrosc 81(1):236–241
31. Nalawade AM, Patil PDV, Mahavidyalaya T (2012) Physicochemical and biological activity studies on Schiff base metal complexes
32. Zehbe K, Kollosche M, Lardong S, Kelling A, Schilde U, Taubert A (2016) Ionogels based on poly (methyl methacrylate) and metal-containing ionic liquids: Correlation between structure and mechanical and electrical properties. Int J Mol Sci 17(3):391
33. Nfor EN, Keenan LL, Nenwa J, Ndifon PT, Njong RN, Dzesse CNT, Offiong OE (2014) A novel mixed ligand dinuclear complex of cobalt(II): synthesis, characterization and magnetic studies. Cryst Struct Theory Appl 3:22
34. Cotton FA, Faut OD, Mague JT (1964) Molecular and electronic structures of some thiourea complexes of Cobalt(II). Inorg Chem 3(1):17–21
35. Nyholm RS (1953) Magnetism and inorganic chemistry. Quart Rev Chem Soc 7(4):377–406
36. Earnshaw A, Figgis BN, Lewis J, Nyholm RS (1957) Magnetic behaviour of the iron-RutheniumOsmium triad. Nature 179(4570):1121–1124
37. Clark RJH, Williams CS (1965) The far-infrared spectra of metal-halide complexes of pyridine and related ligands. Inorg Chem 4(3):350–357

38. Cotton FA, Meyers MD (1960) Magnetic and spectral properties of the spin-free 3d6 systems iron(II) and cobalt(III) in cobalt(III) hexafluoride ion: probable observation of dynamic Jahn-Teller effects. J Am Chem Soc 82(19):5023–5026
39. Dholakiya PP, Patel MN (2004) Metal complexes: preparation, magnetic, spectral, and biocidal studies of some mixed-ligand complexes with Schiff bases containing NO and NN donor atoms. Synth React Inorg Metal-Org Chem 34(3):552–563

Anticancer and Antimicrobial Activities of Schiff Base Ligand and Their Co(II), Cu(II), and Zn(II) Transition Metal Complexes: An Overview

Manisha, Pallavi Jain, and Sunil Kumar Yadav

1 Introduction

Schiff bases are a class of compounds that contain the imine or azomethine functional group (–HC=N–) [1]. In 1864, Schiff, a chemist from Germany, was the first to develop a method for creating these molecules. He achieved this by using a condensation reaction between primary amines and active carbonyls, such as aldehydes or ketones [2]. The primary factor responsible for the biological activity of Schiff bases is the presence of the azomethine group [3]. It is worth mentioning that Schiff bases containing atoms that can bind to different metal ions have important uses in both medical and non-medical fields [4–10]. These compounds form stable complexes with transition metal ions because of their strong ability to chelate [11–13]. Typically, these compounds are created by combining aromatic or aliphatic aldehydes with amines through condensation reactions [14, 15]. Even after extensive research, Schiff bases remain a subject of considerable interest within the realm of inorganic chemistry [16, 17]. These compounds have a wide range of biological uses, including reducing inflammation and pain, fighting against microbes, preventing seizures, treating tuberculosis, combating cancer, protecting against oxidative damage, and killing parasitic worms. As a result, they are highly valuable in the medical and pharmaceutical sectors [18–30]. By judiciously choosing suitable amine precursors, Schiff bases that contain nitrogen and oxygen atoms as donor atoms can increase their effectiveness when they are bonded with transition metal complexes [31–33]. Transition metal complexes of Schiff bases have various biological applications, like antifungal, antibacterial, anticancer, antitumor, antioxidant, and anti-inflammatory activities [34–38]. Transition metal complexes are becoming increasingly important in environmental and biological research, especially in the field of organometallic

Manisha · P. Jain (✉) · S. K. Yadav
Department of Chemistry, SRM Institute of Science and Technology, NCR Campus, Modinagar, India
e-mail: palli24@gmail.com

S. Rattan et al. (eds.), *Emerging Trends in Synthesis and Catalysis in Chemistry*, Springer Proceedings in Materials 51, https://doi.org/10.1007/978-981-97-5169-3_19

chemistry. They provide promising opportunities for the development of new pharmaceuticals with a wide range of mechanisms of action [39, 40]. These complexes have practical uses in the field of therapeutics for the purpose of diagnosing different diseases and are essential in the study of transition metal coordination chemistry [41]. Table 1 displays a list of recently reported complexes. This review aims to provide an informative and up-to-date summary of the existing literature on the biological properties of Schiff bases and their metal complexes.

2 Synthesis of Schiff Bases Metal Complex and Their Biological Applications

The two new SBLs were designated as **3** and **4**, via the reaction of 5-chlorosalicylaldehyde or 5-bromosalicylaldehyde **(2)** with ethyl-2-amino-4,5,6,7-tetrahydrobenzo-1-thiophene-3-carboxylate **(1)**, along with their Co(II) **(5 and 6)**, Cu(II) **(7 and 8)**, and Zn(II) **(9 and 10)** complexes as depicted in Scheme 1 [54]. Co(II) complexes 5 and 6 were obtained by mixing SBL with Co(II) salt in a 2:1 ratio, while Cu(II) complexes **7 and 8** and Zn(II) complexes **9 and 10** were prepared in a 1:1 ratio. Various spectroscopic techniques, including Fourier transform infrared (FT-IR), UV–Visible, mass spectrometry, as well as ^{1}H-NMR and ^{13}C-NMR spectroscopy, were employed to characterize both the SBLs **(3 and 4)** and the metal complexes **(5–10).** The spectroscopic data confirmed the monobasic tridentate nature of the SBLs. It was deduced that complexes **5 and 6** exhibited an octahedral configuration, while complexes **7, 8, 9, and 10** displayed deformed square planar and tetrahedral geometries, respectively. Evaluation of antibacterial activity revealed that the metal chelates exhibited superior antibacterial efficacy compared to the parent ligand against ESBL and MBL pathogens, assessed using the Kirby Bauer method. Notably, Co(II) complex **6** and Cu(II) complexes **7** and **8** demonstrated significant antibacterial activity.

Similarly, a new SBL, named **13**, was obtained by combining quinoline-2-carboxaldehyde **(11)** and 2-aminophenol **(12)** in equal amounts by a condensation reaction. This process is illustrated in Scheme 2 [55]. In addition, two separate cobalt(II) **(15)** and copper(II) **(16)** complexes were synthesized by treating SBLs **13 and 14** with the corresponding metal chlorides in a stoichiometric ratio of 1:1:1. The synthesized compounds **(13, 15, and 16)** were thoroughly characterized using a variety of analytical techniques. The structural configuration of all complexes was determined to be octahedral based on the spectroscopic data and elemental investigations. The complexes were assessed for their antibacterial activity against multiple pathogens, including *S. pneumoniae, B. subtilis, P. aeruginosa, and E. coli.* Additionally, their antifungal activity against *A. fumigatus and C. albicans* was evaluated. The results showed that the complexes had lower antifungal efficacy compared to SBL **13**, as determined by the agar well diffusion method. Furthermore, the assessment of the anticancer efficacy against breast and colon cell lines revealed that compounds

Table 1 Recently reported SBL complexes possessing biological activity

S. no.	Ligand	Metal ion	Complexes	Denticity	Geometry	Biological application	References
1	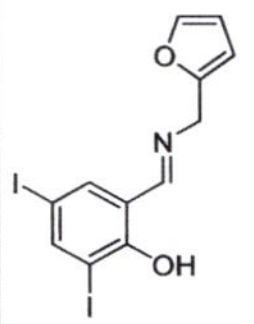	Cu(II)	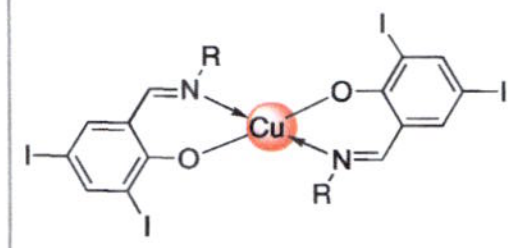	Bidentate	Square planar	Antimicrobial activity	[43]
2	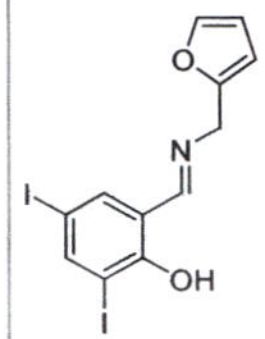	Cu(II)	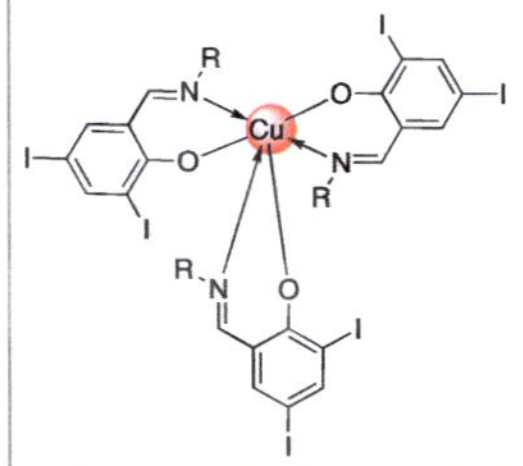	Bidentate	Octahedral	Antimicrobial activity	[43]
3	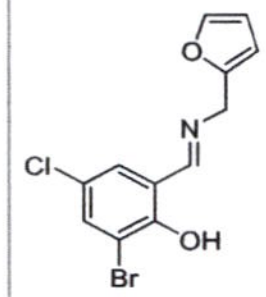	Cu(II)	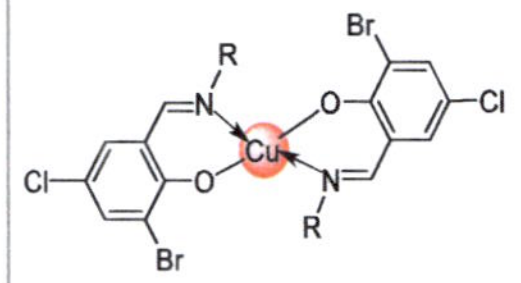	Bidentate	Square planar	Anticancer activity	[44]

(continued)

Table 1 (continued)

S. no.	Ligand	Metal ion	Complexes	Denticity	Geometry	Biological application	References
4	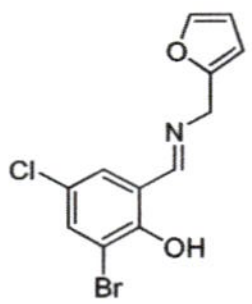	Co(II)	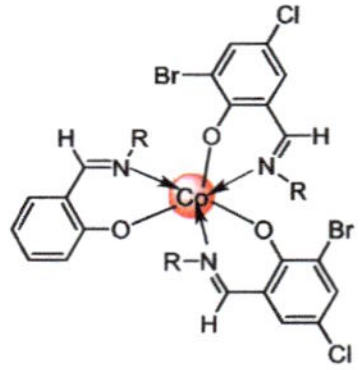	Bidentate	Octahedral	Anticancer activity	[44]
5	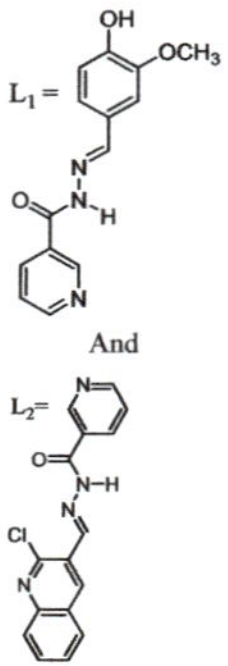	Cu(II)& Co(II)	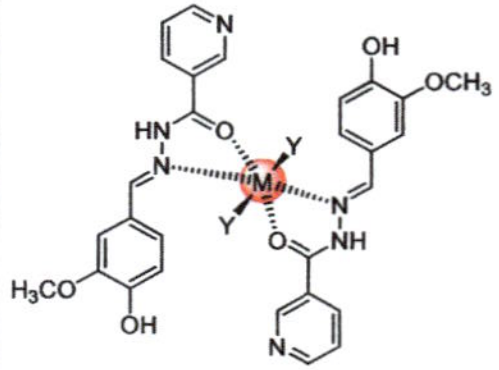 Y = $OCOH_3$, NO_3 M = Co(II), Cu(II)	Bidentate	Octahedral	Antimicrobial activity	[45]

(continued)

Table 1 (continued)

S. no.	Ligand	Metal ion	Complexes	Denticity	Geometry	Biological application	References
6	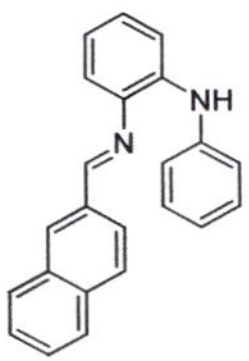	Cu(II)&Zn(II)	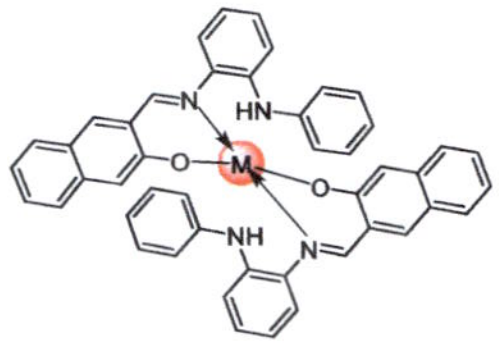 M = Cu(II), Zn(II)	Bidentate	Square planar	Antibacterial activities	[46]
7	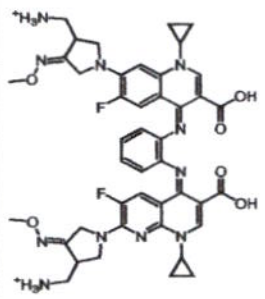	Co(II)&Zn(II)	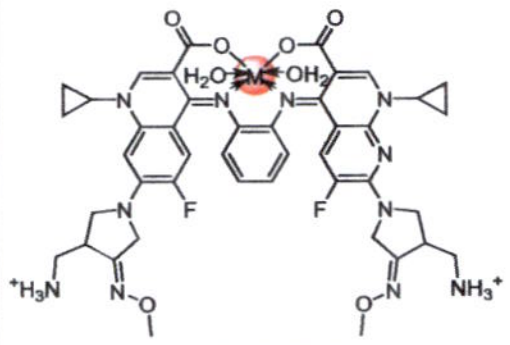 M = Co(II), Zn(II)	Tetradentate	Octahedral	Antimicrobial activity	[47]

(continued)

Table 1 (continued)

S. no.	Ligand	Metal ion	Complexes	Denticity	Geometry	Biological application	References
8	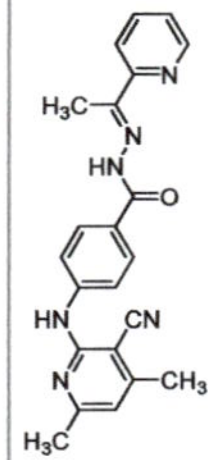	Cu(II)&Co(II)	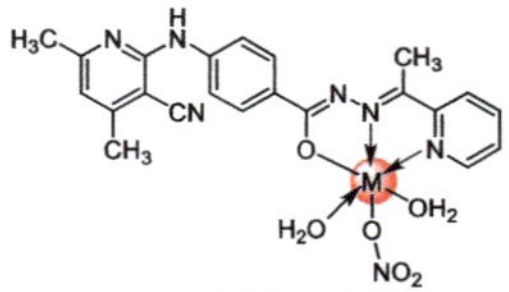 M = Co(II), Cu(II)	Tridentate	Octahedral	Antibacterial activity	[48]
9	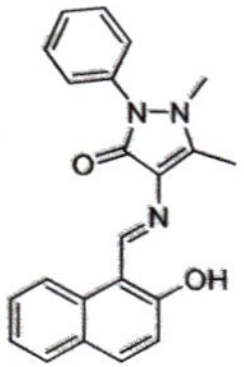	Cu(II)	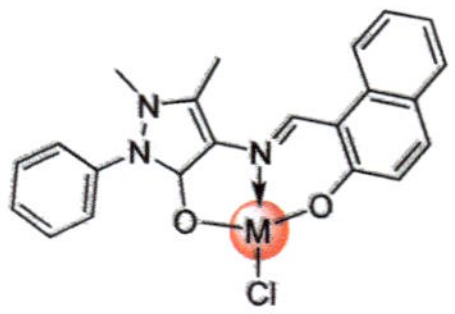	Tridentate	Square planar	Anticancer Activity	[49]
10	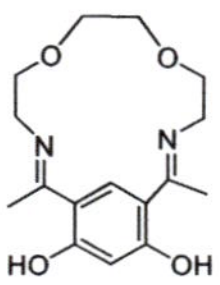	Cu(II),Zn(II)&Co(II)	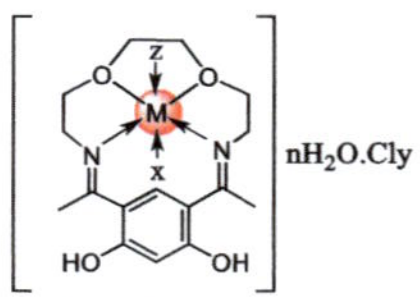 M = Cu(II), X=Z=Cl, n= 4, y= O M = Co(II), X=Z=H_2O, n=y=z M = Zn(II), X=Z=Cl, n=2,y=O	Tetradentate	Octahedral	Antimicrobial activity	[50]

(continued)

Table 1 (continued)

S. no.	Ligand	Metal ion	Complexes	Denticity	Geometry	Biological application	References
11	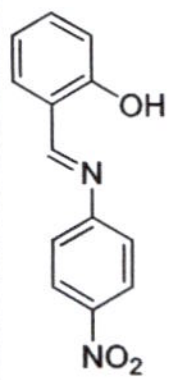	Co(II) &Cu(II)	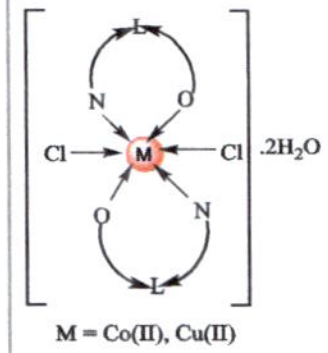	Bidentate	Octahedral	Antimicrobial activity	[51]
12	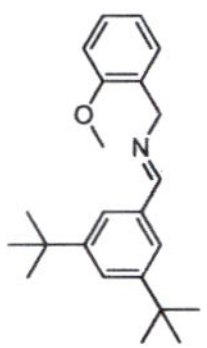	Cu(II)	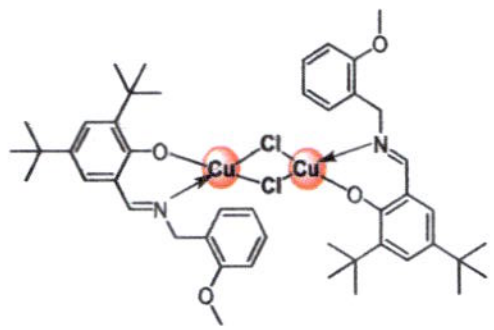	Bidentate	Square planar	Antimicrobial activity	[52]
13	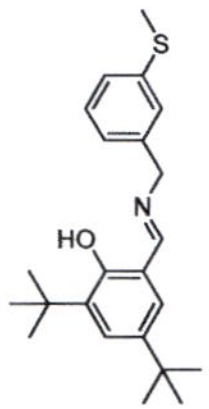	Cu(II)	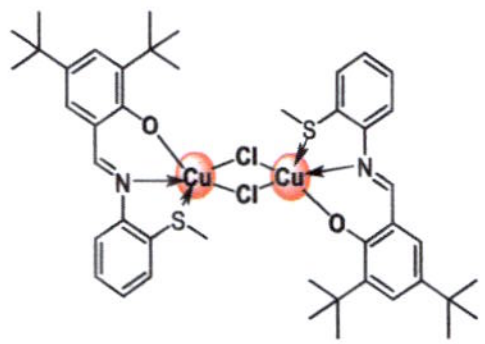	Tridentate	Octahedral	Antimicrobial activity	[52]

(continued)

Table 1 (continued)

S. no.	Ligand	Metal ion	Complexes	Denticity	Geometry	Biological application	References
14	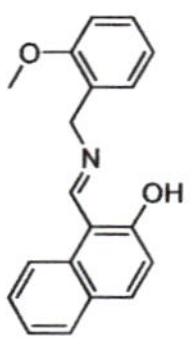	Cu(II)	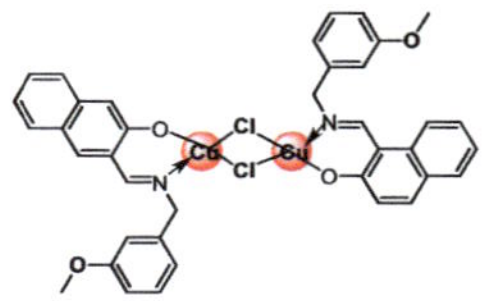	Bidentate	Square planar	Antimicrobial activity	[52]
15	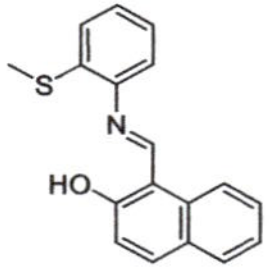	Cu(II)	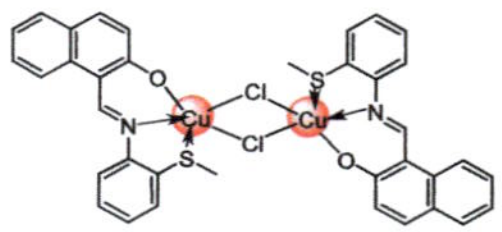	Tridentate	Square pyramidal	Antimicrobial activity	[52]
16	L_1 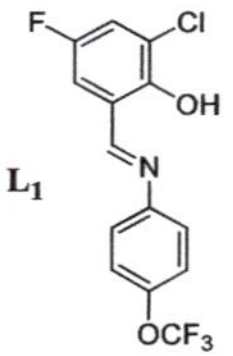	Cu(II)	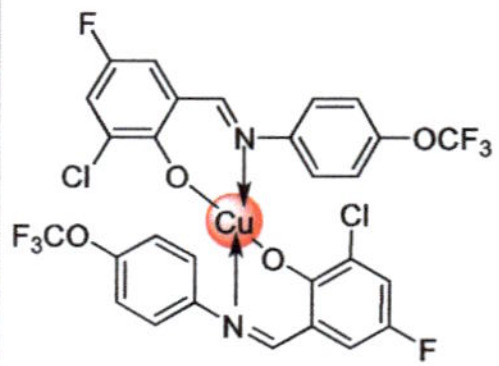	Bidentate	Square planar	Anticancer Activity	[53]

(continued)

Table 1 (continued)

S. no.	Ligand	Metal ion	Complexes	Denticity	Geometry	Biological application	References
17	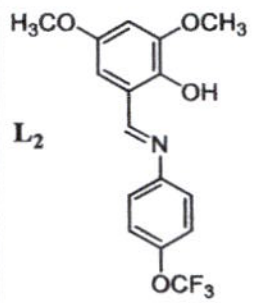 L_2	Cu(II)	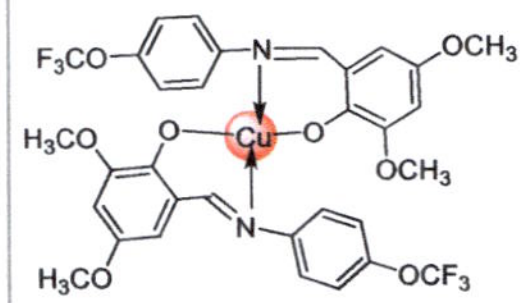	Bidentate	Square planar	Anticancer activity	[53]
18	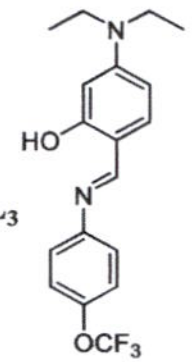 L_3	Cu(II)	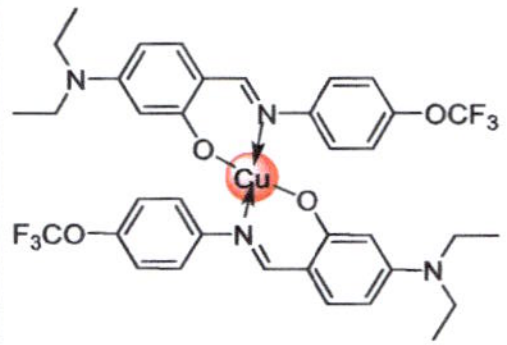	Bidentate	Square planar	Anticancer activity	[53]

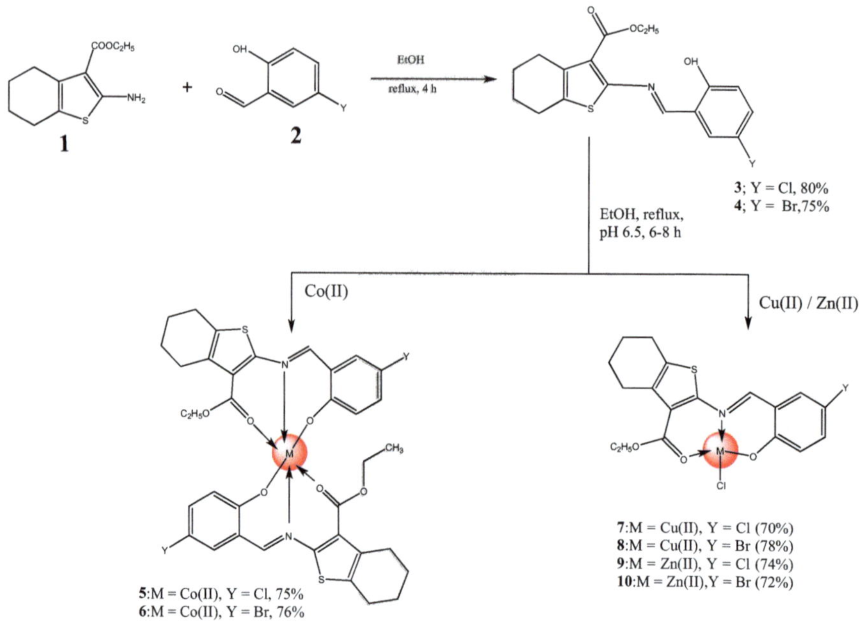

Scheme 1 Synthesis of novel SBL **3, 4** and their metal complexes [54]

15 and **16** exhibited higher IC_{50} values in comparison with compound **13**. The most remarkable finding was that complex **16** had the highest level of effectiveness in inhibiting cancer growth in the breast cell line.

The condensation reaction of terephthalaldehyde(**17**) with ortho-aniline **(18)** derivatives was conducted in two different ratios to synthesize SBLs **19–21**(1:2) and **25–27**(1:1) (Scheme 3) [4]. The resulting ligands were subjected to reflux in methanol with Co(II) salts in a conical flask, yielding six distinct complexes **22–24 and 28–30**. Various analytical techniques were utilized for characterization of all the bidentate ligands **19–21** and **25–27** and the Co(II) complexes **22–24 and 28–30**. The antibacterial activity of compounds **19–30** was assessed against *E. coli, S. marcescens, P. aeruginosa, B. subtilis, and S. aureus* using disk diffusion and broth dilution methods. Notably, SBLs **21 and 27**, featuring hydroxyl groups, exhibited superior antibacterial potential compared to other SBLs **19, 20, 25,** and **26**. Furthermore, the results demonstrated that complexes **22–24** and **28–30** possessed greater antibacterial efficacy than **19–21** and **25–27**. Among these, complex **22** displayed the highest efficiency against all tested bacterial strains, surpassing **23, 24,** and **28–30**.

In 2017, Tadavi et al. successfully synthesized a novel metal complex **34** derived from a newly developed SBL **33**. This SBL was obtained via a straightforward condensation reaction involving 1,2-diaminopropane **32** (1 mol) and 2-hydroxy-6-isopropyl-3-methylbenzaldehyde **31** (1 mol) (Scheme 4) [56]. Both compounds **33 and 34** underwent comprehensive characterization using a variety of spectroscopic techniques. Subsequently, their antibacterial and antifungal properties were

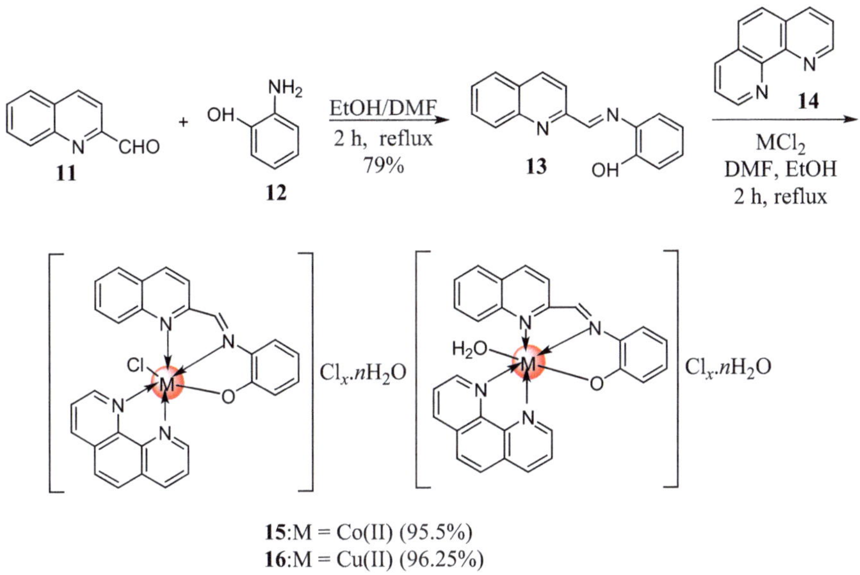

Scheme 2 Synthesis of heterocyclic SBL **13** and its mixed transition metal complexes [55]

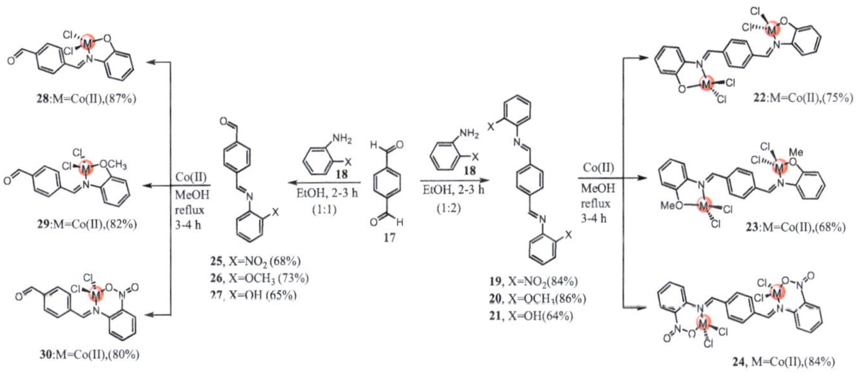

Scheme 3 Synthesis of SBLs **19–21**(1:2) and **25–27**(1:1) and its Co(II) complexes [4]

evaluated. The antibacterial efficacy of compounds **33** and **34** was assessed against *S. aureus, P. aeruginosa, E. coli,* and *B. subtilis* using the colony count method. The results demonstrated an increase in antibacterial activity upon complexation. In particular, SBL **33** exhibited MIC values of 91 g ml^{-1} against *E. coli*, 91 g ml^{-1} against *B. subtilis*, and 111 g ml^{-1} against *S. aureus*. These results highlighted the notable efficacy of SBL compared to Ampicillin, although its activity falls below that of Ciprofloxacin. Interestingly, the Co(II) complex **34** exhibited enhanced MIC values against *B. subtilis* and *S. aureus*, recorded at 165 and 176 $\mu g\ ml^{-1}$, respectively.

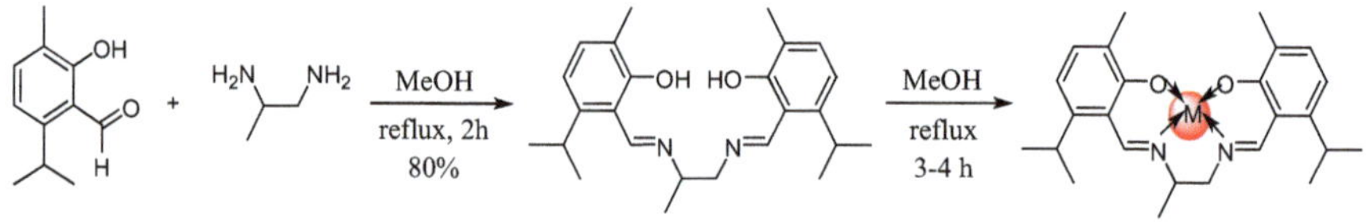

Scheme 4 Synthesis of a new SBL **33** and its Co(II) complex [56]

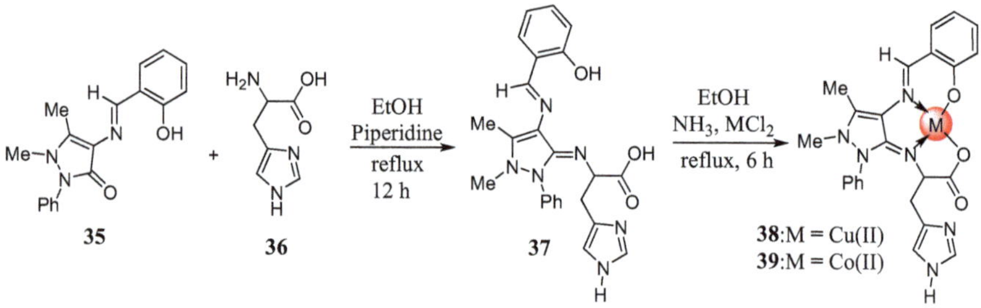

Scheme 5 Synthesis of SBL **37** and its metal complexes [57]

However, these values were considerably higher (500 and 209 $\mu g\ ml^{-1}$, respectively) against *E. coli* and *P. aeruginosa* when compared to standard treatments.

A newly developed tetradentate compound, referred to as SBL **37**, was synthesized by subjecting salicylidene-4-imino-2,3-dimethyl-1-phenyl-3-pyrazolin-5-one **35** and histidine **36** to reflux conditions for a duration of 12 h. Furthermore, complexes involving **37** with Cu(II) **38** and Co(II) **39** were synthesized following the procedure outlined in Scheme 5 [57]. Thorough characterization of all synthesized compounds **(37–39)** was carried out utilizing various spectroscopic techniques to provide a comprehensive understanding of their structures and properties.

The effectiveness of compounds **37–39** in fighting bacteria and fungi was evaluated in a controlled laboratory setting, known as in vitro testing. These compounds underwent testing against a range of bacterial strains. This evaluation utilized a method called the disk diffusion method, where paper disks infused with the compounds were placed on agar plates containing either nutrient agar or potato dextrose agar, which served as the growth medium for the bacteria and fungi, respectively. What stood out in the results was that compounds **38** and **39** demonstrated notably stronger antibacterial and antifungal properties when compared to compound **37** across all tested bacterial strains. This suggests that these particular compounds possess a higher potential for combatting both bacterial and fungal infections.

Sajjad Hussain Sumra synthesized three bidentate SBLs **(44–46)** with N, O donor ligands, along with their corresponding complexes: Co(II) **(47, 50, 53)**, Cu(II) **(48, 51, 54)**, and Zn(II) **(49, 52, 55)** (Scheme 6) [58]. Various spectroscopic techniques employed to characterize each synthesized compound. The efficacy of compounds **44–55** against a range of microorganisms was evaluated in vitro using the agar well diffusion method. Notably, compounds **47–55** exhibited significantly greater antimicrobial activity compared to compounds **44–46.** This indicates their potential for use as antimicrobial agents.

In 2018 Zn(II) complexes **58** were reported via a template mechanism by reacting 3-methoxy-2-hydroxybenzaldehyde **56**, 2,3diamino-cis-2-butenedinitrile **57,** and Zn(II) acetate salt through refluxing in DMSO solvent (Scheme 7) [59]. The synthesized compound 58 underwent analysis using magnetic susceptibilities measurement, molar conductivity, and various spectroscopic techniques. The solid-state structure of **58** was examined using XRD technique, revealing a square pyramidal shape. The antibacterial and antifungal properties of **58** were investigated using agar well diffusion plate methods.

Natarajan Raman synthesized a novel SBL, N,N'-bis (4-chlorobenzylidene)benzene-1,2-diamine **61,** and subsequently synthesized

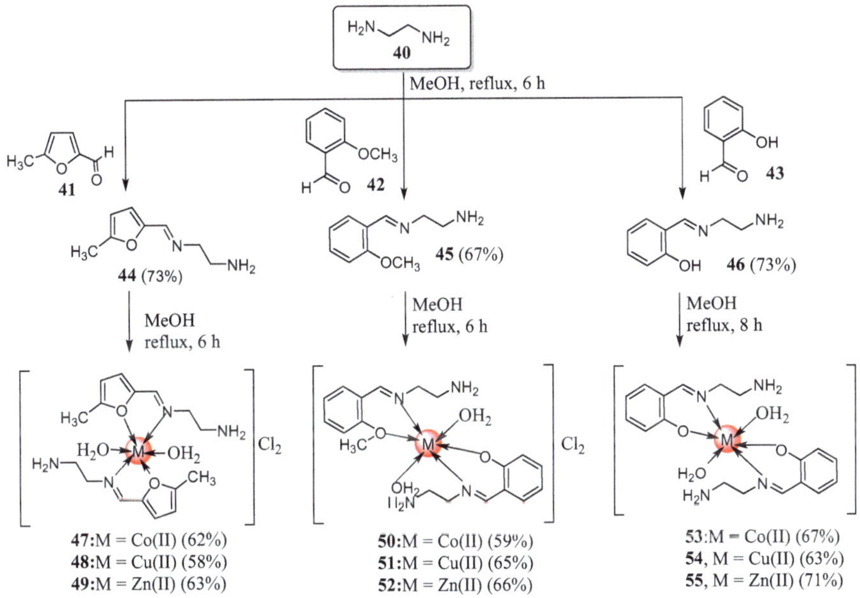

Scheme 6 Synthesis of SBL **44–46** and their metal complexes [58]

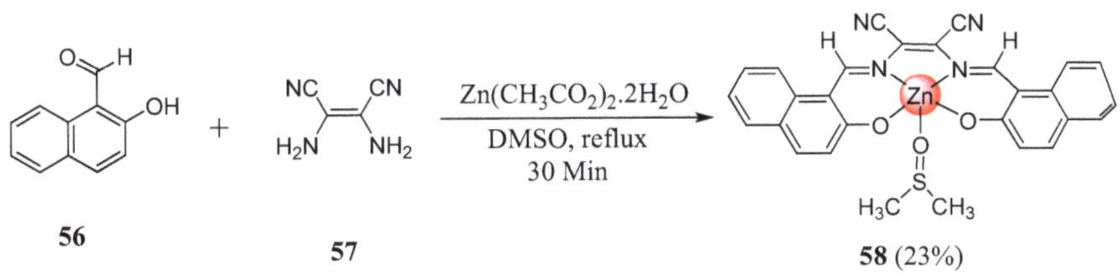

Scheme 7 Synthesis of Zn(II) complex [59]

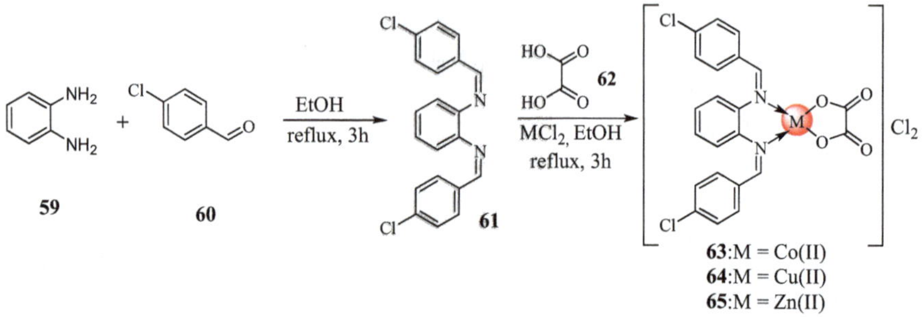

Scheme 8 Synthesis of tetradentate imino-oxalato mixed ligand metal complexes [60]

and characterized its complexes with Co(II) **63**, Cu(II) **64**, and Zn(II) **65** (Scheme 8) [60]. The antibacterial efficacy of these compounds was assessed in vitro against *S. aureus, P. aeruginosa, E. coli, B. subtilis, and S. typhi* using the paper disk method. Results demonstrated that complexes **63–65** exhibited superior antibacterial activity compared to **61** against the same pathogens. The MIC of **61** ranged from 16.7 to 15.4 μg ml^{-1} against all pathogens. Particularly, the Cu(II) complex **64** displayed a higher MIC (1.7–2.8 μg ml^{-1}) compared to complexes **63** and **65** (3.1–4.7 μg ml^{-1}). Furthermore, antifungal activity against *A. niger and C. albicans* revealed enhanced potential of complexes **63–65** (MIC, 2.0–6.7 μg ml^{-1}) compared to **61** (MIC, 11.6–15.8 μg ml^{-1}).

Two novel dinuclear metal complexes, $[Cu_2(HL_2)_2]0.8H_2O$ **(70)** and $[Co_2(HL_1)_2]0.8H_2O$ **(72),** were synthesized from SBLs **69** and **71**, respectively, as illustrated in Scheme 9 [61]. The SB metal complexes were subjected to comprehensive characterization. Both complexes **70** and **72** exhibited a dinuclear structure, with distorted octahedral geometry for **70** and a square pyramidal geometry for **72**. Moreover, the anticancer potential of **70** and **72** was assessed on HeLa and A549 cell lines, showing superior effectiveness compared to standard drug (with an IC50 value of around 7.0) following 48 h of treatment.

The synthesis of a novel SBL **74** was achieved through the condensation reaction between compound **73** and ethylene-1,2-diamine **40** (Scheme 10) [62]. Multiple analytical and spectroscopic techniques were utilized to meticulously analyze **74** and its Co(II) complex **77,** Cu(II) complex **79**, and Zn(II) complex **78**. This comprehensive approach aimed to elucidate the structural and chemical properties of these compounds, providing detailed insights into their molecular structures, bonding patterns, and electronic configurations. By employing a diverse array of analytical techniques, researchers were able to gain a comprehensive understanding of the characteristics and behavior of these compounds and their metal complexes, facilitating further exploration of their potential and diverse applications. Following characterization, **74** and **77–79** were subjected to in vitro assessment of their antibacterial effectiveness against bacterial strains using the disk diffusion method. Remarkably, the Cu(II) complex **79** demonstrated exceptional antimicrobial activity against the tested bacterial strains, comparable to that of ciprofloxacin, highlighting the potential

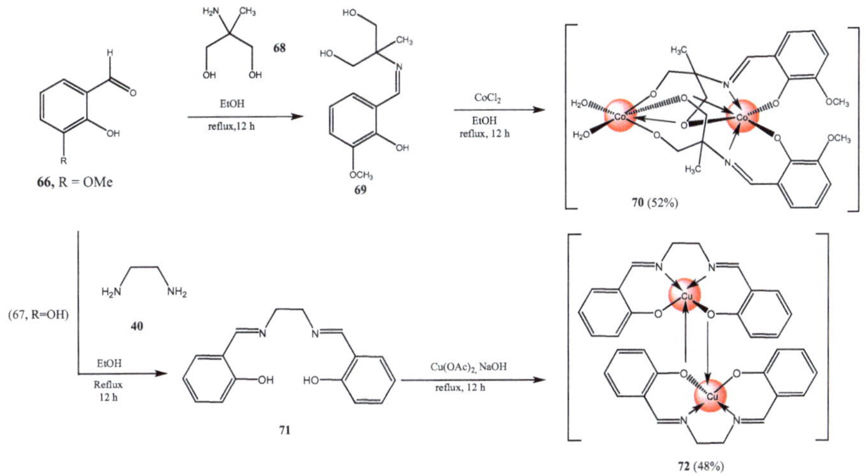

Scheme 9 Synthesis of dinuclear metal complexes [61]

of metal complexes in combating bacterial infections. Furthermore, compounds **76–79** exhibited promising potential as anticancer agents against various human cancer cell lines. The findings suggested that the Cu(II) complex **79** holds promise as a potent agent in cancer therapy, warranting further investigation into its mechanisms of action and potential clinical applications.

A novel SBL **82** derived from the condensation reaction of sulphathiazole **81** and cephalexin antibiotic **80**, along with its complexes with Co(II) **83**, Cu(II) **84**, and Zn(II) **85**, as outlined in Scheme 11 [63]. The synthesized SBL **82**, as well as the mononuclear [ML(OAc)(H_2O)$_2$] complexes **83** and **85**, and the trinuclear Cu(II) complex [$Cu_3L(OH)_5$] were subjected to comprehensive analysis techniques [63].

Similarly, tridentate SBLs **89–92** originating from 2-hydroxynaphthaldehyde/substituted salicylaldehyde **88**, heterocyclic nitrogen base 8-hydroxynaphthaldehyde **86**, and isatinmonohydrazone **87**, alongside their corresponding Co(II) **94**, Cu(II) **95**, and Zn(II) **96** complexes were reported (Scheme 12) [64]. The compounds were characterized and checked for biological action. The findings indicated that **94** and **96** exhibited the highest antimicrobial activity compared to **89–92**, while **95** demonstrated greater efficacy against the same pathogenic strains.

SBLs **100** and **101** derived from salicylaldehyde **97** and 3-t-butyl salicylaldehyde **98**, in combination with morpholine **99**, along with their corresponding Cu(II) complexes **102** and **103** (Scheme 13) were reported [6]. Similar results were obtained, when compounds **100–103** were evaluated for antibacterial potential, i.e., complexes are more potent than SBLs. Furthermore, **103** demonstrated lower MIC values than **102**, indicating enhanced antibacterial potency due to two highly hydrophobic t-butyl groups present in **103**. Moreover, the in vitro anticancer potential of compounds **100–103** was assessed using the MTT colorimetric assay on human breast cancer cell lines (MCF-7), reporting significant cytotoxic effects of **100–103**.

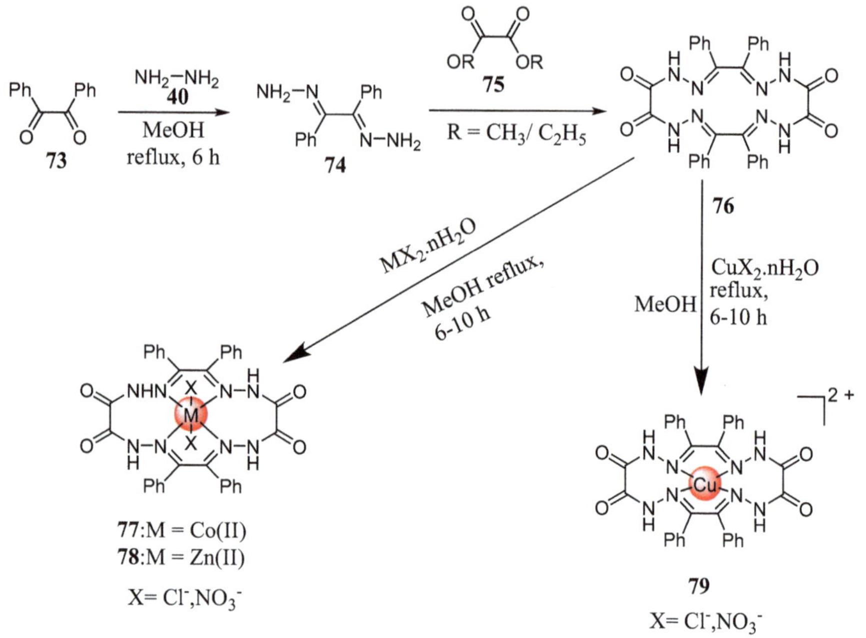

Scheme 10 Synthesis of Schiff base octa azamacrocyclic metal complexes [62]

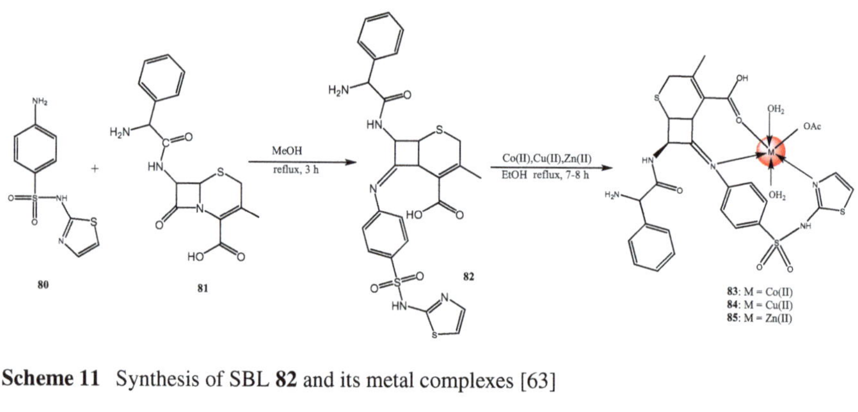

Scheme 11 Synthesis of SBL **82** and its metal complexes [63]

In another study, the complexes **107–109** of tetradentate ligand **106** displayed greater antibacterial activity compared to **106**. The compound **106** was achieved through the condensation reaction of malonyl **104** and dehydroacetic acid **105** in an ethanolic solvent (Scheme 14) [65]. SBL **112** was synthesized by same technique involving pyridine-3-carboxamide **110** and 1,7-bis-(4-hydroxy-3-methoxphenty)-1,6-heptadiene-3,5-dione **111** (Scheme 15). The octahedral complexes **113, 114** of **112** were examined for antimicrobial efficacy using the disk diffusion method.

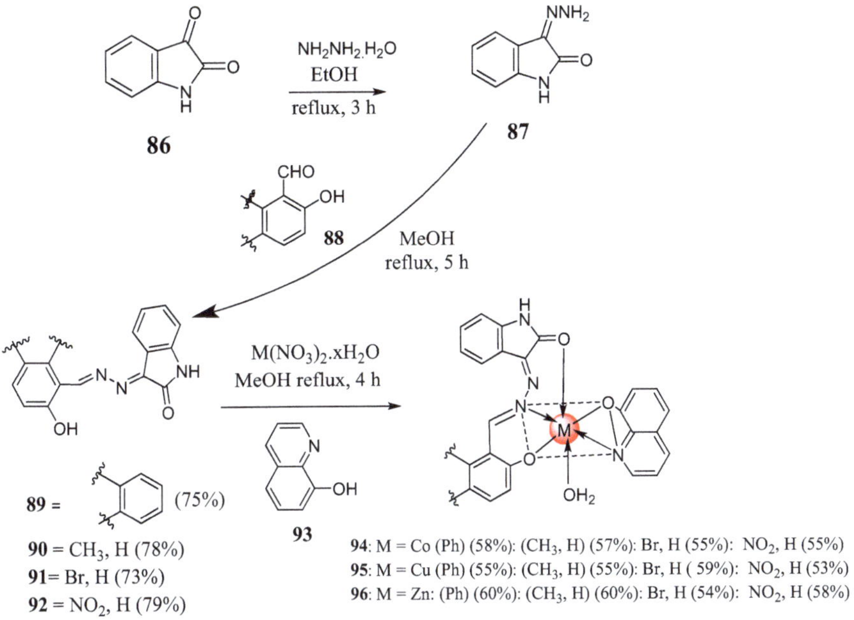

Scheme 12 Synthesis of mixed SBLs **89–92** and their complexes [64]

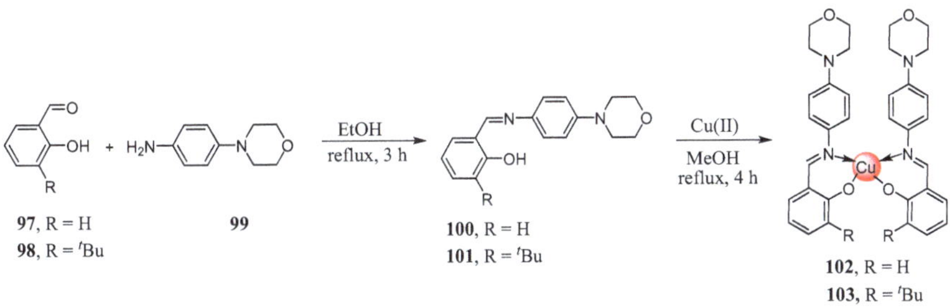

Scheme 13 Synthesis of SBL and complexes **100–103** [6]

Results revealed a distinct order of antimicrobial potency against *S. aureus, S. pyogenes, and E. coli*, with Cu(II) **114** exhibiting the highest efficacy followed by Co(II) **113**, and the ligand **112** displaying the least activity. This investigation underscores their metal complexes as antimicrobial agents, with promising implications for combating bacterial infections [66].

A tridentate SBL, namely 2-acetyl-5-methylfuranthiosemicarbazone **117,** and its Co(II) **118** and Cu(II) **119** complexes (Scheme 16) were evaluated against *E. coli and S. typhi* bacteria and found superior biological activity of **118** and **119** compared to **117** [67], whereas SBL 1,10-dithia-3,8,12,17-tetraoxo-4,4,13,16-tetraazacyclooctadecane **121 and** its Co(II) and Cu(II) complexes **122–127**

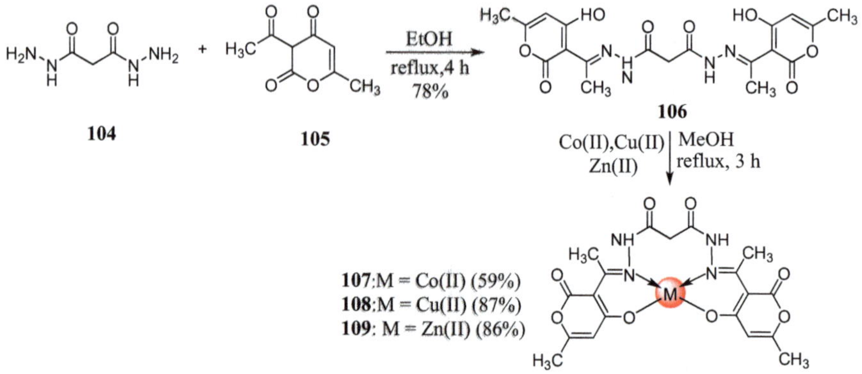

Scheme 14 Synthesis of a heterocyclic SBL **106** and its complexes [65]

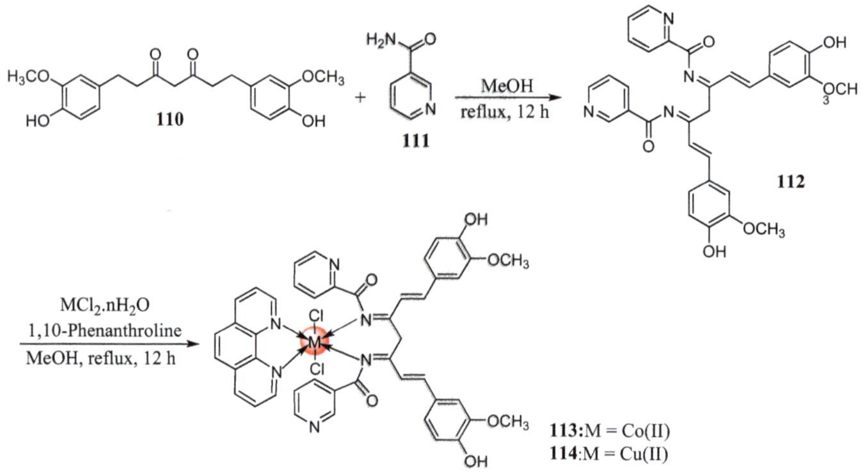

Scheme 15 Synthesis of SBL **112** and its complexes [66]

(Scheme 17) were tested using the broth dilution method against various bacterial and fungal strains, i.e., *P. aeruginosa, E. coli, C. tropicalis, C. glabrata, C. kruesi, and C. albicans* and reported same enhanced antibacterial efficacy of complexes **122–127** than **121** [68].

A sequence of Cu(II) **133**, Co(II) **135**, and Zn(II) **132** and **134** was synthesized through the reaction of SBLs **130** and **131** (as outlined in Scheme 18) [69]. The SBLs and resultant complexes underwent thorough characterization employing spectroscopic methods. The structural elucidation of ZnL_2 complexes was accomplished via X-ray crystallography. As per the crystallographic data, the metal centers within these complexes exhibited tetrahedral geometry. Notably, complex **134** displayed a distorted tetrahedral geometry. In terms of antibacterial activity, SBL **130** exhibited comparatively lower efficacy against bacterial strains than **132–135**.

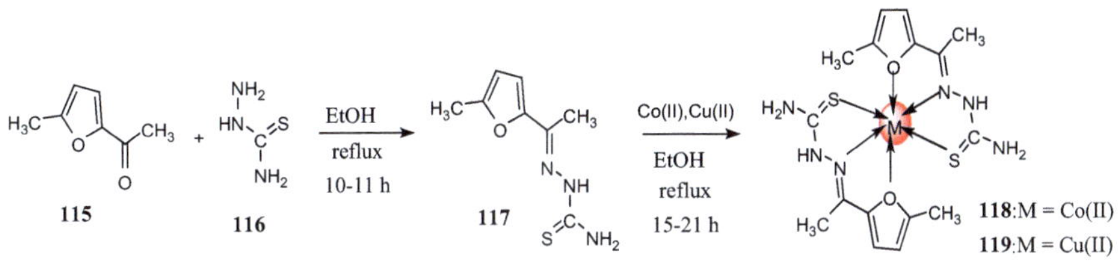

Scheme 16 Synthesis of 2-acetyl-5methyl-furanthiosemicarbazone and its complexes [67]

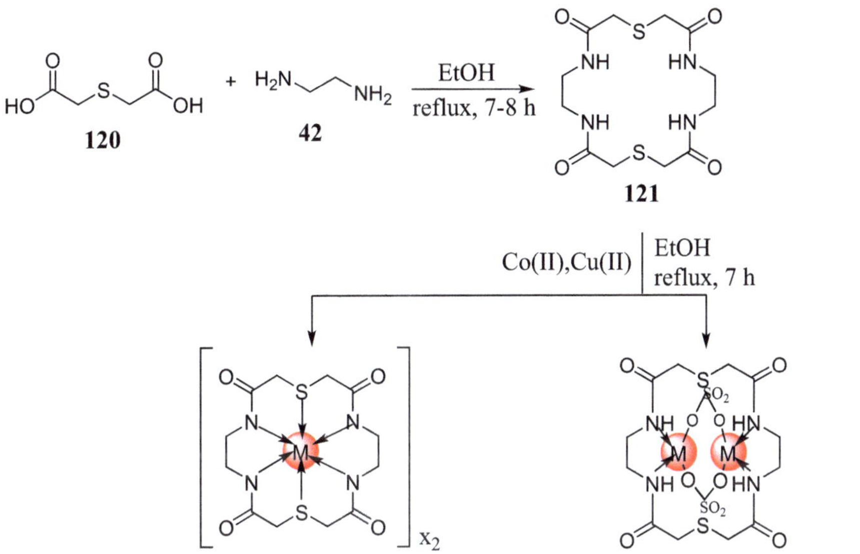

122:M = [CoL]Cl_2 (62%)
123:M = [CoL]$(NO_3)_2$ (60%)
124:M = [CuL]Cl_2 (64%)
125:M = [CuL]$(NO_3)_2$ (60%)

126:M = [$Co_2L(SO_4)_2$] (58%)
127:M = $Cu_2L(SO_4)_2$] (58%)

Scheme 17 Synthesis of macrocyclic SBL **121** and its complexes **122–127** [68]

Another series of compounds featuring the bidentate ligand **138**, scientifically denoted as (E)-N-(2-methoxy benzylidene) anthracene-2-amine, together with its metal complexes **139–141**, has been detailed in scientific literature (Scheme 19). The coordination mechanism between SBL **138** and the central metal atom was elucidated, highlighting the involvement of its azomethine nitrogen and oxygen groups in the bonding process. The investigation was not stopped at mere characterization rather it extended to evaluate the biological activities of compounds **138–141**. This evaluation focused on assessing their potential as antimicrobial agents using well-established disk diffusion method. The results of these biological assays

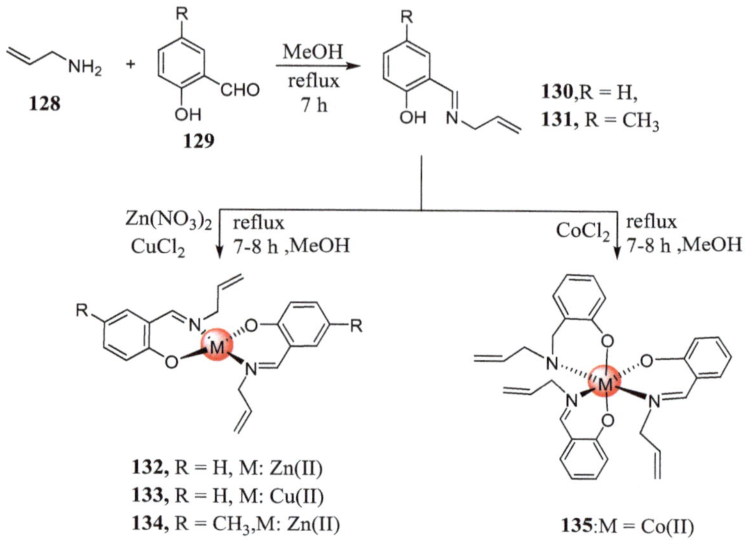

Scheme 18 Synthesis of SBLs **130, 131** and its metal complexes [69]

unveiled intriguing findings confirming significantly heightened activities of **139–141** compared to the parent compound **138**. This enhancement was attributed to the chelation effects facilitated by the coordination of **138** to the metal ions, suggesting promising avenues for further exploration in the realm of medicinal chemistry.

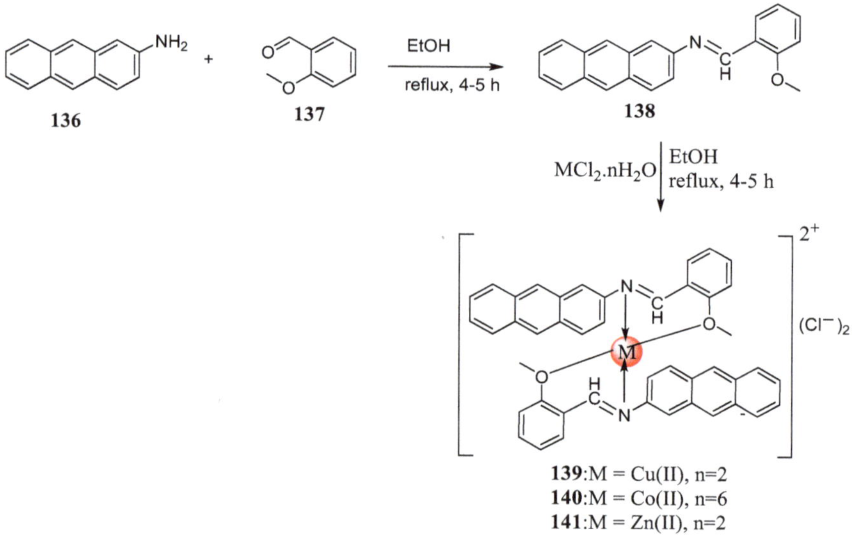

Scheme 19 Formation of a SBL **138** and its complexes [70]

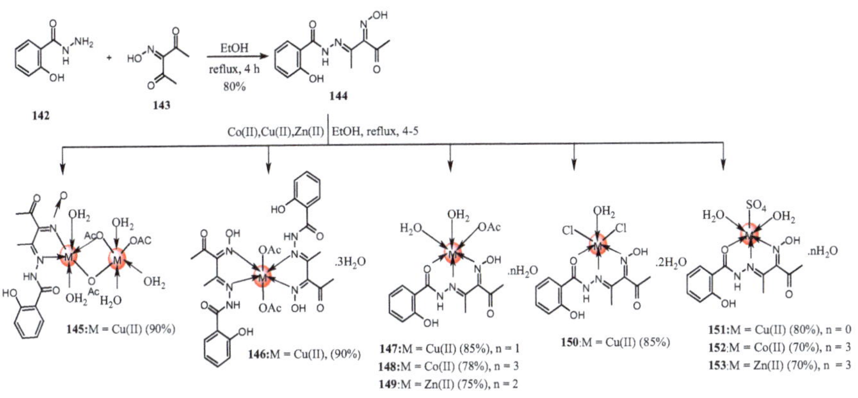

Scheme 20 Synthesis of SBL **144** and its complexes [71]

Nine distorted octahedral complexes **145–153**, derived from SBL, 2-hydroxy-N-((Z)-3(hydroxyimino)-4oxopentan-2-ylidene)benzohydrazide **144**, have been documented (Scheme 20) [71]. Notably, in comparison with the standard medication Sorafenib (Nexavar), the investigated complexes **145–153** and SBL **144** displayed significantly heightened cytotoxicity against human liver cancer HepG2 tumor cell lines. The results substantiated the superior effectiveness of complexes **145–153** over SBL **144**. Particularly, Cu(II) complex **146** demonstrated the highest cytotoxicity ($IC_{50} = 2.24\ \mu M$), followed by complex **145** ($IC_{50} = 2.67\ \mu M$), and then complex 147 ($IC_{50} = 6.49\ \mu M$).

A similar reason as mentioned for **139–141** can be quoted to explain the better antifungal potency of complexes **157–165** of SBL **156.** The ligand **156** was synthesized via an equimolar condensation reaction involving 3-amino-1-H-1,2,4-triazole **154** and pyrrol-2-carboxaldehyde **155**. The complexes **162, 164,** and **165** exhibited significant potential against all tested pathogens, while complexes **159** and **160** showed noteworthy activity against *F. solani* and *C. glabrata* fungal strains (Scheme 21).

Some other Cu(II) **(171 and 173)** and Co(II) **(172 and 174)** transition metal complexes of two novel bidentate ligands, namely (2-benzo[d]thiazol-6-ylimino)-4,6-dichlorophenol **169** and (1-benzo[d]thiazol-6-ylimino)-6-bromo-4-chlorophenol **170** (Scheme 22), were meticulously characterized and examined for the antibacterial activity against *E. coli and B. amyloliquefaciens* and antifungal activity against *S. rolfsii and M. phaseolina.* Comparative analysis with conventional antibacterial (Streptomycin) and antifungal (Mancozeb) drugs revealed that complexes **171–174** displayed superior efficacy compared to compounds **169 and 170** [73].

The metal complexes of Cu(II) **178**, Co(II) **179**, and Zn(II) **180** of bidentate ligand 2-((1H-benzoimidazole-4-yl)imino)methylphenol **177** are shown in Scheme 23. The identification and characterization of **177–180** involved a comprehensive approach integrating elemental analysis alongside an array of spectroscopic techniques. Through these methods, the molecular structure and composition were elucidated, affirming the octahedral configuration of complexes **178–180**. Additionally, CV was

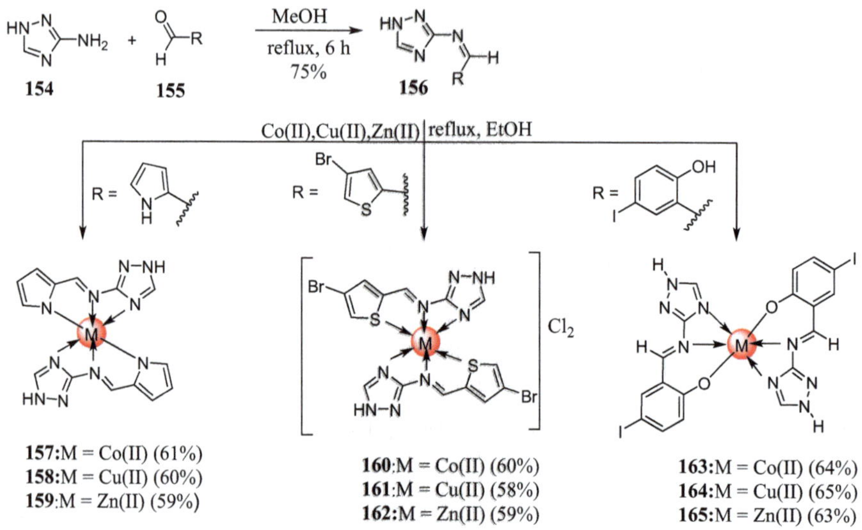

Scheme 21 Synthesis of SBL **156** and its metal complex [72]

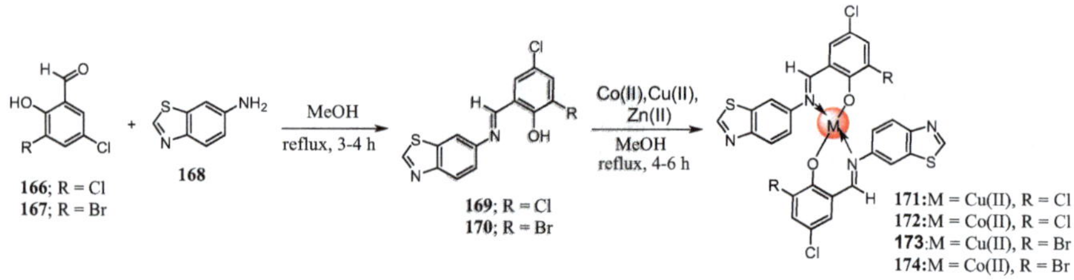

Scheme 22 Synthesis of SBLs **169, 170** with Co(II) and Cu(II) complexes [73]

utilized to delve into the electrochemical behaviors of complexes **178–180,** providing insights into their redox properties.

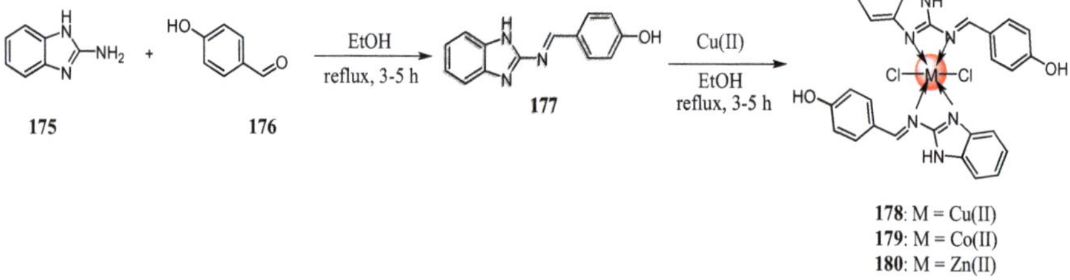

Scheme 23 Synthesis of Cu(II), Co(II), and Zn(II) complexes from SBL **177** [74]

Furthermore, the antimicrobial efficacy of compounds **177–180** was rigorously assessed against a panel of pathogenic microorganisms. Notably, these evaluations unveiled remarkable potency across the tested compounds. Of particular interest, the Cu(II) complex **178** exhibited markedly superior antimicrobial activity compared to its counterparts **177, 179, and 180**. This comprehensive investigation not only confirmed the structural characteristics and electrochemical nature of complexes **178–180** but also highlighted the significant antimicrobial potential, with complex **178** emerging as a particularly promising candidate for further exploration in antimicrobial applications.

Similar methods as reported earlier were used to synthesize Co(II) **184**, Cu(II) **185**, and Zn(II) **186** complexes of 4-nitrobenzaldehydeglycylglycine **183** (through the refluxing of glycylglycine **182** and 4-nitrobenzaldehyde (4-NBA) **181**) as outlined in Scheme 24. The ligand **183** displayed tridentate monobasic donor nature, with the azomethine nitrogen, carboxylate oxygen, and deprotonated peptide nitrogen atoms contributing to this property. Moreover, complexes **184–186** displayed water solubility, with octahedral geometry. In terms of biological activity, compounds **183, 185, and 186** demonstrated anticancer effects against HeLa and HCT114 cancer cell. Notably, the Cu(II) complex **185** exhibited the highest anticancer potential, with IC_{50} values of approximately 36.5 μM for HeLa cells and 26.5 μM for HCT116 cells, surpassing the activity of **184, 186, and 183**. Complexes **185 and 186** demonstrated interaction with *E. coli* DNA and effective destruction of both cancer cell types. Furthermore, compounds **183–186** displayed antibacterial and antifungal potential against *E. coli, S. aureus, B. subtilis, P. aeruginosa, C. albicans, A. flavus, and A. niger* using the disk diffusion method, with complex **185** showing superior activity compared to **184, 186, and 183**.

A mononuclear species denoted as complex **190** and a bi-nuclear species denoted as complex **191**, employing the ligand SBL 6,6-dimethoxy-2,2′-[O-phenylenebis(nitrilomethylidyne)]diphenolato **189**. This ligand was obtained through the refluxing of ortho-vanillin **187** and ortho-phenylenediamine **188** (Scheme 25) [76]. Exhibiting a paramagnetic nature, both complexes **190 and 191** were assessed for their potential in anticancer activity against colorectal cancer in

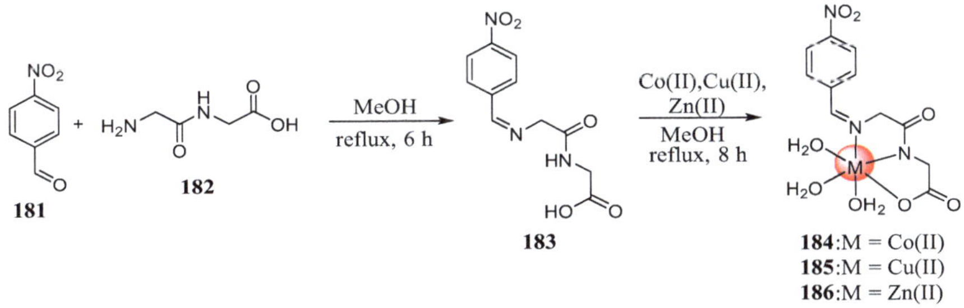

Scheme 24 Synthesis of 4-nitrobenzaldehydeglycylycine and its metal complexes [75]

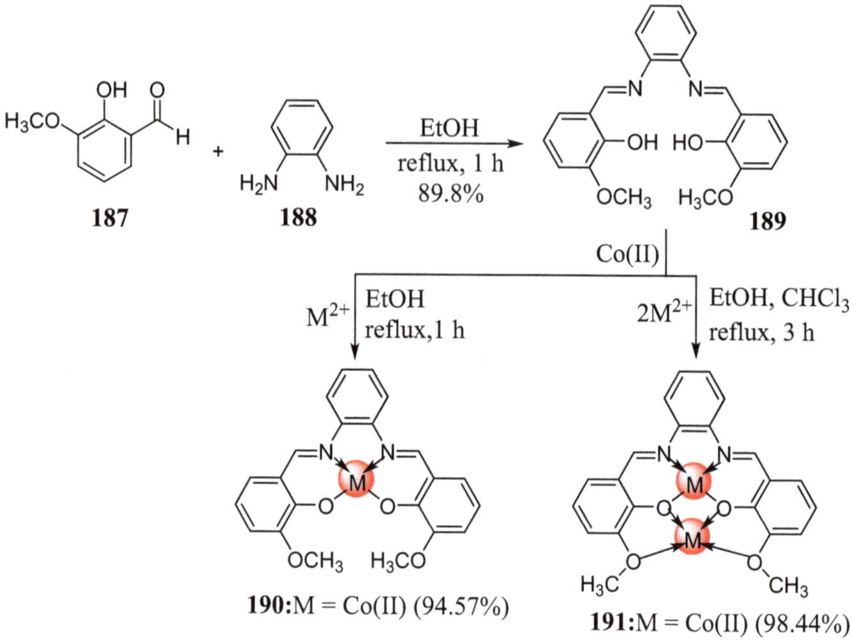

Scheme 25 Synthesis of SBL **189** and its Co(II) complexes [76]

human subjects. Remarkably, dinuclear complex **191** demonstrated higher efficacy compared to mononuclear complex **190** in this context.

In 2019, SBL **194** derived from 6-salicyaldehyde **193** and 3,4-toluenediamine **192** as well as Cu(II) **195** and Zn(II) **196** complexes was reported (Scheme 26). Prepared compounds **194–196** were characterized by different spectroscopic technique. *E. coli, S. dysenteriae,* and *S. Sonnei* bacteria were used to study the antibacterial potential of **194–196** against streptomycin and ciprofloxacin standard drugs [77]. In the same year, H. Guzin Aslan and colleagues embarked on a scientific journey that culminated in the birth of a novel compound, labeled as SBL 1-((2-hydroxynaphthalen-1-yl)methyleneamino)-5-(4-methybenzoyl)-4-tolyrimidin-2(1H)-one **199.** This ground-breaking achievement was realized through a meticulous synthesis process involving the refluxing of 1-amino-5-(4-methylbenzoy)-4-(4-methylphenyl)pyrimidin-2(1H)one **197** with 2-hydroxy-1-naphthaldehyde **198**. Furthermore, the research unveiled the creation of Co(II) **200** and Cu(II) **201** complexes derived from the compound **199**, as delineated in Scheme 27. The elucidation of the synthesized compounds **199–201** was meticulously conducted employing an array of spectroscopic techniques. This thorough characterization provided crucial insights into the structural and chemical properties of the compounds. Upon evaluation, compounds **199–201** exhibited promising potential as anticancer agents against A549 cell line. This revelation opens new avenues

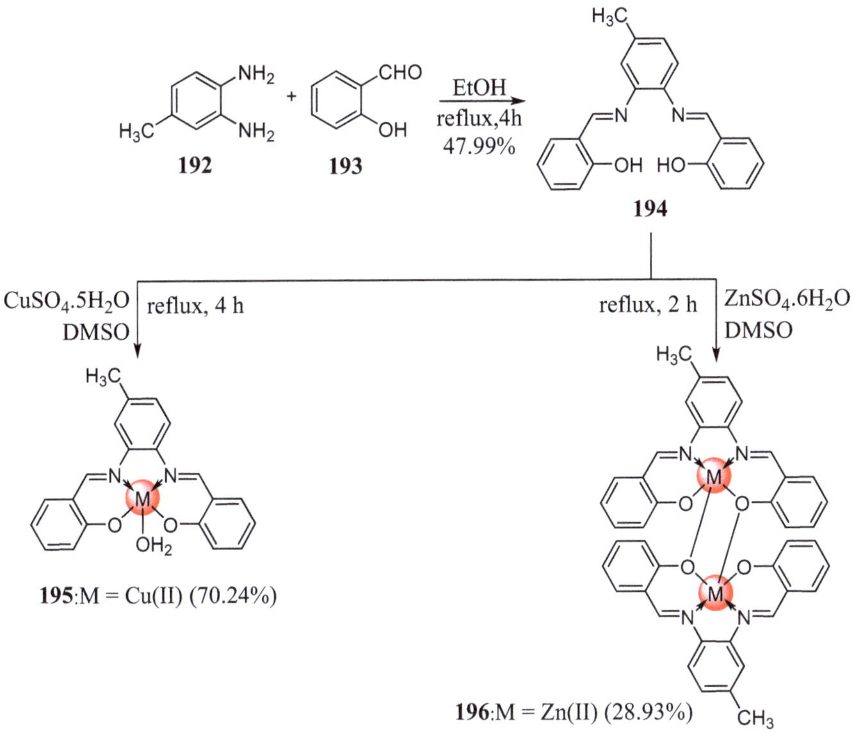

Scheme 26 Synthesis of SBL 194 and Cu(II) and Zn(II) complexes [77]

in the pursuit of effective treatments for cancer, showcasing the significance of the synthesized compounds in the realm of medical science.

Compound **206**, a benzimidazole derivative was synthesized alongside its corresponding Cu(II) **207** and Zn(II) complex **208** as outlined in Scheme 28. The synthesis process involved meticulous steps to ensure the formation of each compound. Following synthesis, a comprehensive characterization of compounds **206–208** was undertaken utilizing a range of spectroscopic techniques. These analyses provided valuable insights into the molecular structure and composition of the synthesized compounds. To further validate their structural integrity, single crystal XRD analysis was employed, offering a detailed three-dimensional view of their arrangement at the atomic level. In addition to structural elucidation, the antibacterial properties of compounds **206–208** were investigated against various bacterial strains including *M. lutenus, E. coli, and E. aerogenes*. This assessment aimed to gauge their potential as antimicrobial agents. The evaluation involved rigorous testing procedures to ascertain the efficacy of the compounds in inhibiting bacterial growth. This research sheds light on the synthesis, characterization, and antimicrobial activity of the SBL benzimidazole derivative and its metal complexes, providing valuable insights for potential applications in combating bacterial infections.

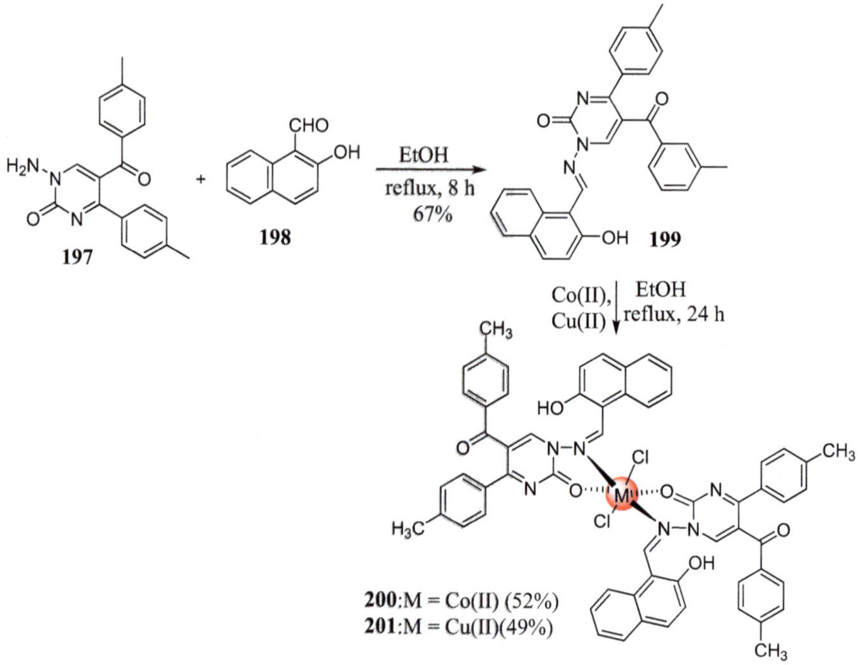

Scheme 27 Synthesis of SBL **199** and metal complexes [78]

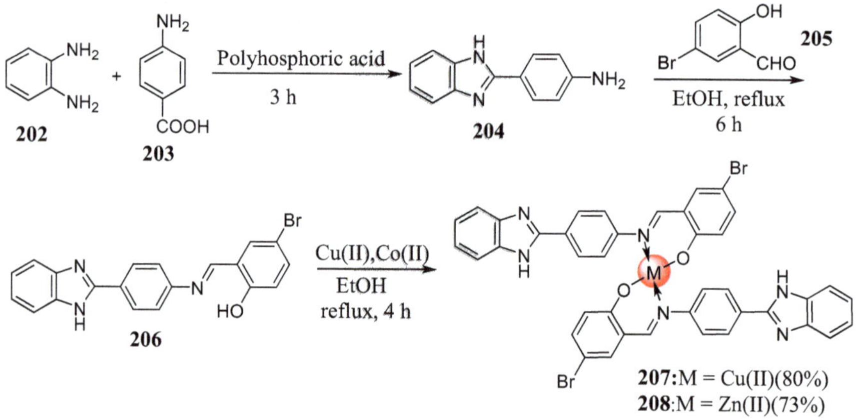

Scheme 28 Synthesis of SBL **206** and its metal complexes [79]

In 2021, HadiKargar and his team boarded on a comprehensive project involving the creation of four distinct Schiff base ligands. These ligands were shown in Scheme 29. These compounds served as the foundation for the subsequent synthesis of Cu(II) complexes, a process outlined in Scheme 29 of their study. The characterization of these complexes involved a multifaceted approach, encompassing analyses utilizing molar conductivity, elemental examination, and an array of spectroscopic methods. Notably, XRD analysis unveiled a distinctive distorted tetrahedral geometry for complexes **215–218**, shedding light on their structural intricacies. Beyond structural elucidation, the research delved into the practical implications of these compounds by evaluating their antibacterial efficacy. Specifically, complexes **211–218** underwent rigorous testing against two common bacterial strains. The results were revealing, demonstrating a notable disparity in activity levels among the tested complexes. In particular, complexes **215–218** exhibited significantly heightened antibacterial activity when compared to their counterparts, **211–214**, showcasing their potential for future therapeutic applications in combating bacterial infections.

Musa A. Said undertook a ground-breaking synthesis involving complexes **222–224**. This innovative process employed a newly developed Schiff base ligand, represented by a complex chemical formula: [[6,6′(1E,1′E)(4,4′methylenebis(cyclohexane-4,1-diyl)bis(azan-1-yl-1-ylidene)bis (methan-1-yl-1ylidene)bis(2,4di-tert-butylphenol)] **221.** This ligand was derived from a sophisticated condensation reaction between 3,5-di-tert-butyl-2-hydroxybenzaldehyde **220** and 4,4′methylenedicyclohexan-amine **219,** as outlined in Scheme 30. This synthesis represents a significant advancement in the field, demonstrating the potential for novel ligands to enable the creation of complex metal

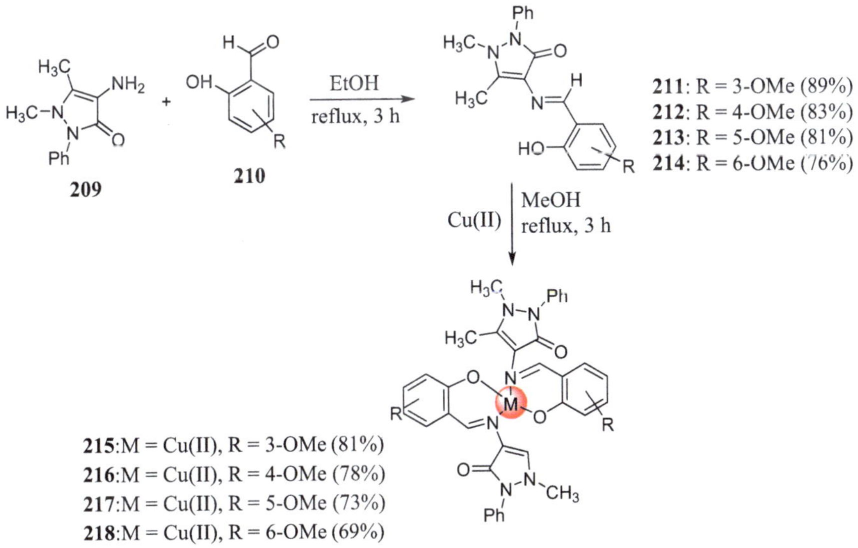

Scheme 29 Synthesis of SBL **211–214** and Cu(II) complexes [80]

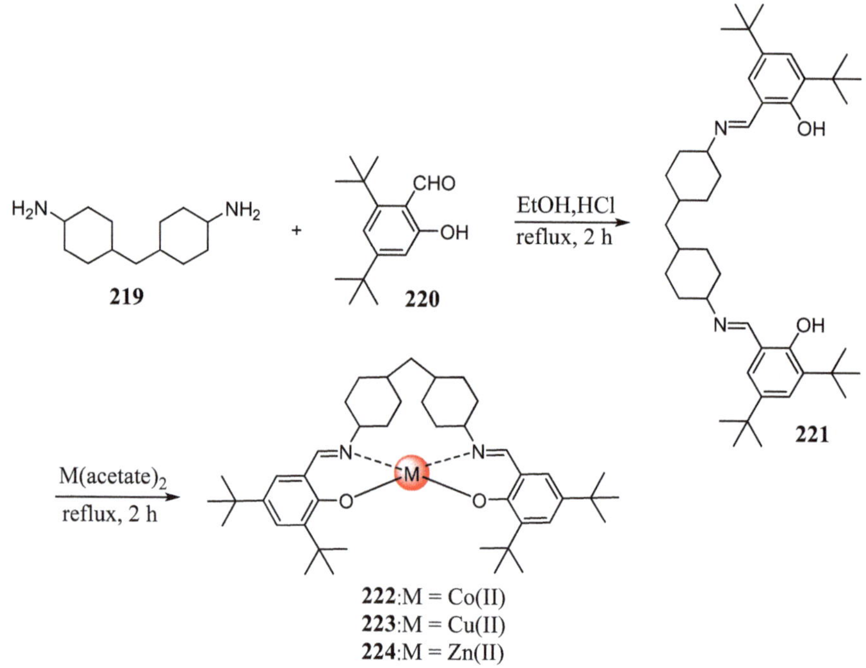

Scheme 30 Synthesis of SBL **221** and metal complexes [81]

complexes with diverse applications. The anticancer efficacy of compounds **221–224** was assessed utilizing A549 and H1299 lung cancer cell lines. Notably, complex **224** exhibited significant activity at a concentration of 20 μM, demonstrating suppression of A548 and H1297 lung cancer cell lines by 82.61 and 87.51%, respectively.

Some complexes **230–233** of same metal ions incorporating two newly developed SBL **228, 229** derived from 1H-Indole 2,3-dione (isatin) **226** and N,N-dimethyl amino benzaldehyde **227** in conjunction with cefotaxime sodium antibiotic **225** (Scheme 31) [82] were assessed for antibacterial efficacy against *E. coli and S. aureus.* Results demonstrated that complexes **231 and 233** exhibited superior activity against *S. aureus* compared to **230 and 232.**

Similar metal complexes **238** and **239** of SBL 1-((2,4-dihydroxybenzylidene)amino)quinoline-2(1H)-one **237** were successfully synthesized through a condensation pathway (Scheme 32) [83]. In the biological assessment of **237–239**, it was claimed that complex **238** demonstrated the superior potential of over **239** and **237.**

A set of copper **243** and zinc **244** complexes derived from SBL **242 is** outlined in Scheme 33. Complex 243 was found to possess a square planar geometry, while a tetrahedral structure was hypothesized for the complex 244. The characterization process enhances our understanding of the structural and chemical properties of these

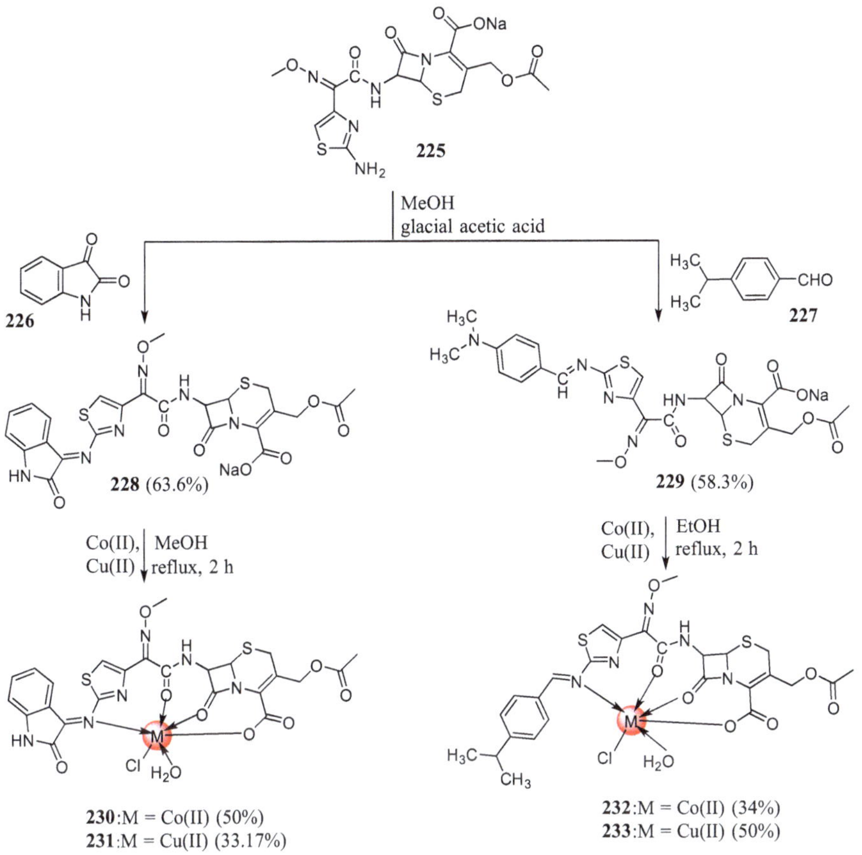

Scheme 31 Synthesis of SBLs **228, 229** and their metal complexes [83]

compounds, providing valuable insights into their potential applications. Antimicrobial activity screening of compounds **242–244** against different strains confirmed superior antimicrobial efficacy of complex **243** compared to both the complex **244** and the parent SBL **242** [84].

The synthesis of complexes **248** and **249** of 1,4-bis[(2-hydroxybenzaldehyde)propyl]piperazine **247** was done via the condensation reaction of 1,4-bis(3-aminopropyl)piperazine **246** and salicylaldehyde **245** (Scheme 34). The antibacterial activities of compounds **247–249** were evaluated utilizing the Kirby–Bauer disk diffusion method. Notably, the complex **248** exhibited superior antibacterial activity compared to the complex **249** [85].

Likewise, A. Varada Reddy and colleagues synthesized a novel SBL benzloxybenzaldehyde-4-phenyl-3-thiosemicarbazone **252**, derived from the reaction between 4-phenyl-3-thiosemicarbazide **251** and benzyloxy benzaldehyde 250, along with its Cu(II) complex **253** (Scheme 35). The antibacterial activity of **252**

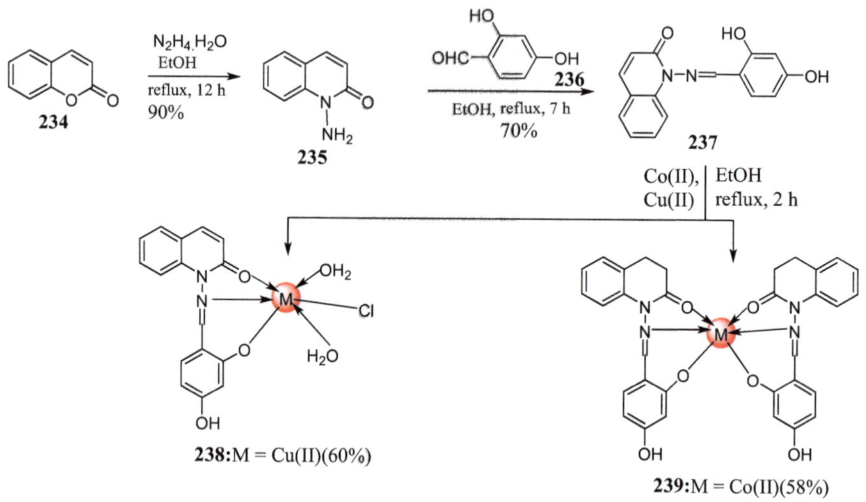

Scheme 32 Synthesis of SBL **237** and its Co(II), Cu(II) metal complexes [83]

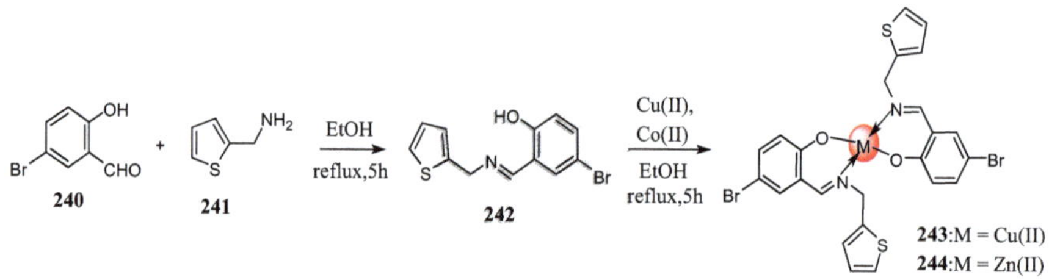

Scheme 33 Synthesis of SBL **242** and its metal complexes [84]

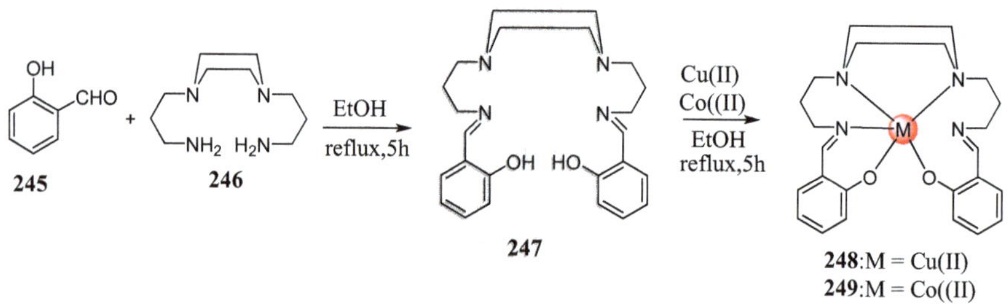

Scheme 34 Synthesis of polydentate SBL **247** and its cobalt and copper metal complexes [85]

and 253 was evaluated using the agar disk diffusion method. Notably, complex **253** exhibited superior antibacterial efficacy compared to **252** [86].

Two SBL compounds, namely **256 and 257**, along with their corresponding Cu(II) complexes, designated as **258 and 259** are shown in Scheme 36. This multifaceted

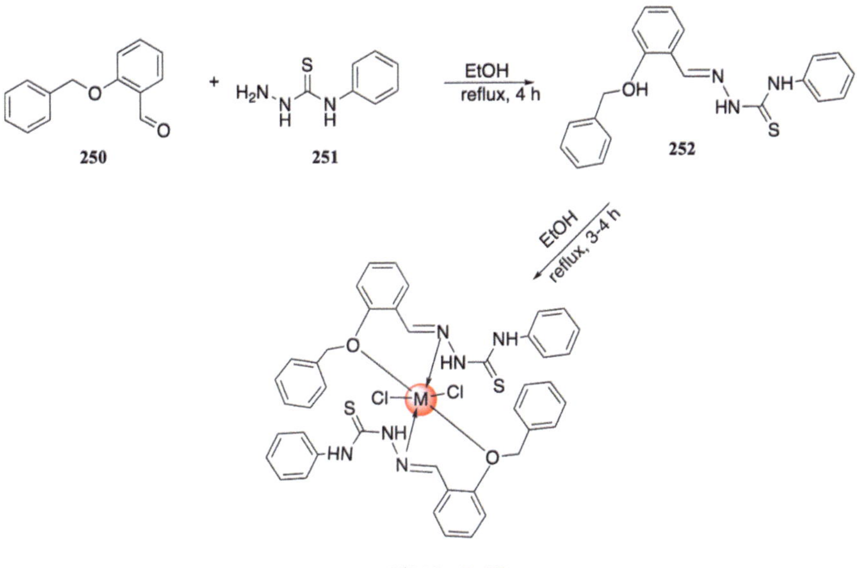

Scheme 35 Synthesis of SBL **252** and metal complex [86]

analysis provided a detailed understanding of their molecular structures and properties. Moreover, the crystal structures of complexes **258 and 259** were meticulously confirmed through XRD analysis, ensuring the accuracy and reliability of the findings. This analytical method allowed for a precise depiction of the spatial arrangement of atoms within these complexes. In addition to elucidating their chemical composition and structure, the biological activities of compounds **256–259** were thoroughly investigated. Their antibacterial potential was assessed against a spectrum of bacterial strains. This assessment aimed to gauge their effectiveness in combating bacterial infections, providing valuable insights into their pharmacological potential. Furthermore, the cytotoxicity of these compounds was evaluated against MDAMB-231 and MCF-7 human breast adenocarcinoma cancer cell lines. This investigation delved into their ability to inhibit cancer cell growth, offering prospects for potential therapeutic applications in oncology. Remarkably, complexes **258 and 259** demonstrated remarkable cytotoxic effects surpassing those of compounds **256 and 257**. This observation underscores their promising potential as anticancer agents, warranting further exploration and development in the realm of cancer therapy [87].

A study by J. Wu et al. led to the synthesis of a SBL denoted as **262**, which was derived from the reaction between 4-(bis(4-methoxyphenyl)amino)benzaldehyde **260** and hydrazine carbodithioate **261**. Furthermore, the study extended to the preparation of its Cu(II) **263** and Zn(II) **264** complexes (Scheme 37). Characterization of compounds **262–264** was conducted using a comprehensive suite of analytical techniques including elemental analysis, TGA, molar conductivity measurements,

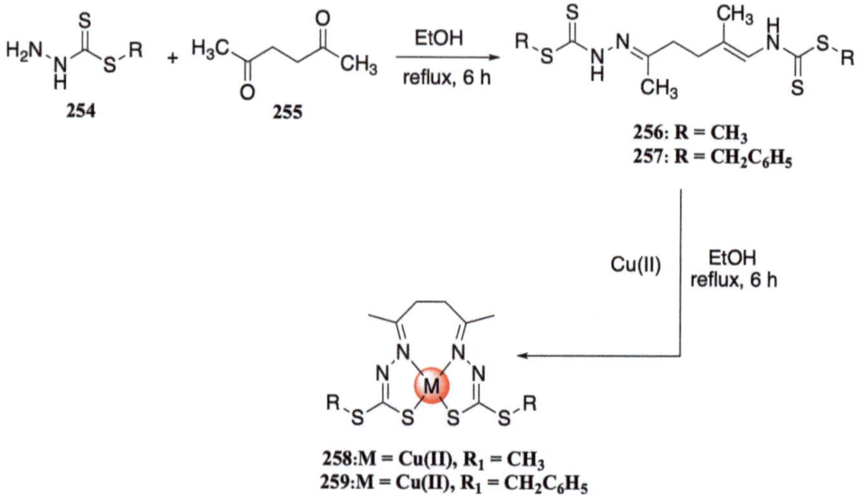

Scheme 36 Synthesis of SBL **256, 257** and macrocyclic Cu(II) complex [87]

and various spectroscopic methods. Additionally, the structure elucidation of these compounds was achieved through XRD analysis [88].

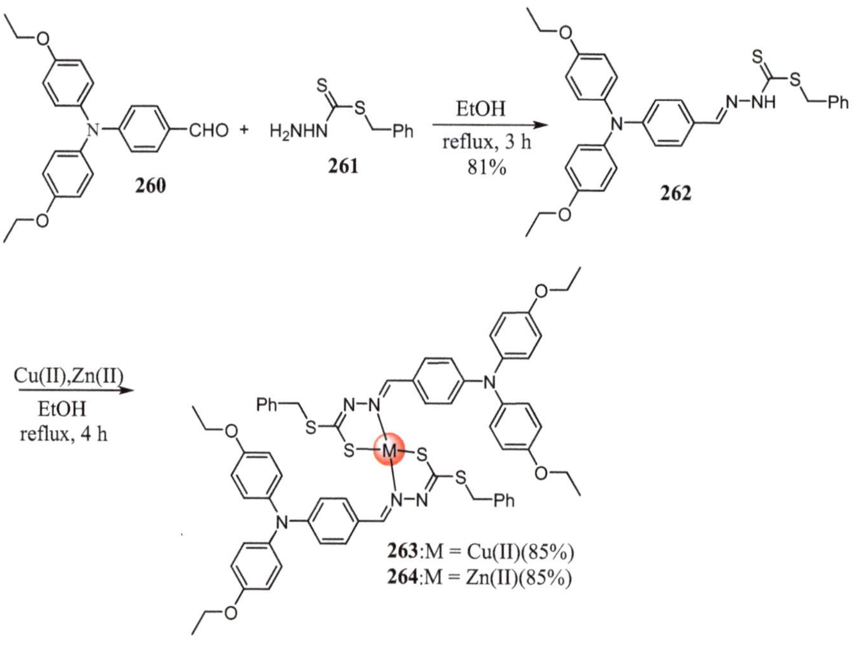

Scheme 37 Synthesis of SBL **262** and its Cu(II) and Zn(II) complexes [88]

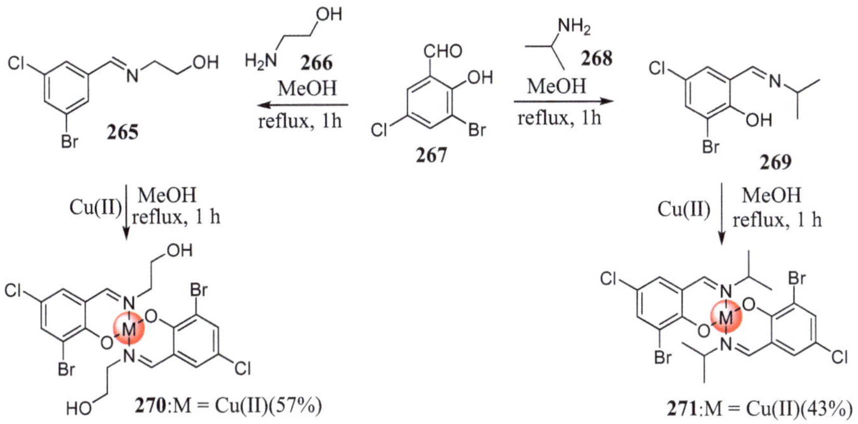

Scheme 38 Synthesis of SBL **265, 269** and its Cu(II) complexes [89]

Cai presented findings on two novel SBL **269** and 265, along with their corresponding Cu(II) complexes **270** and **271** (Scheme 38). These ligands and complexes underwent thorough characterization, confirming the coordination of two phenolate oxygen and two imino nitrogen groups with Cu ions resulted in the formation of a square planar geometry. The in vitro antibacterial efficacy of compounds **265, 269–271** was evaluated using Mueller–Hinton medium [89].

SBLs **272, 276** and its Zn(II) **277, 278** complexes (Scheme 39). The SBLs **272, 276** and **277, 278** complexes demonstrated enhanced antibacterial action against *E. coli* and *Pseudomonas fluorescens* (*P. fluorescens*) than Penicillin, but were less effective against *S. aureus* and *B. subtilis*. The complexes **277, 278** showed better potential than **272** and **276** against all microorganisms [90].

Mahmoud et al. [34] conducted the synthesis of Cu(II) **282**, Co(II) **284**, and Zn(II) **283** complexes with SBL 281 (Scheme 40). The characterization of these synthesized compounds (281–284) encompassed elemental analysis, molar conductivity measurement, and various spectroscopic techniques. Antimicrobial screening was performed using the diffusion agar technique for compounds **281–284**. Zn(II) **283** complex exhibited superior antimicrobial activity compared to **281, 282, and 284**. The assessment of anticancer potential against the MCF-7 human breast cancer cell line was conducted for compounds **281–284**. The results indicated that complex **282** displayed enhanced anticancer efficacy compared to **283 and 284**.

In 2022, a novel Schiff base ligand, designated as 2-((E)-((4-((E)benzylidene)amino)phenyl)amino)methyl)-naphthalene-1-ol **288,** along with its corresponding complexes **289–291**, was synthesized by Alorini et al. [91] (Scheme 41). The SBL coordinated with metal ions through its azomethine nitrogen and phenolic oxygen atoms in a bidentate manner, exhibiting an octahedral geometry for complexes **289–291**. Subsequent assessment of compounds **288–291** focused on their antibacterial and antifungal activities using the hole plate diffusion method. Furthermore, in vitro anticancer evaluations of **288–291** were performed against

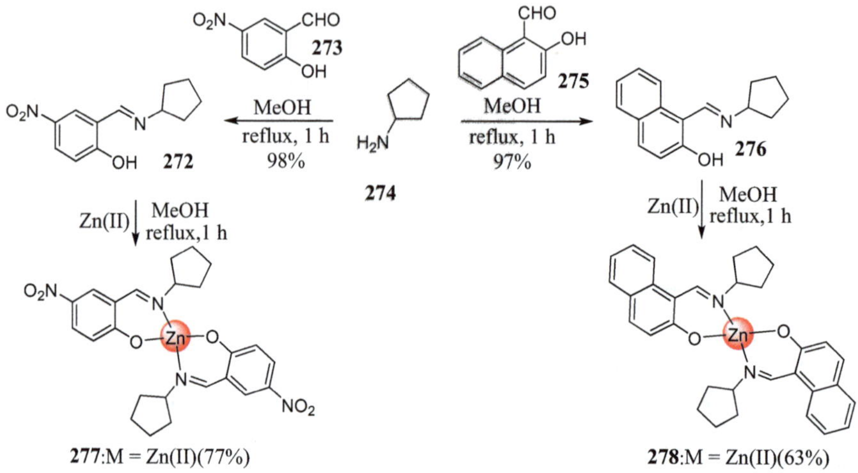

Scheme 39 Synthesized SBLs **272, 276** and its Zn(II) complexes [90]

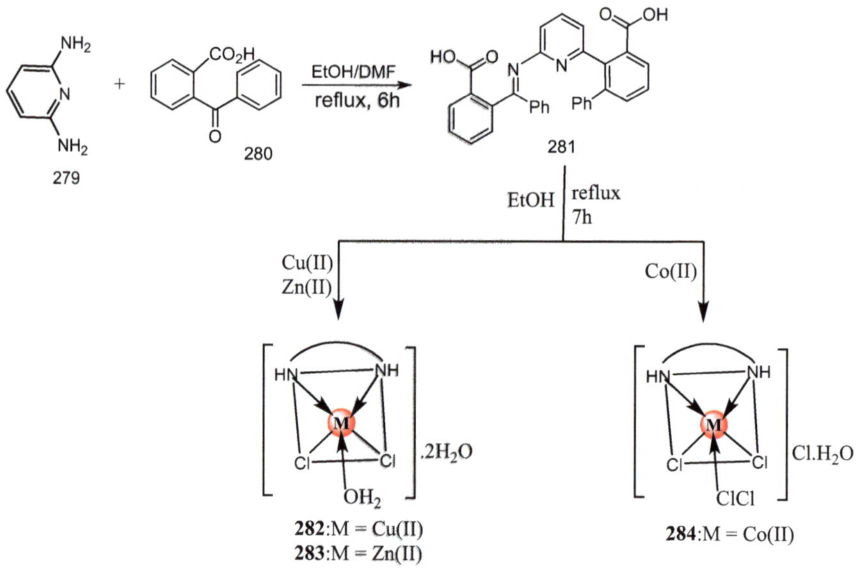

Scheme 40 Synthesis of new SBL **281** and its metal complexes [34]

HeLa tumor cell lines and PC-3, SKOV3 cell lines. Notably, complex **290** exhibited superior antitumor activity compared to clinically utilized medications such as cisplatin, estramustine, and etoposide against these human cell lines.

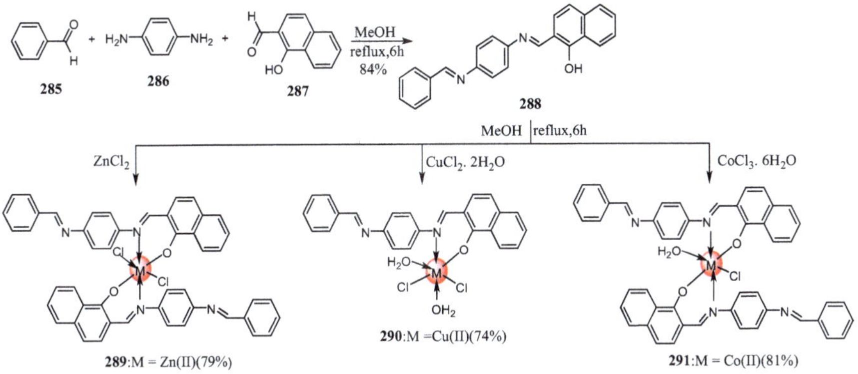

Scheme 41 Synthesis of a novel SBL **288** and metal complexes [92]

SBL **294** was obtained through the reaction between o-phthaldehyde**292** and 2-aminobenzoic acid **293** (Scheme 42) [92]. Following the synthesis of SBL **294**, subsequent steps involved the formation of complexes with metal complexes, namely **295–298**. The spectral analyses unequivocally affirmed the tetradentate donation capability of SBL **294.** To evaluate their potential antimicrobial properties, compounds **294–298** were subjected to antimicrobial assays against a panel of microorganisms. The results indicated that complexes **295–298** exhibited superior antimicrobial activity compared to SBL **294.**

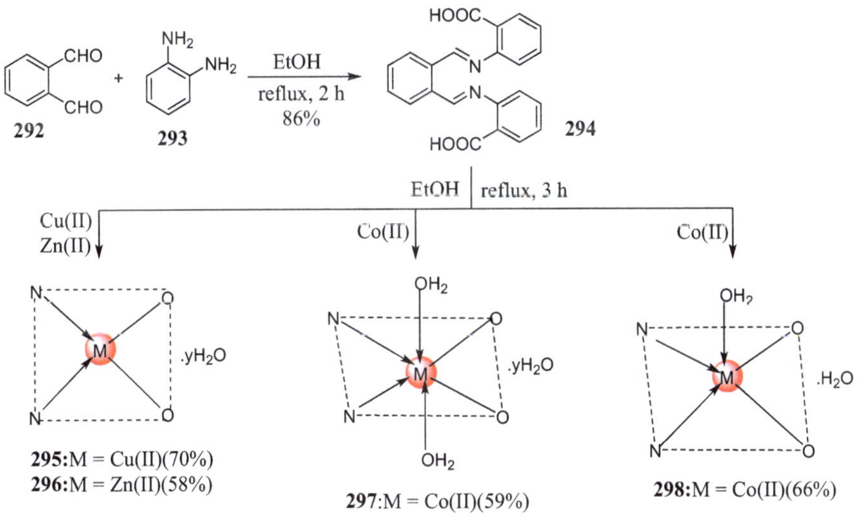

Scheme 42 Synthesis of SBL **294** and metal complexes [92]

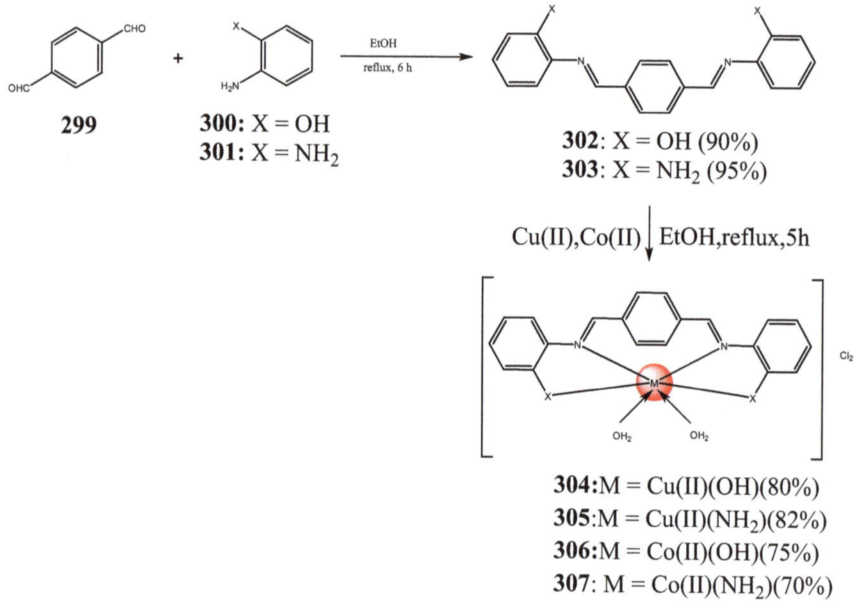

Scheme 43 Synthesis of SBLs **302, 303** and metal complexes [93]

In the pursuit of novel compounds, two distinct SBLs, **302** and 303, were synthesized through the condensation reaction involving 2-aminophenol/o-phenylenediamine **300/301** and terephthaldehyde **299**. Additionally, the copper **304, 305** and Co^{balt} **306, 307** complexes of **302** and **303** were prepared and investigated (Scheme 43). Evaluation of the antibacterial and antifungal properties of 302–307 was performed against various bacterial strains. Remarkably, compounds 304–307 exhibited enhanced antimicrobial activity compared to **302 and 303**. Specifically, the MIC values of complexes **304–307** against *Aspergillus niger* were found to be 250, 125, 500, and 250 μg/ml, respectively.

Two SBLs, namely (2-hydroxybenzylidenehydrazinocarbonyl)-3-methylquinoxaline-4-oxide **312** and (3-methyl-2(pyridine-2-ylmethylenehydrazinocarbonyl)quinoxaline-4-oxide **313 and its complexes 114–116** (Scheme 44), were investigated [94]. The compounds **312–316** were prepared and characterized utilizing elemental analyses, magnetic susceptibility measurements, and various spectroscopic techniques. The antifungal efficacy of compounds **310–314** was evaluated against a bacterial strain. Complex **313** exhibited the maximum inhibition zone against *A. niger*, albeit less than that of the standard drug.

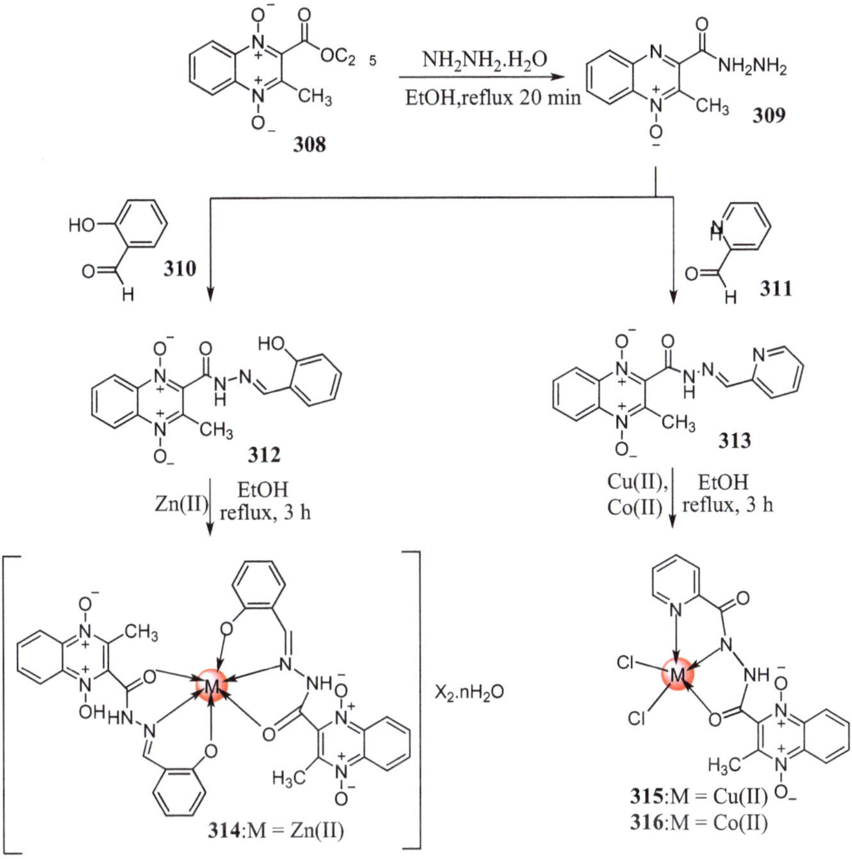

Scheme 44 Synthesis of SBLs **312, 313** and their metal complex [94]

3 Conclusion

In summary, Schiff base ligands (SBLs) hold significant importance in chemistry due to their ability to form metal complexes, along with their therapeutic and industrial relevance. However, further research is warranted given their demonstrated biological activity. Analysis of existing literature strongly indicates that Schiff base metal complexes exhibit enhanced biological activity compared to their parent SBLs. This heightened activity can be attributed to the π-electron delocalization during chelation, leading to increased lipophilicity of the complexes. Moreover, various factors including the type of donor atoms or metal ions, complex geometry, total charge, solubility, conductivity, and metal–ligand bond length influence the enhanced activity of Schiff base metal complexes. This review aims to present a comprehensive and current overview of the literature on the biological activities of Schiff bases and their metal complexes.

References

1. Cozzi PG (2004) Metal-Salen Schiff base complexes in catalysis: practical aspects. Chem Soc Rev 33:410–421
2. Schiff H (1864) Mittheilungen aus dem Universitätslaboratorium in Pisa: Eine neue Reihe organischer Basen. Ann Chem Pharm 131:118–119
3. Sahan F, Kose M, Hepokur C, Karakas D, Kurtoglu M (2019) New azo-azomethine-based transition metal complexes: synthesis, spectroscopy, solid-state structure, density functional theory calculations and anticancer studies. Appl Organomet Chem 33:1124
4. Shaygan S, Pasdar H, Foroughifar N, Davallo M, Motiee F (2018) Cobalt(II) complexes with Schiff base ligands derived from terephthalaldehyde and ortho-substituted anilines: synthesis, characterization and antibacterial activity. Appl Sci 8:385
5. Al-Zoubi W, Al-Hamdani AAS, Kaseem M (2016) Synthesis and antioxidant activities of Schiff bases and their complexes: a review. Appl Organomet Chem 30:810–817
6. Dhahagani K et al (2018) Crystal structure, optical properties, DFT analysis of new morpholine based Schiff base ligands and their copper(II) complexes: DNA, protein docking analyses, antibacterial study and anticancer evaluation. Mater Sci Eng C90:119–130
7. Goel P, Kumar D, Chandra S, Kumar A, Misra N (2020) Experimental and theoretical studies of Mn(II) and Co(II) metal complexes of a tridentate Schiff's base ligand and their biological activities. Appl Organomet Chem 34:e5371
8. Jain P, Pandey G, Kumar D, Chandra S (2019) Prospects of biologically active Schiff's base ligand and metal complexes in drug discovery. Sci Am 11:144–154
9. Goel P, Kumar D, Chandra S (2019) Antibacterial screening of nitrogen and sulphur donor atom containing methylcarbamatethiosemicarbazone and its Mn(II) and Co(II) complexes: synthesis, spectroscopic approach, molecular modeling. Iran J Sci Technol Trans Sci 43:477–487
10. Goel P, Kumar D, Chandra S, Kumar A (2018) Synthesis and spectroscopic study of biologically active tridentate Schiff's base ligand 2-acetyl-5-methyl-furanthiosemicarbazone and its Mn(II), Co(II), Ni(II), and Cu(II) complexes. Iran J Sci Technol Trans Sci 42:557–565
11. Priya NP, Arunachalam S, Manimaran A, Muthupriya D, Jayabalakrishnan C (2009) Mononuclear Ru(III) Schiff base complexes: synthesis, spectral, redox, catalytic and biological activity studies. Spectrochim Acta A Mol Biomol Spectrosc 72:670–676
12. Kaur M et al (2021) Post-synthesis modification of metal-organic frameworks using Schiff base complexes for various catalytic applications. Chem Eng J 423:130230
13. Mahmoud WH, Mohamed GG, El-Sayed OY (2018) Coordination compounds of some transition metal ions with new Schiff base ligand derived from dibenzoyl methane: structural characterization, thermal behavior, molecular structure, antimicrobial, anticancer activity and molecular docking studies. Appl Organomet Chem 32:48
14. Gupta KC, Sutar AK (2008) Catalytic activities of Schiff base transition metal complexes. Coord Chem Rev 252:1420–1450
15. Rakhtshah J (2022) A comprehensive review on the synthesis, characterization, and catalytic application of transition-metal Schiff-base complexes immobilized on magnetic Fe_3O_4 nanoparticles. Coord Chem Rev 467:214614
16. Liu X, Hamon J-R (2019) Recent developments in penta-, hexa- and heptadentate Schiff base ligands and their metal complexes. Coord Chem Rev 389:94–118
17. Brooker S (2002) Some copper and cobalt complexes of schiff-base macrocycles containing pyridazine head units. Eur J Inorg Chem 2002:2535–2547
18. Alka S, Vishvakarma VK, Yadav S, Singh P, Jain P (2023) Pharmacological aspects of Co(II), Ni(II) and Cu(II) schiff base complexes: an insight. Results Chem 5:100849
19. Alka S, Vishvakarma VK, Yadav S, Singh P, Jain P (2023) In vitro, in silico, ADME and theoretical analysis of Mn(II) and Co(II) complexes derived from methyl-(Z)-N′-carbamothioylcarbamohydrazonate Schiff base ligand. Chem Biodivers 12:e202300042. https://doi.org/10.1002/cbdv.202300042
20. Jain P, Singh P (2023) Structural and biological applications of Schiff base metal complexes. CRC Press, New York

21. da Silva CM et al (2011) (2011) Schiff bases: a short review of their antimicrobial activities. J Adv Res 2:1–8
22. Jin X, Wang J, Bai J (2009) Synthesis and antimicrobial activity of the Schiff base from chitosan and citral. Carbohydr Res 344:825–829
23. Rather B et al (2010) Anticonvulsant activity of schiff bases of 3-amino-6,8-dibromo-2-phenyl-quinazolin-4(3H)-ones. Indian J Pharm Sci 72:375
24. Hearn MJ et al (2009) Preparation and antitubercular activities in vitro and in vivo of novel Schiff bases of isoniazid. Eur J Med Chem 44:4169–4178
25. Malik MA, Dar OA, Gull P, Wani MY, Hashmi AA (2018) Heterocyclic Schiff base transition metal complexes in antimicrobial and anticancer chemotherapy. MedChemComm 9:409–436
26. Tadele KT, Tsega TW (2019) Schiff bases and their metal complexes as potential anticancer candidates: a review of recent works. Anticancer Agents Med Chem 19:1786–1795
27. Alam MS, Choi J-H, Lee D-U (2012) Synthesis of novel Schiff base analogues of 4-amino-1,5-dimethyl-2-phenylpyrazol-3-one and their evaluation for antioxidant and anti-inflammatory activity. Bioorg Med Chem 20:4103–4108
28. Manjunath M, Kulkarni AD, Bagihalli GB, Malladi S, Patil SA (2017) Bio-important antipyrine derived Schiff bases and their transition metal complexes: synthesis, spectroscopic characterization, antimicrobial, anthelmintic and DNA cleavage investigation. J Mol Struct 1127:314–321
29. Hossain S, Zakaria CM, Kudrat-E-Zahan S (2017) Structural and biological activity studies on metal complexes containing thiosemicarbzone and isatin based schiff base: a review. Asian J Res Chem 10:6
30. Adibi H, Khodaei MM, Pakravan P, Abiri R (2010) Synthesis, characterization, and in vitro antimicrobial evaluation of hydrazone and bishydrazone derivatives of isatin. Pharm Chem J 44:219–227
31. Saddam-Hossain M (2017) Selected pharmacological applications of 1st row transition metal complexes: a review. Clin Med Res 6:177
32. Tandon SS, Larkworthy LF (1984) Synthesis and characterization of Nickel(II) complexes with cyclic and linear Schiff-base ligands incorporating sulphur, nitrogen, and oxygen donor atoms. J Chem Soc Dalt Trans 11:2389–2393
33. Chakraborty J, Samanta B, Pilet G, Mitra S (2006) Synthesis, structure and spectral characterisation of a hydrogen-bonded polymeric manganese(III) Schiff base complex. Struct Chem 17:585–593
34. Mahmoud WH, Deghadi RG, Mohamed GG (2016) Novel Schiff base ligand and its metal complexes with some transition elements: synthesis, spectroscopic, thermal analysis, antimicrobial and in vitro anticancer activity. Appl Organomet Chem 30:221–230
35. Devi J, Batra N, Malhotra R (2012) Ligational behavior of Schiff bases towards transition metal ion and metalation effect on their antibacterial activity. Spectrochim Acta A Mol Biomol Spectrosc 97:397–405
36. Ejidike IP, Ajibade PA (2015) Transition metal complexes of symmetrical and asymmetrical Schiff bases as antibacterial, antifungal, antioxidant, and anticancer agents: progress and prospects. Rev Inorg Chem 35:191–224
37. Bursal E et al (2021) Transition metal complexes of a multidentate Schiff base ligand containing pyridine: synthesis, characterization, enzyme inhibitions, antioxidant properties, and molecular docking studies. Biometals 34:393–406
38. Parashar RK, Sharma RC, Kumar A, Mohan G (1988) Stability studies in relation to IR data of some schiff base complexes of transition metals and their biological and pharmacological studies. Inorg Chim Acta 151:201–208
39. Cheng G-J, Zhang X, Chung LW, Xu L, Wu Y-D (2015) Computational organic chemistry: bridging theory and experiment in establishing the mechanisms of chemical reactions. J Am Chem Soc 137:1706–1725
40. Ko C-N, Li G, Leung C-H, Ma D-L (2019) Dual function luminescent transition metal complexes for cancer theranostics: the combination of diagnosis and therapy. Coord Chem Rev 381:79–103

41. Shebl M, Saif M, Nabeel AI, Shokry R (2016) New non-toxic transition metal nanocomplexes and Zn complex-silica xerogel nanohybrid: synthesis, spectral studies, antibacterial, and antitumor activities. J Mol Struct 1118:335–343
42. Pervaiz M et al (2019) Synthesis, spectral and antimicrobial studies of amino acid derivative Schiff base metal (Co, Mn, Cu, and Cd) complexes. Spectrochim Acta A Mol Biomol Spectrosc 206:642–649
43. Venkateswarlu K, Rambabu A, Shankar DS, Lakshmi S, Shivaraj PVA (2022) A treatise on furan cored Schiff base Cu(II), Ni(II) and Co(III) complexes accentuating their biological efficacy: synthesis, thermal and spectroscopic characterization, DNA interactions, antioxidant and antibacterial activity studies. Chem Biodivers 19:e202100686
44. Venkateswarlu K, Anantha Lakshmi PV, Shivaraj S (2022) Synthesis, spectroscopic and thermal studies of Cu^{2+}, Ni^{2+} and Co^{3+} complexes of Schiff base containing furan moiety: antitumor, antioxidant, antibacterial and DNA interaction studies. Appl Organomet Chem 36:e6530
45. Hashem HE, Nath A, Kumer A (2022) Synthesis, molecular docking, molecular dynamic, quantum calculation, and antibacterial activity of new Schiff base-metal complexes. J Mol Struct 1250:131–915
46. Gurusamy S, Krishnaveni K, Sankarganesh M, Nandini Asha R, Mathavan A (2022) Synthesis, characterization, DNA interaction, BSA/HSA binding activities of VO(IV), Cu(II) and Zn(II) Schiff base complexes and its molecular docking with biomolecules. J Mol Liq 345:117045
47. Elshafie HS, Sadeek SA, Camele I, Mohamed AA (2022) Biochemical characterization of new gemifloxacin Schiff base (GMFX-o-phdn) metal complexes and evaluation of their antimicrobial activity against some phyto- or human pathogens. Int J Mol Sci 23:2110
48. El-Gammal OA, El-Bindary AA, Mohamed FS, Rezk GN, El-Bindary MA (2022) Synthesis, characterization, design, molecular docking, anti COVID-19 activity, DFT calculations of novel Schiff base with some transition metal complexes. J Mol Liq 346:117850
49. Qi J, Luo Y, Zhou Q, Su G, Zhang X, Nie X, Lv M, Li W (2022) Synthesis, structure and anticancer studies of Cu(II) and Ni(II) complexes based on 2-hydroxy-1-naphthaldehyde-4-aminoantipyrine Schiff-base. J Mol Struct 1255:132458
50. Ahmed YM, Mohamed GG (2022) Synthesis, spectral characterization, antimicrobial evaluation and molecular docking studies on new metal complexes of novel Schiff base derived from 4,6-dihydroxy-1,3-phenylenediethanone. J Mol Struct 1256:132496
51. Sekhar EV, Rangaswamy J, Bhat M, Naik N (2022) 2-(4-nitrophenyl)iminomethyl phenol Schiff base metal complexes: Synthesis, spectroscopic characterization, anticancer and antimicrobial studiesInorg. Nano-Metal Chem 52:1289–1298
52. Hossain AMS, Méndez-Arriaga JM, Xia C, Xie J, Gómez-Ruiz S (2022) Metal complexes with ONS donor Schiff bases: a review. Polyhedron 217:115692
53. Phang YL, Zheng C, Xu H (2022) Structural diversity and biological activities of caged *Garcinia xanthones*: recent updates. Acta Mater Med 12:72–95
54. More G, Bootwala S, Shenoy S, Mascarenhas J, Aruna K (2018) Synthesis, characterization and in vitro antitubercular and antimicrobial activities of new aminothiophene Schiff bases and their Co(II), Ni(II), Cu(II) and Zn(II) metal complexes. Orient J Chem 34:800–812
55. Abd El-Halim HF, Mohamed GG, Anwar MN (2018) Antimicrobial and anticancer activities of Schiff base ligand and its transition metal mixed ligand complexes with heterocyclic base. Appl Organomet Chem 32:e3899
56. Tadavi SK, Yadav AA, Bendre RS (2018) Synthesis and characterization of a novel schiff base of 1,2-diaminopropane with substituted salicyaldehyde and its transition metal complexes: single crystal structures and biological activities. J Mol Struct 1152:223–231
57. Palanimurugan A, Dhanalakshmi A, Selvapandian P, Kulandaisamy A (2019) Electrochemical behavior, structural, morphological, Calf Thymus-DNA interaction and in-vitro antimicrobial studies of synthesized Schiff base transition metal complexes. Heliyon 5:e02039
58. Sumrra SH et al (2014) Synthesis, spectral characterization, and biological evaluation of transition metal complexes of bidentate N, O donor Schiff bases. Bioinorg Chem Appl 2014:1–10

59. Sheikhshoaie I, Lotfi N, Sieler J, Krautscheid H, Khaleghi M (2018) Synthesis, structures and antimicrobial activities of nickel(II) and zinc(II) diaminomaleonitrile-based complexes. Transit Met Chem 43:555–562
60. Pravin N, Raman N (2013) DNA interaction and antimicrobial activity of novel tetradentate imino-oxalato mixed ligand metal complexes. Inorg Chem Commun 36:45–50
61. Naqi Ahamad M et al (2020) Anticancer properties, apoptosis and catecholase mimic activities of dinuclear cobalt(II) and copper(II) Schiff base complexes. Bioorg Chem 95:103561
62. Zafar H et al (2015) Synthesis and characterization of Schiff base octaazamacrocyclic complexes and their biological studies. J Photochem Photobiol B Biol 142:8–19
63. Anacona JR, Rodriguez JL, Camus J (2014) Synthesis, characterization and antibacterial activity of a Schiff base derived from cephalexin and sulphathiazole and its transition metal complexes. Spectrochim Acta A Mol Biomol Spectrosc 129:96–102
64. Devi J, Batra N (2015) Synthesis, characterization and antimicrobial activities of mixed ligand transition metal complexes with isatin monohydrazone Schiff base ligands and heterocyclic nitrogen base. Spectrochim Acta A Mol Biomol Spectrosc 135:710–719
65. Saini RP, Kumar V, Gupta AK, Gupta GK (2014) Synthesis, characterization, and antibacterial activity of a novel heterocyclic Schiff's base and its metal complexes of first transition series. Med Chem Res 23:690–698
66. Kareem A, Laxmi C, Arshad M, Nami SAA, Nishat N (2016) Herbo-mineral based Schiff base ligand and its metal complexes: synthesis, characterization, catalytic potential and biological applications. J Photochem Photobiol B Biol 160:163–171
67. Goel P, Kumar D, Chandra S, Kumar A (2018) Synthesis and spectroscopic study of biologically active tridentate Schiff's base ligand 2-acetyl-5-methyl-furanthiosemicarbazone and its Mn(II), Co(II), Ni(II), and Cu(II) complexes. Iran J Sci Technol Trans A Sci 42:557–565
68. Rayees Ahmad Shiekh IR, Malik MA, Luddin N, Masudi SM, Al-Thabaiti S (2013) A ion metal complexes with mixed nitrogen–sulphur (N_S) donor macrocyclic Schiff base ligand: synthesis, spectral, electrochemical and antimicrobial studies. Int J Electrochem Sci 8:6972–6987
69. Amiri Rudbari H et al (2016) Synthesis, characterization, X-ray crystal structures and antibacterial activities of Schiff base ligands derived from allylamine and their vanadium(IV), cobalt(III), nickel(II), copper(II), zinc(II) and palladium(II) complexes. J Mol Struct 1125:113–120
70. Ashraf MA, Karamat Mahmood AW (2011) Synthesis, characterization and biological activity of Schiff Bases. Int Conf Chem Chem Process 10:1–7
71. El-Tabl AS, Mohamed Abd El-Waheed M, Wahba MA, Abd El-Halim Abou El-Fadl N (2015) Synthesis, characterization, and anticancer activity of new metal complexes derived from 2-hydroxy-3-(hydroxyimino)-4-oxopentan-2-ylidene)benzohydrazide. Bioinorg Chem Appl 2015:1–14
72. Hanif M, Chohan ZH (2013) Design, spectral characterization and biological studies of transition metal(II) complexes with triazole Schiff bases. Spectrochim Acta A Mol Biomol Spectrosc 104:468–476
73. Daravath S et al (2017) Design, synthesis, spectral characterization, DNA interaction and biological activity studies of copper(II), cobalt(II) and nickel(II) complexes of 6-amino benzothiazole derivatives. J Mol Struct 1144:147–158
74. Kalarani R, Sankarganesh M, Kumar GGV, Kalanithi M (2020) Synthesis, spectral, DFT calculation, sensor, antimicrobial and DNA binding studies of Co(II), Cu(II) and Zn(II) metal complexes with 2-amino benzimidazole Schiff base. J Mol Struct 1206:127725
75. Shiju C, Arish D, Kumaresan S (2020) Novel water soluble Schiff base metal complexes: synthesis, characterization, antimicrobial-, DNA cleavage, and anticancer activity. J Mol Struct 1221:128770
76. Bahron H, Khaidir SS, Tajuddin AM, Ramasamy K, Yamin BM (2019) Synthesis, characterization and anticancer activity of mono- and dinuclear Ni(II) and Co(II) complexes of a Schiff base derived from o-vanillin. Polyhedron 161:84–92
77. Bhowmick A et al (2019) Synthesis and structure determination of some Schiff base metal complexes with investigating antibacterial activity. Am J Chem 9:21–25

78. Aslan HG, Akkoç S, Kökbudak Z (2020) Anticancer activities of various new metal complexes prepared from a Schiff base on A549 cell line. Inorg Chem Commun 111:107645
79. Mahmood K et al (2019) Synthesis, DNA binding and antibacterial activity of metal(II) complexes of a benzimidazole Schiff base. Polyhedron 157:326–334
80. Kargar H et al (2021) Synthesis, crystal structure, theoretical calculation, spectroscopic and antibacterial activity studies of copper(II) complexes bearing bidentate schiff base ligands derived from 4-aminoantipyrine: influence of substitutions on antibacterial activity. J Mol Struct 1230:129908
81. Ali I, Mahmood LMA, Mehdar YTH, Aboul-Enein HY, Said MA (2020) Synthesis, characterization, simulation, DNA binding and anticancer activities of Co(II), Cu(II), Ni(II) and Zn(II) complexes of a Schiff base containing o-hydroxyl group nitrogen ligand. Inorg Chem Commun 118:108004
82. Abdulghani AJ, Hussain RK (2015) Synthesis and characterization of Schiff base metal complexes derived from cefotaxime with 1-indole-2,3-dione (Isatin) and 4-N, N-dimethylaminobenzaldehyde. Open J Inorg Chem 05:83–101
83. Abid KK, Redha H, Al-Bayati AAF (2016) Transition metal complexes of new N-amino quinolone derivative; synthesis, characterization, thermal study and antimicrobial properties. Am J Chem 6:29–35
84. El-Sherif AA, Eldebss TMA (2011) Synthesis, spectral characterization, solution equilibria, in vitro antibacterial and cytotoxic activities of Cu(II), Ni(II), Mn(II), Co(II) and Zn(II) complexes with Schiff base derived from 5-bromosalicylaldehyde and 2-aminomethylthiophene. Spectrochim Acta A Mol Biomol Spectrosc 79:1803–1814
85. El-Sherif AA, Shehata MR, Shoukry MM, Barakat MH (2012) Synthesis, characterization, equilibrium study and biological activity of Cu(II), Ni(II) and Co(II) complexes of polydentate Schiff base ligand. Spectrochim Acta A Mol Biomol Spectrosc 96:889–897
86. Prathima B, Subba Rao Y, Adinarayana Reddy S, Reddy YP, Varada Reddy A (2010) Copper(II) and nickel(II) complexes of benzyloxybenzaldehyde-4-phenyl-3-thiosemicarbazone: synthesis, characterization and biological activity. Spectrochim Acta A Mol Biomol Spectrosc 77:248–252
87. Low ML et al (2016) New insight into the structural, electrochemical and biological aspects of macroacyclic Cu(II) complexes derived from S-substituted dithiocarbazate schiff bases. Eur J Med Chem 120:1–12
88. Liu H et al (2017) Syntheses, characterizations and third-order NLO properties of a series of Ni(II), Cu(II) and Zn(II) complexes using a novel S-benzyldithiocarbazate ligand. Polyhedron 121:53–60
89. Cai LF (2017) Copper(II) complexes of mono-condensed N, O-donor Schiff base ligands: synthesis, crystal structures, and antibacterial activity. Russ J Coord Chem 43:535–539
90. Guo YN (2012) Synthesis, crystal structures, and antibacterial activities of Schiff base zinc(II) complexes [Zn(L1)2] and [Zn(L2)2]. Russ J Coord Chem 38:121–125
91. Alorini TA, Al-Hakimi AN, El-Sayed Saeed S, Alhamzi EHL, Albadri AEAE (2022) Synthesis, characterization, and anticancer activity of some metal complexes with a new Schiff base ligand. Arab J Chem 15:103559
92. Abdallah SM, Zayed MA, Mohamed GG (2010) Synthesis and spectroscopic characterization of new tetradentate Schiff base and its coordination compounds of NOON donor atoms and their antibacterial and antifungal activity. Arab J Chem 3:103–113
93. Saranya J et al (2020) Tetradentate Schiff base complexes of transition metals for antimicrobial activity. Arab J Sci Eng 45:4683–4695
94. Mahal A, Abu-El-Halawa RSAZ, Ibrahim M, Al-Refai M, Kaimari T (2015) Synthesis, characterization and antifungal activity of some metal complexes derived from quinoxaloylhydrazone. World J Org Chem 3:1–8

Design, Synthesis, and Characterization of Some Oxadiazole Derivatives and Their *S*-Glucosides Bearing Thiazole Scaffold

Shital M. Dekate, Kishor M. Hatzade, and Ajay M. Ghatole

1 Introduction

Modern research has proven a considerable fascinate in thiazole derivatives, because of far-reaching biological activities, such as anticonvulsant activity [1], anti-HIV [2], antidiabetic [3], anti-Alzheimer [4], antimalarial [5], antimicrobial [6], anti-inflammatory [7], anti-proliferative towards MiaPaCa-2 cell line [8], anti-proliferative towards gastric carcinoma cells [9], anti-proliferative towards diffuse malignant peritoneal mesothelioma cell lines, a completely competitive shape of cancer [10], and CDK1 inhibitory activity of thiazole [11]. The majority of pharmaceutical medications, such as Fanetizole, Meloxicam, Tiazofurin [12] and Ritonavir [13] contain thiazole ring.

On the other hand, there are a bunch of reports on 1,3,4-oxadiazoles exhibiting various biological activities, like anti-diabetic, antihypertension [14], analgesic [15], antiviral [16], anticonvulsant [17], antifungal [18] antibacterial [19], anticancer [20], anti-glycation [21], anti-inflammatory [22], antimicrobial, and ulcerogenic [23]. Currently, clinical medications such as Nesapidil, Furamizole, and Zibotentan use compounds containing oxadiazole units.

Thioglycosides are with cytotoxic action of interest, such as thioglycosides of pyridine [24], pyrimidine [25], imidazole [26], oxadiazole [27], thiophene [28], quinoline [29], and thienopyrazole [30].

We were particularly interested in *S*-glycosylation of 5{[4-(substituted phenylthiazol[49]-2-yl)]-thio methyl}-1,3,4- oxadiazole-2-thiol (**7a-j**). A unique series

S. M. Dekate (✉) · K. M. Hatzade
Department of Chemistry, Dhote Bandhu Science College, Gondia, Maharashtra, India
e-mail: snimje04@gmail.com

A. M. Ghatole (✉)
Department of Chemistry, J. M. Patel College, Bhandara, Maharashtra, India
e-mail: ajay.ghatole5@gmail.com

S. Rattan et al. (eds.), *Emerging Trends in Synthesis and Catalysis in Chemistry*, Springer Proceedings in Materials 51, https://doi.org/10.1007/978-981-97-5169-3_20

of 5-{[4-(substituted phenylthiazol-2-yl)]-thio methyl}-1,3,4-oxadiazole-2-thiols and its S-glucoside are synthesized in the most recent work.

2 Results and Discussion

Our research work starts with the synthesis of 2-amino-4-(substituted phenyl)-1,3-thiazoles **(2a-j)** [31] synthesized by the solid phase reaction of thiourea, substituted acetophenone **(1a-j),** and iodine which on further undergo the reaction with sodium nitrite, cuprous chloride, and HCl under cold condition (Sandmeyer reaction) yielded 2-chloro-4-(substituted phenyl)-1,3-thiazoles **(3a-j). 3a-j** were further refluxed with thiourea and ethanol produced 4-(substituted phenyl)-1,3-thiazol-2–thiols **(4a-j)**. [4-(substituted phenyl) thiazol-2-yl]-ethyl-2-thioacetates **(5a-j)** were obtained by refluxing with ethyl bromoacetate, sodium carbonate, and ethanol which undergo further reaction with hydrazine hydrate and ethanol yielded [4-(substituted phenyl) thiazol-2-yl)-2-thioacetohydrazides **(6a-j)** [32]. Finally, **7a-j** were obtained by refluxing with CS_2, KOH with **(6a-j)** [33] as shown in Scheme 1.

The 5-{[4-(substituted phenyl) [18]thiazol-2-yl]-thio methyl}-1,3,4-oxadiazole-2-thiols **7a-j** reacts [45]with 2,3,4,6-tetraO-acetyl-α-D-glucopyranosyl bromide and acetone gives **8a-j** which undergo demethylation using methanolic ammonia to give **9a-j** [27] as shown in Scheme 2.

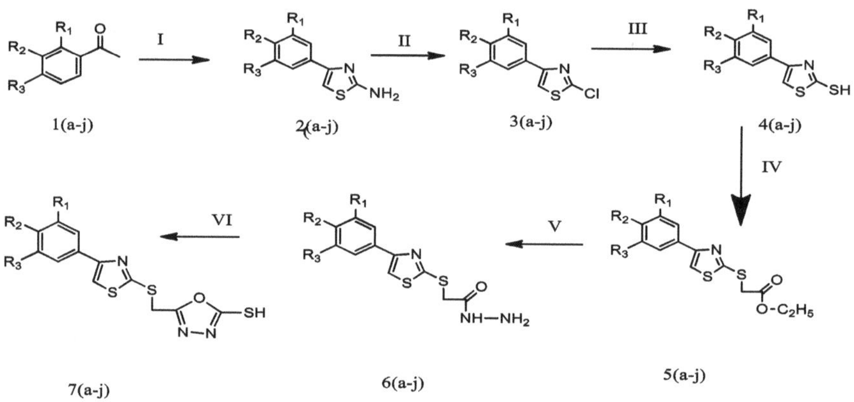

Scheme 1 Reaction Conditions and Reagents: where I = thiourea, I_2, ethanol, reflux on 100 °C for 24 h; II = $NaNO_2$, HCl, CuCl, cool at 0–5 °C; III = thiourea, ethanol, reflux for 5 h; IV = Na_2CO_3, ethyl bromoacetate, ethanol, reflux for 5 h; V = hydrazine hydrate, ethanol, reflux for 4 h; and VI = CS_2, KOH, water, reflux for 3 h

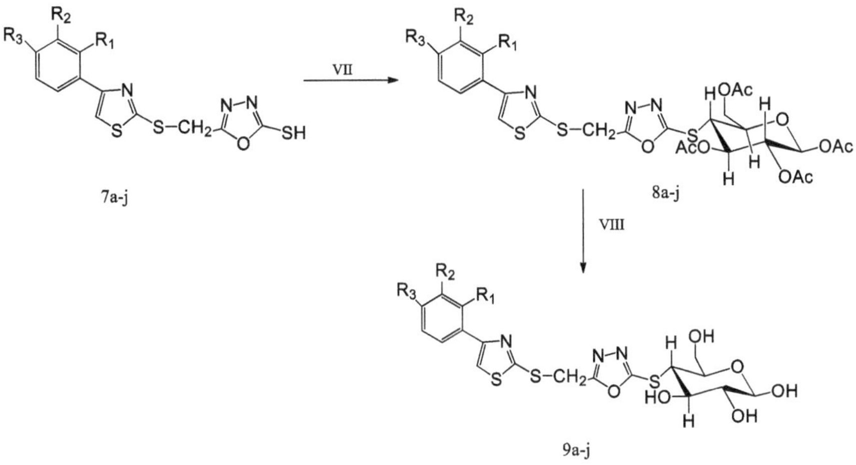

Scheme 2 Reaction Conditions and Reagents: Where VII = 2,3,4,6-tetra-O-acetyl-α-D-glucopyranosyl bromide, acetone, KOH. VIII = methanolic ammonia solution, diethyl ether

3 Experimental

All solvents were distilled before being used. All melting points were determined in an open capillary tube on a liquid 6-paraffin bath without being adjusted. TLC was used on silica gel 60 F254 (Merck). From Himedia and Merck Chemicals, all of the chemicals and solvents were purchased. The IR spectra were captured using the Bruker FT-IR spectrometer. Using the trimethylsilane standard as the reference, ^{1}H NMR and ^{13}C NMR spectra were recorded on a Brucker AC 400 (MHz) apparatus. Elemental studies were carried out using the Perkin-Elmer 2400 CHN analyzer.

General Procedure for Synthesis of 2-chloro-4-(Substituted Phenyl)-1, 3-Thiazoles (3a-j) [31]

The reaction was maintained at 0 °C by adding cold solutions of $NaNO_2$ (0.192 mol) in 40 mL water dropwise to a well-stirred solution of 2-amino-4-(substituted phenyl)-1,3-thiazoles (0.0512 mol) in 35% aqueous HCl (100 mL). Before adding cuprous chloride (0.128 mol) in a 35% aq. HCl solution (30 mL), the diazonium salt solution was stirred at ambient temperature for 4 h to allow for the release of N_2. TLC was used to confirm that the reaction had ended. To dissolve any remaining copper, the suspension was chilled and then basified with concentrated ammonia solution. After filtering, water washing, and drying, the precipitate was formed. Recrystallization using ethanol allowed for the crude product's purification.

General Procedure for Synthesis of 4-(Substituted Phenyl)-1,3-Thiazol-2-Thiols (4a-j)

5 h long refluxing in the water bath was done with a mixture of 2-chloro-4-(substituted phenyl)-1, 3-thiazoles (10 mmol), and thiourea (10 mmol) in ethanol.

TLC kept an eye on the reaction to see if it vanished when the reaction was finished (3 h). Extra ethanol was evaporated using a Rota evaporator, and the residue was then added to 30–40 mL of ice water. The separated solid was filtered and dried, by recrystallizing the crude products from ethyl acetate.

General Procedure for Synthesis of [4-(Substituted Phenyl) Thiazol-2-yl]-Ethyl-2-Thioacetates (5a-j)

A 100 mmol solution of the mercapto heterocyclic compound **(4a-j)** in 50 mL of acetone was mixed with anhydrous sodium carbonate (55 mmol). Slowly and while stirring, 100 mmol of ethyl bromoacetate was added to the reaction mixture. Thin layer chromatography was used to track the reaction's development while employing ethyl acetate and n-hexane (3:7) as the eluent. By using filtration, the sodium bromide byproduct was eliminated. The acetone was taken from the product's mother liquor by concentration under vacuum, and any leftover acetone was then simply eliminated with methanol. Without purification, the residue was employed for the following step.

General Procedure for Synthesis of [4-(Substituted Phenyl) Thiazol-2-yl)-2-Thioacetohydrazides (6a-j) [32]

In 30 mL of methanol, the previously obtained residue (5a-j) was dissolved. The clear solution received 200 mmol of 100% hydrazine hydrate, which was heated via reflux. The reaction was monitored using thin-layer chromatography, with a 9:1 solution of chloroform and methanol serving as the eluent. The mass of the reaction was cooled to 5 °C to crystallize the product. On filtration and washing with cold methanol, it produced an acylated hydrazine derivatives of mercapto heterocyclic compounds **(6a-j).**

General Procedure for Synthesis of 5-{[4-(Substituted Phenyl) Thiazol-2-yl]-Thio methyl}-1, 3, 4-oxadiazole-2-thiols 7a-j [33]

The above-mentioned residue (0.01 mol) in ethanol (25 mL) was combined with carbon disulfide (0.02 mol) and potassium hydroxide (0.01 mol) at 0 °C. The solvent was removed after the combination had refluxed for 7 h, and the residual solid was then dissolved in cold water before being acidified with hydrochloric acid. Filtered, dried, and then recrystallized from ethanol was the process used to obtain the white precipitate.

5-{[(4-Phenyl) Thiazol-2-yl]-Thiomethyl}-1, 3, 4-Oxadiazole-2-Thiol (7a)

Yield 65%, m.p. 160 °C. IR (KBr, 1): 2670, 2626 (SH-atom str), 1632 (C = N), 1064 (C–O–C). H^1-NMR δ_H (CDC13): 13.09 (1H, s, SH), 8.10 (s, 1H, thiazole-H), 7.40–7.80 (5H, m, Ar–H), 4.22 (2H, s, CH_2). ^{13}C NMR δc (DMSO- d6): 167.4 (N = C–S), 163.5 (O–C = N), 155.2 (C = C–N), 127–135 (aromatic 6C-atom), 34.9 (CH_2). EI-MS: m/z (%) 309(2%), 308 (14%), 307 (16%), 306 (100%) base peak. Anal. Calcd for $C_{12}H_9N_3OS_3$: C, 46.88; H, 2.95; N, 13.67; O, 5.20; S, 31.29. Found: C, 46.98; H, 2.99; N, 13.76; O, 5.24; S, 31.35(%).

5-{[(4-(4-Bromophenyl) Thiazol-2-yl)]-Thiomethyl}-[4]1, 3, 4-Oxadiazole-2-Thiol (7b)

Yield 60%, m.p. 200 °C. IR (KBr, 1):2680, 2636 (SH-atom str), 1637 (C = N), 1076 (C–O–C). H^1-NMR δ_H. (CDCl3): 13.07 (1H, s, SH), 8.11 (s, 1H, thiazole[20] –H), 7.70–7.85 (4H, m, Ar H). C^{13} NMR δc (DMSO-d6):167.5 (^{3}N = C–S), 163.4 (O–C = N), 155.1 (C = C–N), 123–135 (aromatic 6C-atom), 34.5 (CH_2). EI-MS: m/z (%) 389 (1%), 388 (12%), 388 (12%), 386 (100%) base peak, 385 (14%), 384 (89%). Anal. Calcd for $C_{12}H_8BrN_3OS_3$: C, 37.31; H, 2.09; Br, 20.68; N, 10.88; O, 4.14; S, 24.90. Found: C, 37.38; H, 2.19; Br, 20.72; N, 10.86; O, 4.24; S, 24.95(%).

5-{[(4-(4-Chlorophenyl) Thiazol-2-yl)]-Thio methyl}-1, 3, 4-Oxadiazole-2-Thiol (7c)

Yield 62%, m.p. 270 °C. IR (KBr, 1): 2671, 2631 (SH-atom str), 1622 (C = N), 1065 (C–O–C). H^1-NMR δ_H. (CDCl3): 13.02 (1H, s, SH), 8.08 (s, 1H, thiazole-H), 7.60–8.05 (4H, Ar –H, m), 4.22 (2H, s, CH). C^{13} NMR δc (DMSO-d6): 167.3 (N = C–S), 163.3 (O–C = N), 155.3 (C = C–N), 129–135 (aromatic 6C-atom), 34.3 (CH_2). EI-MS: m/z (%) 344 (5%), 343 (7%), 342 (46%), 341 (13%), 341 (3%), 340 (100%) base peak. Anal. Calcd for $C^7{}_{12}H_8ClN_3OS_3$: C, 42.16; H, 2.36; Cl, 10.37; N, 12.29; O, 4.68; S, 28.14. Found: C, 42.28; H, 2.39; Cl, 10.42; N, 12.26; O, 4.74; S, 28.15(%).

5-{(4-(4-Methoxyphenyl) Thiazol-2-yl)]-Thio methyl}-1, 3, 4-Oxadiazole-2-Thiol (7d)

Yield 65%, m.p. 252 °C. IR (KBr, 1): 2677, 2630 (SH-atom str), 1630 (C = N), 1070 (C–O–C). H1-NMR δH (CDCl3): 13.08 (1H, s, SH), 8.09 (s, 1H, thiazole-H), 7.10–7.65 (4H, m, Ar–H), 4.24 (2H, s, CH2), 3.87 (3H, s, OCH3). C13 NMR δc (DMSO-d6): 167.1 (N = C–S), 163.4 (O–C = N), 155.4 (C = C–N), 114–160 (aromatic 6C-atom), 112.7 (C = C–S) (thiazole), 55.10 (OCH3), 34.3 (CH2). EI-MS: m/z (%) 340 (2%), 339 (1%), 339 (14%), 338 (17%), 337 (100%) base peak. Anal. 10 Calcd for C13H11N3O2S3: C, 46.27; H, 3.29; N, 12.45; O, 9.48; S, 28.51. Found: C, 46.28; H, 3.39; N, 12.46; O, 9.54; S, 28.55(%).

5-{[(4-(4-Methylphenyl) Thiazol-2-yl)]-Thio methyl}-1, 3, 4-Oxadiazole-2-thiol (7e)

Yield 62%, m.p. 280 °C. IR (KBr, 1): 2667, 2621 (SH-atom str), 1629 (C = N), 1066 (C–O–C). H1-NMR δH (CDCl3): 13.05 (1H, s, SH), 8.11 (s, 1H, thiazole-H), 7.30–7.85 (4H, m, Ar–H), 4.22 (2H, s, CH2), 2.37 (3H, s, CH3). C13 NMR δc (DMSO-d6): 163.8 (O–C = N), 155.7 (C = C–N), 125–135 (aromatic 6C-atom), 112.6 (C = C–S) (thiazole), 34.3 (CH2), 21.10 (CH3). EI-MS: m/z (%) 324 (2%), 323 (1%), 323 (13%), 322 (16%), 322 (1%), 321 (100%) base peak. Anal. Calcd for C13H11N3OS3: C, 48.57; H, 3.45; N, 13.07; O, 4.98; S, 29.93. 20 Found: C, 48.58; H, 3.49; N, 13.16; O, 4.94; S, 98.97(%).

5-{[(4-(4-nitrophenyl) thiazol-2-yl)]-thio methyl}-1, 3, 4-oxadiazole-2-thiol (7f)

Yield 68%, m.p. 248 °C. IR (KBr, 1): 2666, 2622 (SH-atom str), 1629 (C = N), 1061 (C–O–C). H1-NMR δH. (CDCl3): 13.08 (1H, s, SH), 8.12 (s, 1H, thiazole-H), 8.10–8.35 (4H, Ar–H), 4.20 (2H, CH2). C13 NMR δc (DMSO-d6): 163.4 (O–C = N), 155.6 (C = C–N), 124–150 (aromatic 6C-atom), 112.7 (C = C–S) (thiazole), 34.2 (CH2), 21.10 (CH3). EI-MS: m/z (%) 354 (1%), 353 (1%), 353 (13%), 352 (15%), 352 (1%), 351 (100%) base peak. Anal. Calcd for C12H8N4O3S3: C, 40.90; H, 2.29; N, 15.90; O, 13.62; S, 27.30. Found: C, 40.98; H, 2.39; N, 15.96; O, 13.64; S, 27.37(%).

5-{[(4-(4-hydroxyphenyl) thiazol-2-yl)]-thio methyl}-1, 3, 4-oxadiazole-2-thiol (7g)

Yield 70%, m.p. 225 °C. IR (KBr, 1): 2671, 2625 (SH-atom str), 1628 (C = N), 1055 (C–O–C). H1-NMR δH (CDCl3): 13.04 (1H, s, SH), 8.11 (s, 1H, thiazole-H), 6.90–7.55 (4H, Ar–H), 5.43 (1H, s, OH), 4.23 (2H, s, CH2). C13 NMR δc (DMSO-d6): 167.7 (N = C–S), 163.5 (O–C = N), 155.5 (C = C–N), 116–160 (aromatic 6C-atom), 34.9 (CH2). EI-MS: m/z (%) 325 (2%), 324 (1%), 324 (13%), 323 (15%), 323 (1%), 322 (100%) base peak. Anal. Calcd for C12H9N3O2S3: C, 44.56; H, 2.80; N, 12.99; O, 9.89; S, 29.74. Found: C, 44.58; H, 2.89; N, 12.96; O, 9.84; S, 29.77(%).

5-{[(4-(3,4-dimethoxyphenyl) thiazol-2-yl)]-thio methyl}-1, 3, 4-oxadiazole-2-thiol (7h)

Yield 64%, m.p. 260 °C. IR (KBr, ν, cm^{-1}): 2667, 2628 (SH-atom str), 1612 (C = N), 1067 (C–O–C). H1-NMR δH (CDCl3): 13.06 (1H, s, SH), 8.07 (s, 1H, thiazole-H), 6.95–7.35 (3H, m, Ar–H), 3.90 (6H, s, OCH3), 4.21 (2H, s, CH2). C13 NMR δc (DMSO-d6): 167.1 (N = C–S), 163.8 (O–C = N), 155.3 (C = C–N), 113–150 (aromatic 6C-atom), 112.7 (C = C–S) (thiazole), 56.3 (OCH3), 34.7 (CH2). EI-MS: m/z (%) 370 (2%), 369 (1%), 369 (14%), 368 (15%), 368 (3%), 367 (100%) base peak. Anal. Calcd for C14H13N3O3S3: C, 45.76; H, 3.57; N, 11.44; O, 13.06; S, 26.18. Found: C, 45.78; H, 3.59; N, 11.46; O, 13.14; S, 26.17(%).

5-{[(4-(3-nitrophenyl) thiazol-2-yl)]-thio methyl}-1, 3, 4-oxadiazole-2-thiol (7i)

Yield 60%, m.p. 240 °C. IR (KBr, ν, cm cm^{-1}): 2687, 2642 (SH-atom str), 1644 (C = N), 1082 (C–O–C). H1-NMR δH (CDCl3): 13.03 (1H, s, SH), 8.12 (s, 1H, thiazole-H), 6.90–7.55 (4H, m, Ar–H), 4.21 (2H, s, CH2). C13 NMR δc (DMSO-d6): 167.8 (N = C–S), 163.4 (O–C = N), 155.4 (C = C–N), 116–160 (aromatic 6C-atom), 112.5 (C = C–S) (thiazole), 34.8 (CH2). EI-MS: m/z (%) 325 (2%), 324 (1%), 324 (13%), 323 (15%), 323 (1%), 322 (100%) base peak. Anal. Calcd for C12H9N3O2S3: C, 44.56; H, 2.80; N, 12.99; O, 9.89; S, 29.74. Found: C, 44.59; H, 2.88; N, 12.97; O, 9.85; S, 29.78(%)0.95.

5-{[(4-(2-hydroxyphenyl) thiazol-2-yl)]-thio methyl}-1, 3, 4-oxadiazole-2-thiol (7j)

Yield 68%, m.p. 185 °C. IR (KBr, ν, cm^{-1}): 2695, 2635 (SH-atom str), 1630 (C = N), 1072 (C–O–C). H1-NMR δH (CDCl3): 13.09 (1H, s, SH), 8.54 (s, 1H, thiazole-H), 6.90–7.55 (4H, m, Ar–H), 5.30 (1H, s, OH), 4.15 (2H, s, CH2). C13 NMR δc (DMSO-d6): 166.7 (N = C–S), 161.2 (O–C = N), 150.4 (C = C–N), 114–161 (aromatic 6C-atom), 111.5 (C = C–S) (thiazole), 34.8 (CH2). EI-MS: m/z (%) 325 (2%), 324 (1%), 324 (13%), 323 (15%), 323 (1%), 322 (100%) base peak. Anal. Calcd for C12H9N3O2S3: C, 44.56; H, 2.80; N, 12.99; O, 9.89; S, 29.74. Found: C, 44.59; H, 2.88; N, 12.97; O, 9.85; S, 29.78(%).

General procedure for synthesis of 5-{[(4-(4-substituted phenyl)thiazol-2-yl]-thio methyl}-2-(2, 3, 4, 6-tetra-O-acetyl-*S*-β-D-glucopyranosyl)-1, 3, 4-oxadiazoles (8a-j)

A solution of 2,3,4,6-tetra-O-acetyl-D-glucopyranosyl bromide (10 mmol) in acetone (20 mL) was added portion by portion to a stirred suspension of oxadiazole thiol derivative **(7a-j)** (10 mmol) in aqueous potassium hydroxide (10 mmol) in distilled water (2 mL). TLC results showed that the reaction was finished after 6 h of stirring at room temperature. The solvent was withdrawn under reduced pressure, and the remaining material was stirred for an hour while ice-cold water was added. After being dried and crystallized from the resulting substance, compound **(8a-j)** was created.

5-{[4-phenyl thiazol-2-yl]-thio methyl}-2-(2, 3, 4, 6-tetra-O-acetyl-*S*-β-D-glucopyranosyl)-1, 3, 4-oxadiazole (8a)

Yield 60%, m.p. 170 °C. IR (KBr, ν, cm^{-1}): 3045 (C–H aromatic), 2865 (glucosidic C–H str), 1765 (C = O of O acetyl gps of aglycone moiety), 1632 (C = N). H1-NMR δH (CDCl3): 13.09 (1H, s, SH), 8.10 (s, 1H, thiazole-H), 7.40–7.80 (5H, m, Ar–H), 4.22 (2H, s, CH2). C13 NMR δc (DMSO-d6): 167.4 (N = C–S), 163.5 (O–C = N), 155.2 (C = C–N), 127–135 (aromatic 6C-atom), 34.9 (CH2). EI-MS: m/z (%) 516 (3%), 515 (4%), 515 (13%), 514 (24%), 514 (3%), 513 (100%) base peak. Anal. Calcd for C22H15N3O6S3: C, 51.45; H, 2.94; N, 8.18; O, 18.69; S, 18.73. Found: C, 51.48; H, 2.99; N, 8.26; O, 18.74; S, 18.75(%).

5-{[(4-(4-bromophenyl)thiazol-2-yl)]-thio methyl}-2-(2, 3, 4, 6-tetra-O-acetyl-*S*-β-D-glucopyranosyl)-1, 3, 4oxadiazole (8b)

Yield 62%, m.p. 180 °C. IR (KBr, H str), 1732 (C = O of O-acetyl gps of aglycone moiety), 1623 (C = N). H1-NMR δH (CDCl3): 13.07 (1H, s, SH), 8.11 (s, 1H, thiazole-H), 7.70–7.85 (4H, m, Ar–H), 4.21 (2H, s, CH2). C13 NMR δc (DMSO-d6): 167.5 (N = C–S), 163.4 (O–C = N), 155.1 (C = CN), 123–135 (aromatic 6C-atom), 34.5 (CH2). EI-MS: m/z (%) 595 (3%), 594 (3%), 594 (12%), 593 (27%), 592 (4%), 592 (100%) base peak, 591 (21%), 591 (3%), 590(90%). Anal. Calcd for C22H14BrN3O6S3: C, 44.60; H, 2.38; Br, 13.49; N, 7.09; O, 16.20; S, 16.24. Found: C, 44.68; H, 2.39; Br, 13.52; N, 7.06; O, 16.24; S, 16.25(%).

5-{[(4-(4-chlorophenyl) thiazol-2-yl)]-thio methyl}-2-(2, 3, 4, 6-tetra-O-acetyl-*S*-β-D-glucopyranosyl)-1, 3, 4- oxadiazole (8c)

Yield 64%, m.p. 185 °C. IR (KBr, H aromatic), 1620 (C = N), 1782 (C = O of O-acetyl gps of aglycone moiety), 1612 (C = N). H1-NMR δH (CDCl3): 13.02 (1H, s, SH), 8.08 (s, 1H, thiazole-H), 7.60–8.05 (4H, m, Ar–H), 4.22 (2H, s, CH2). C13 NMR δc (DMSO-d6): 167.3 (N = C–S), 163.3 (OC = N), 155.3 (C = C–N), 129–135 (aromatic 6C-atom), 34.3 (CH2). EI-MS: m/z (%) 551 (1%), 550 (1%), 550 (5%), 549 (12%), 548 (4%), 548 (45%), 547 (24%), 547 (3%), 546 (100%) base peak. Anal. Calcd for C22H14ClN3O6S3: C, 48.22; H, 2.57; Cl, 6.47; N, 7.67; O, 17.52; S, 17.55. Found: C, 48.28; H, 2.59; Cl, 6.49; N, 7.66; O, 17.54; S, 17.57(%).

5-{[4-(4-methoyphenyl) thiazol-2-yl)]-thio methyl}-2-(2, 3, 4, 6-tetra-O-acetyl-*S*-β-D-glucopyranosyl)-1, 3, 4- oxadiazole (8d)

Yield 65%, m.p. 195 °C. IR (KBr, ν, cm^{-1}): 3053 (C–H aromatic), 1634 (C = N), 1774 (C = O of O- acetyl gps of aglycone moiety), 1628 (C = N). H1-NMR δH (CDCl3): 13.08 (1H, s, SH), 8.09 (s, 1H, thiazole-H), 7.10–7.65 (4H, m, Ar–H), 4.24 (2H, s, CH2), 3.87 (3H, s, OCH3). C13 NMR δc (DMSO-d6): 167.1 (N = C–S), 163.4 (O–C = N), 155.4 (C = C–N), 114–160 (aromatic 6C-atom), 112.7 (C = C–S) (thiazole), 55.10 (OCH3), 34.3 (CH2). EI-MS: m/z (%) 546 (3%), 545 (5%), 545 (13%), 544 (25%), 544 (3%), 543 (100%) base peak. Anal. Calcd for C23H17N3O7S3: C, 50.82; H, 3.15; N, 7.73; O, 20.60; S, 17.70. Found: C, 50.88; H, 3.19; N, 7.76; O, 20.64; S, 17.75(%).

5-{[(4-(4-methylphenyl) thiazol-2-yl)]-thio methyl}-2-(2, 3, 4, 6-tetra-O-acetyl-*S*-β-D-glucopyranosyl glucopyranosyl)-1, 3, 4- oxadiazole (8e)

Yield 60%, m.p. 200 °C. IR (KBr, ν, cm^{-1}): 3041 (C–H aromatic), 1643 (C = N), 1747 (C = O of O-acetyl gps of aglycone moiety), 1631 (C = N). H1-NMR δH (CDCl3): 13.05 (1H, s, SH), 8.11 (s, 1H, thiazole-H), 7.30–7.85 (4H, m, Ar–H), 4.22 (2H, s, CH2), 2.37 (3H, s, CH3). C13 NMR δc (DMSO-d6): 163.8 (O–C = N), 155.7 (C = C–N), 125–135 (aromatic 6C-atom), 112.6 (C = C–S) (thiazole), 34.3 (CH2), 21.10 (CH3). EI-MS: m/z (%) 530 (3%), 529 (5%), 529 (13%), 528 (27%), 528 (1%), 527 (100%) base peak. Anal. Calcd for C23H17N3O6S3: C, 52.36; H, 3.25; N, 7.96; O, 18.20; S, 18.23. Found: C, 52.38; H, 3.29; N, 7.98; O, 18.24; S, 18.27(%).

5-{[(4-(4-nitrophenyl) thiazol-2-yl)]-thio methyl}-2-(2, 3, 4, 6-tetra-O-acetyl-*S*-β-D-glucopyranosyl)-1, 3, 4- oxadiazole (8f)

Yield 64%, m.p. 280 °C. IR (KBr, ν, cm^{-1}): 3059 (C–H aromatic), 1652 (C = N), 1753 (C = O of-acetyl gps of aglycone moiety), 1628 (C = N). H1-NMR δH (CDCl3): 13.08 (1H, s, SH), 8.12 (s, 1H, thiazole-H), 8.10–8.35 (4H, m, Ar–H), 4.20 (2H, s, CH2). C13 NMR δc (DMSO-d6):163.4 (OC = N), 155.6 (C = C–N), 124–150 (aromatic 6C-atom), 112.7 (C = C–S) (thiazole), 34.2 (CH2), 21.10 (CH3). EI-MS: m/z (%) 561 (3%), 560 (5%), 559 (13%), 559 (26%), 558 (1%), 558 (100%)

base peak. Anal. Calcd for C22H14N4O8S3; C, 47.31; H, 2.53; N, 10.03; O, 22.92; S, 17.22. Found: C, 47.38; H, 2.59; N, 10.06; O, 22.94; S, 17.27(%).

5-{[(4-(4-hydroyphenyl) thiazol-2-yl)]-thio methyl}-2-(2, 3, 4, 6-tetra-O-acetyl-*S*-β-D-glucopyranosyl)-1, 3, 4- oxadiazole (8g)

Yield 62%, m.p. 220 °C. IR (KBr, ν, cm^{-1}): 3039 (C–H aromatic), 1637 (C = N), 1741 (C = O of O-acetyl gps of aglycone moiety), 1633 (C = N). H1-NMR δH (CDCl3): 13.04 (1H, s, SH), 8.11 (s, 1H, thiazole-H), 6.90–7.55 (4H, m, Ar–H), 5.43 (1H, s, OH), 4.23 (2H, s, CH2). C13 NMR δc (DMSO-d6): 167.7 (N = CS), 163.5 (O–C = N), 155.5 (C = C–N), 116–160 (aromatic 6C-atom), 34.9 (CH2). EI-MS: m/z (%) 532 (3%), 531 (5%), 531 (13%), 530 (26%),530 (1%), 529 (100%) base peak. Anal. Calcd for C22H15N3O7S3: C, 49.90; H, 2.86; N, 7.93; O, 21.15; S, 18.16. Found: C, 49.98; H, 2.89; N, 7.96; O, 21.14; S, 18.17(%).

5-{[(4-(3,4-dimethoyphenyl) thiazol-2-yl)]-thio methyl}-2-(2, 3, 4, 6-tetra-O-acetyl-*S*-β-D glucopyranosyl)-1, 3, 4-oxadiazole (8h)

Yield 60%, m.p. 180 °C. IR (KBr, H aromatic), 1649 (C = N), 2854 (glucosidic C6 –H str), 1723 (C = O of O- acetyl gps of aglycone moiety), 1643 (C = N). H1-NMR δH (CDCl3): 13.06 (1H, s, SH), 8.07 (s, 1H, thiazole-H), 6.95–7.35 (3H, m, Ar–H), 3.90 (6H, s, OCH3), 4.21 (2H, s, CH2). C13 NMR δc (DMSO-d6): 163.8 (O–C = N), 155.3 (C = C–N), 113–150 (aromatic 6C-atom), 112.7 (C = C–S) (thiazole), 56.3 (OCH3), 34.7 (CH2). EI-MS: m/z (%) 577 (1%), 576 (4%), 575 (5%), 575 (13%), 574 (26%), 574 (3%), 573 (100%) base peak. Anal. 8 Calcd for C24H19H19N3O8S3: C, 50.25; H, 3.34; N, 7.33; O, 22.31; S, Found: C, 50.28; H, 3.39; N, 7.36; O, 22.34; S, 16.78(%).

5-{[(4-(3-nitrophenyl) thiazol-2-yl)]-thio methyl}-2-(2, 3, 4, 6-tetra-O-acetyl-*S*-β-D-glucopyranosyl)-1, 3, 4-oxadiazole (8i)

Yield 65%, m.p. 250 °C. IR (KBr, H aromatic), 1631 (C = N), 1744 (C = O of O-acetyl gps of aglycone moiety), 1654 (C = N). H1-NMR δH (CDCl3): 13.08 (1H, s, SH), 8.06 (s, 1H, thiazole-H), 7.80–8.70 (4H, m, Ar–H), 4.25 (2H, s, CH2). C13 NMR δc (DMSO-d6): 163.5 (O–C = N), 155.7 (C = C–N), 122–150 (aromatic 6C-atom), 34.9 (CH2). EI-MS: m/z (%) 561 (3%), 560 (5%), 559 (13%), 559 (26%), 558 (1%), 558 (100%) base peak. Anal. Calcd for C22H14N4O8S3: C, 47.31; H, 2.53; N, 10.03; O, 22.92; S, 17.22. Found: C, 47.38; H, 2.51; N, 10.06; O, 22.94; S, 17.27(%). 1

5-{[(4-(2-hydroyphenyl) thiazol-2-yl)]-thio methyl}-2-(2, 3, 4, 6-tetra-O-acetyl-*S*-β-D-glucopyranosyl)-1, 3, 4- oxadiazole (8j)

Yield 60%, m.p. 200 C. IR (KBr, ν, cm^{-1}): 3044 (C–H aromatic), 1621 (C = N), 1751 (C = O of O-acetyl gps of aglycone moiety), 1641 (C = N). H1-NMR δH (CDCl3): 13.03 (1H, s, SH), 8.12 (s, 1H, thiazole-H), 6.90–7.55 (4H, m, Ar–H), 5.40 (1H, s, OH), 4.21 (2H, s, CH2). C13 NMR δc (DMSO-d6): 167.8 (N = C–S), 163.4 (O–C = N), 155.4 (C = C–N), 116–160 (aromatic 6C-atom), 34.8 (CH2). EI-MS: m/z (%) 532 (3%), 531 (5%), 531 (13%), 530 (26%), 530 (1%), 529 (100%) base peak. Anal.

Calcd for C22H15N3O7S3: C, 49.90; H, 2.86; N, 7.93; O, 21.15; S, 18.16. Found: C, 49.92; H, 2.88; N, 7.97; O, 21.17; S, 18.18(%).

5-{[(4-(substituted phenyl) thiazol-2-yl)]-thio methyl}-2-(2-*S*-β-D-glycopyranosyl)-1,3,4-oxadiazoles (9a-j)

The acetylated thio-glycoside (8a-j) (5 mmol) was treated with a saturated methanolic ammonia solution (15 mL) for 10 min at 0 °C with stirring. The reaction mixture was stirred at room temperature for a further 7 h. The reaction was finished when the solvent was withdrawn under reduced pressure to yield yellowish residue, as indicated by TLC (petroleum ether: hexane, 2:1). After that, it was treated with 20 mL of cold diethyl ether while being agitated to create a solid product, which was then filtered, dried, and recrystallized from methanol (9a-j).

5-{[(4-(phenyl) thiazol-2-yl)]-thio methyl}-2-(2-*S*-β-D-glycopyranosyl)-1, 3, 4-oxadiazole (9a)

Yield 79%, m.p. 170 °C. IR (KBr, 1):3450–3387 (OH), 3049 (CH aromatic), 2490 (CH aliphatic), 1632 (C = N). H^1-NMR δH (CDCl3): 13.09 (1H, s, SH), 8.10 (s, 1H, thiazole-H), 7.40–7.80 (5H, m, Ar–H), 4.22 (2H, s, CH_2). C13 NMR δc (DMSO-d6): 167.4 (N = C–S), 163.5 (O–C = N), 155.2 (C = C–N), 127–135 (aromatic 6C-atom), 34.9 (CH_2). EI-MS: m/z 472% (3%), 471 (3%), 471 (13%), 470 (19%), 470 (3%), 469 (100%) base peak. Anal. Calcd for $C_{18}H_{19}N_3O_6S_3$: C, 46.04; H, 4.08; N, 8.95; O, 20.44; S, 20.49. Found: C, 46.08; H, 4.09; N, 8.96; O, 20.74; S, 20.55(%).

5-{[(4-(4-bromophenyl) thiazol-2-yl)]-thio methyl}-2-(2-*S*-β-D-glycopyranosyl)-1, 3, 4-oxadiazole (9b)

Yield 62%, m.p. 180 °C. IR (KBr, 1): 3480–3395 (OH), 3123 (CH aromatic), 3010 (CH aliphatic), 1643 (C = N).

H1-NMRδH (CDCl3): 13.07 (1H, s, SH), 8.11 (s, 1H, thiazole-H), 7.70–7.85 (4H, m, Ar–H), 4.21 (2H, s, CH2). C13 NMR δc (DMSO-d6):163.4 (O–C = N), 155.1 (C = C–N), 123–135 (aromatic 6C-atom), 34.5 (CH2). EI-MS: m/z (%) 551 (2%), 550 (3%), 550 (13%), 549 (17%), 549 (5%), 548 (3%), 548 (100%) base peak, 547 (17%), 546 (90%). Anal. Calcd for C18H18BrN3O6S3: C, 39.42; H, 3.31; Br, 14.57; N, 7.66; O, 17.50; S, 17.54. Found: C, 39.48; H, 3.39; Br, 14.58; N, 7.69; O, 17.54; S, 17.55(%).

5-{[(4-(4-chlorophenyl) thiazol-2-yl)]-thio methyl}-2-(2-*S*-β-D-glycopyranosyl)-1, 3, 4-oxadiazole (9c)

Yield 64%, m.p. 185 °C. IR (KBr, 1): 3392–3325 (OH), 3133 (CH aromatic), 3043 (CH aliphatic), 1649 (C = N). H1-NMR δH (CDCl3): 13.02 (1H, s, SH), 8.08 (s, 1H, thiazole-H), 7.60–8.05 (4H, m, Ar–H), 4.22 (2H, s, CH2). C13 NMR δc (DMSO-d6): 167.3 (N = C–S), 163.3 (O–C = N), 155.3 (C = C–N), 129–135 (aromatic 6C-atom), 34.3 (CH2). EI-MS: m/z (%) 508 (1%), 507 (1%), 507 (5%), 506 (6%), 506 (4%), 505 (3%), 505 (45%), 504 (19%), 504 (3%), 503 (100%) base peak. Anal. Calcd for C18H18ClN3O6S3: C, 42.90; H, 3.60; Cl, 7.03; N, 8.34; O, 19.05; S, 19.09 Found: C, 42.98; H, 3.63; Cl, 7.09; N, 8.36; O, 19.04; S, 19.07(%).

5-{[(4-(4-methoxyphenyl) thiazol-2-yl)]-thio methyl}-2-(2-*S*-β-D-glycopyranosyl)- 1, 3, 4-oxadiazole (9d)

Yield 65%, m.p. 195 °C. IR (KBr, 1): 3482–3315 (OH), 3139 (CH aromatic), 3052 (CH aliphatic), 1654 (C = N).

H1 NMR δH (CDCl3): 13.08 (1H, s, SH), 8.09 (s, 1H, thiazole-H), 7.10–7.65 (4H, m, Ar–H), 4.24 (2H, s, CH2), 3.87 (3H, s, OCH3). C13 NMR δc (DMSO-d6): 163.4 (O–C = N), 155.4 (C = C–N), 114–160 (aromatic 6C-atom), 112.7 (C = C–S) (thiazole), 55.10 (OCH3), 34.3 (CH2). EI-MS: m/z (%) 502 (3%), 501 (4%), 501 (13%), 500 (21%), 500 (3%), 499 (100%) base peak. Anal. Calcd for C19H21N3O7S3: C, 45.68; H, 4.24; N, 8.41; O, 22.42; S, 19.26. Found: C, 45.70; H, 4.29; N, 8.46; O, 22.44; S, 19.25(%).

5-{[(4-(4-methylphenyl) thiazol-2-yl)]-thio methyl}-2-(2-*S*-β-D-glycopyranosyl)- 1, 3, 4-oxadiazole (9e)

Yield 60%, m.p. 210 °C. IR (KBr, 1): 3390–3343 (OH), 3145 (CH aromatic), 3076 (CH aliphatic), 1658 (C = N).

H1-NMRδH (CDCl3): 13.05 (1H, s, SH), 8.11 (s, 1H, thiazole-H), 7.30–7.85 (4H, m, Ar–H), 4.22 (2H, s, CH2), 2.37 (3H, s, CH3). C13 NMR δc (DMSO-d6): 163.8 (O–C = N), 155.7 (C = C–N), 125–135 (aromatic 6C-atom), 112.6 (C = C–S) (thiazole), 34.3 (CH2), 21.10 (CH3). EI-MS: m/z (%) 486 (2%), 485 (2%), 485 (15%), 484 (24%), 483 (100%) base peak. Anal. 17 Calcd for C19H21N3O6S3: C, 47.19; H, 4.38; N, 8.69; O, 19.85; S, 19.89. Found: C, 47.18; H, 4.39; N, 8.68; O, 19.84; S, 19.87(%).

5-{[(4-(4-nitrophenyl) thiazol-2-yl)]-thio methyl}-2-(2-*S*-β-D-glycopyranosyl)- 1, 3, 4-oxadiazole (9f)

Yield 60%, m.p. 200 °C. IR (KBr, ν, cm^{-1}): 3472–3355 (OH), 3154 (C–H aromatic), 3050 (C–H aliphatic), 1645 (C = N). H1-NMRδH (CDCl3): 13.08 (1H, s, SH), 8.12 (s, 1H, thiazole-H), 8.10–8.35 (4H, m, Ar–H), 4.20 (2H, s, CH2). C13 NMR δc (DMSO-d6): 167.2 (N = C–S), 163.4 (O–C = N), 155.6 (C = C–N), 124–150 (aromatic 6C-atom), 112.7 (C = C–S) (thiazole), 34.2 (CH2), 21.10 (CH3). EI-MS: m/z (%) 517 (2%), 516 (1%), 516 (2%), 515 (23%), 514 (100%) base peak. Anal. Calcd for C18H18N4O8S3: C, 42.02; H, 3.53; N, 10.89; O, 24.88; S, 18.69. Found: C, 42.08; H, 3.59; N, 10.86; O, 24.89; S, 18.71 (%).

5-{[(4-(4-hydroxyphenyl) thiazol-2-yl)]-thio methyl}-2-(2-*S*-β-D-glycopyranosyl)-1, 3, 4-oxadiazole (9g)

Yield 65%, m.p. 260 °C. IR (KBr, ν, cm^{-1}): 3477–3315 (OH), 3155 (C–H aromatic), 3032 (C–H aliphatic), 1652 (C = N). H1-NMR δH (CDCl3): 13.04 (1H, s, SH), 8.11 (s, 1H, thiazole-H), 6.90–7.55 (4H, m, Ar–H), 5.43 (1H, s, OH), 4.23 (2H, s, CH2). C13 NMR δc (DMSO-d6): 167.7 (N = C–S), 163.5 (O–C = N), 155.5 (C = C–N), 116–160 (aromatic 6C-atom), 34.9 (CH2). EI-MS: m/z (%) 488 (2%), 487 (1%), 487 (2%), 486 (23%), 485 (100%) base peak. Anal. Calcd for C18H19N3O7S3: C,

44.52; H, 3.94; N, 8.65; O, 23.07; S, 19.81. Found: C, 44.58; H, 3.99; N, 8.66; O, 23.09; S, 19.87(%).

5-{[(4-(3,4-dimethoxyphenyl) thiazol-2-yl)]-thio methyl}-2-(2-*S*-β-D-glycopyranosyl)-1, 3, 4-oxadiazole (9h)

Yield 60%, m.p. 245 °C. IR (KBr, ν, cm^{-1}): 3490–3345 (OH), 3147 (C–H aromatic), 3058 (C–H aliphatic), 1667 (C = N). H1-NMR δH (CDCl3): 13.06 (1H, s, SH), 8.07 (s, 1H, thiazole-H), 6.95–7.35 (3H, m, Ar–H), 3.90 (6H, s, OCH3), 4.21 (2H, s, CH2). C13 NMR δc (DMSO-d6): 163.8 (O–C = N), 155.3 (C = C–N), 113–150 (aromatic 6C-atom), 112.7 (C = C–S) (thiazole), 56.3 (OCH3), 34.7 (CH2). EI-MS: m/z (%) 532 (3%), 531 (4%), 531 (13%), 530 (22%), 530 (3%), 529 (100%) base peak. Anal. Calcd for C20H23N3O8S3: C, 45.36; H, 4.38; N, 7.93; O, 24.17; S, 18.16. Found: C, 45.38; H, 4.39; N, 7.96; O, 24.14; S, 18.18(%).

5-{[(4-(3-nitrophenyl) thiazol-2-yl)]-thio methyl}-2-(2-*S*-β-D-glycopyranosyl)-1, 3, 4-oxadiazole (9i)

Yield 60%, m.p. 200 °C. IR (KBr, ν, cm^{-1}): 3472–3375 (OH), 3112 (C–H aromatic), 3043 (C–H aliphatic), 1620 (C = N). H1-NMR δH (CDCl3): 13.08 (1H, s, SH), 8.06 (s, 1H, thiazole-H), 7.80–8.70 (4H, m, Ar–H), 4.25 (2H, s, CH2). C13 NMR δc (DMSO-d6): 167.7 (N = C–S), 163.5 (O–C = N), 155.7 (C = C–N), 122–150 (aromatic 6C-atom), 34.9 (CH2). EI-MS: m/z (%) 517 (2%), 516 (1%), 516 (2%), 516 (13%), 515 (23%), 514 (100%) base peak. Anal. Calcd for C18H18N4O8S3: C, 42.02; H, 3.53; N, 10.89; O, 24.88; S, 18.69. Found: C, 42.08; H, 3.55; N, 10.86; O, 24.90; S, 18.67(%).

5-{[(4-(2-hydroxyphenyl) thiazol-2-yl)]-thio methyl}-2-(2-*S*-β-D-glycopyranosyl)-1, 3, 4-oxadiazole (9j)

Yield 64%, m.p. 185 °C. IR (KBr, ν, cm^{-1}): 3423–3320 (OH), 3093 (C–H aromatic), 3063 (CH-aliphatic), 1612 (C = N). H1-NMR δH (CDCl3): 13.03 (1H, s, SH), 8.12 (s, 1H, thiazole-H), 6.90–7.55 (4H, m, Ar–H), 5.40 (1H, s, OH), 4.21 (2H, s, CH2). C13 NMR δc (DMSO-d6): 163.4 (O–C = N), 155.4 (C = C–N), 116–160 (aromatic 6C-atom), 34.8 (CH2). EI-MS: m/z (%) 488 (2%), 487 (1%), 487 (2%), 487 (13%), 486 (23%), 485 (100%) base peak. Anal. Calcd for C18H19N3O7S3: C, 44.52; H, 3.94; N, 8.65; O, 23.07; S, 19.81. Found: C, 44.59; H, 3.98; N, 8.67; O, 23.05; S, 19.88(%).

4 Concluding Remarks

A wide range of techniques have been reported on the synthetic pathway for of oxadiazole derivatives and their S-glucosides incorporating thiazole scaffold. The high-yield synthesis of thiazole-oxadaizole derivatives and their S-glucosides can be achieved by developing a novel practical approach. In order to enhance the biological activities, various novel compounds were synthesized for use in biological applications by changing the substituent at the thiazole ring with the appropriate substituent groups. Due to its unique characteristics and promise, research on thiazole derivatives and S-glucosides may, therefore, be the focus of a further investigation.

References

1. Siddiqui WAN (2011) Synthesis, anticonvulsant and toxicity screening of thiazolylethiadiazole derivatives. Med Chem Res 20:261–268
2. Rawal RK, Tripathi R, Katti S, Pannecouque C, Clercq ED (2008) Design and synthesis of 2(2,6dibromophenyl)-3-heteroaryl-1,3-thiazolidin-4-ones as anti-HIV agents. Eur J Med Chem 43(12):2800–2806
3. Iino T, Tsukahara D, Kamata K, Sasaki K, Ohyama S, Hosaka H, Hasegawa T, Chiba M, Nagata Y, Eiki J-I, Nishimura T (2009) Potent and Discovery of potent and orally active 3-alkoxy-5-phenoxy-Nthiazolyl benzamides as novel allosteric glucokinase activators. Bioorg Med Chem 17(7):2733–2743
4. Shiradkar MR, Akula KC, Dasari V, Baru V, Chiningiri B, Gandhi S, Kaur R (2007) Clubbed thiazoles by MAOS: a novel approach to cyclin-dependent kinase 5/p25 inhibitors as a potential treatment for Alzheimer's disease. Bioorg Med Chem 15(7):2601–2610
5. Parameshwar M, Kumar TP, Tharanikkarasu K (2014) In vitro and in silico antimalarial activity of 2(2hydrazinyl)thiazole derivatives. Eur J Pharm Sci 52:138–145
6. DNC, BN, SH, TA, Desai NC, Bhatt N, Somani H, Trivedi A (2013) Synthesis, antimicrobial and cytotoxic activities of some novel thiazole clubbed 1,3,4-oxadiazoles. Euro J Med Chem 67:54–59
7. Helal M, Salem M, El-Gaby M, Aljahdali M (2013) Synthesis and biological evaluation of some novel thiazole compounds as potential anti-inflammatory agents. Eur J Med Chem 65:517–526
8. Diana P, Carbone A, Barraja P, Montalbano A, Parrino B, Lopergolo A, Pennati M, Zaffaroni N, Cirrincione G (2011) Synthesis and antitumor activity of 3-(2-Phenyl thiazol-4-yl)-1H-indoles and 3-(2Phenyl-1,3thiazol-4-yl)-1H-7-azaindoles. Chem Med Chem 6(7):1300–1309
9. Zhang B-L, Song L-X, Li Y-F, Li Y-L, Guo Y-Z, Zhang E, Liu H-M (2014) Synthesis and biological evaluation of dehydroepiandrosterone-fused thiazole, imidazo[2,1-b]thiazole, pyridine steroidal analogues. Steroids 80:92–101
10. Carbone A, Pennati M, Parrino B, Lopergolo A, Barraja P, Montalbano A, Spanò V, Sbarra S, Doldi V, De Cesare M, Cirrincione G, Diana P, Zaffaroni N (2013) Novel 1H-pyrrolo[2,3-b]pyridine derivatives nortopsentin analogues: synthesis and antitumor activity in peritoneal mesothelioma experimental models. J Med Chem 56(17):7060–7072
11. Parrino B, Attanzio A, Spanò V, Cascioferro S, Montalbano A, Barraja P, Tesoriere L, Diana P, Cirrincione G, Carbone A (2017) Synthesis, antitumor activity and CDK1 inhibiton of new thiazole nortopsentin analogues. Eur J Med Chem 138:371–383
12. Jayarama HN, Pillwein K, Nichols CR, Hoffman R, Weber G (1985) Selective sensitivity to tiazofurin of human leukemic cells. Biochem Pharmacol 35(12):2029–2032
13. Izawa K, Onishi T (2006) Industrial syntheses of the central core molecules of HIV protease inhibitors. Chem Rev 106(7):2811–2827
14. Bankar GR, Nandakumar K, Nayak PG, Thakur A, Chamallamudi MR, Nampurath GK (2009) Vasorelaxant effect in rat aortic rings through calcium channel blockage: a preliminary in vitro assessment of a 1,3,4-oxadiazole derivative. Chem Biol Interact 181(3):377–382
15. SAKNS, Gilani SJ (2010) Synthesis and pharmacological evaluation of condensed heterocyclic 6substituted 1,2,4-triazolo-[3,4-b]-1,3,4-thiadiazole and 1,3,4-oxadiazole derivatives of isoniazid. Bioorgan Med Chem Lett 20(16):4762–4765

16. AZRZ, Hajimahdi MRAZ (2013) Synthesis, biological evaluation, and molecular modeling studies of new 1,3,4-oxadiazole- and 1,3,4-thiadiazole-substituted 4-oxo-4H-pyrido[1,2-a]pyrimidines as anti-HIV-1 agents. Med Chem Res 22:2467–2475
17. SATMFAAPNVZ, Afshin Zarghi AS (2005) Synthesis and anticonvulsant activity of new 2substituted-5-(2-benzyloxyphenyl)-1,3,4-oxadiazoles. Bioorgan Med Chem Lett 15(7):1863–1865
18. Li Y, Liu J, Zhang H, Yang X, Liu Z (2006) Stereoselective synthesis and fungicidal activities of (E)α(methoxyimino)-benzeneacetate derivatives containing 1,3,4-oxadiazole ring. Bioorg Med Chem Lett 16(8):2278–2282
19. Madhu Sekhar M, Nagarjuna U, Padmavathi V, Padmaja A, Reddy N, Vijaya T (2018) Synthesis and antimicrobial activity of pyrimidinyl 1,3,4-oxadiazoles, 1,3,4-thiadiazoles and 1,2,4-triazoles. Euro J Med Chem 145(10):1–10
20. Yadav M, Shirude S, Puntambekar D, Patel P, Prajapati H, Parmar A, Balaraman R, Giridhar R (2007) Studies in 3,4-diaryl-1,2,5-oxadiazoles and their N-oxides: search for better COX-2 inhibitors. Acta Pharm 57(1):13–30
21. NHIWJSIFRSMK, Muhammad Taha MZ (2016) Synthesis of 2-(2-methoxyphenyl)-5phenyl1,3,4-oxadiazole derivatives and evaluation of their antiglycation potential. Med Chem Res 25:225–234
22. Palaska E, Şahin G, Kelicen P, Durlu N, Altinok G (2002) Synthesis and anti-inflammatory activity of 1acylthiosemicarbazides, 1,3,4-oxadiazoles, 1,3,4-thiadiazoles and 1,2,4-triazole-3-thiones. Il Farmaco 57(2):101–107
23. Abd-Ellah HS, Abdel-Aziz M, Shoman ME, Beshr EA, Kaoud T, Ahmed A-SF (2017) New 1,3,4oxadiazole/oxime hybrids: design, synthesis, anti-inflammatory, COX inhibitory activities and ulcerogenic liability. Bioorg Chem 74:15–29
24. Elgemeie GH, Mahdy EM, Elgawish MA, Ahmed MM, Shousha WG, Eldin ME (2010) A new class of antimetabolites: pyridine thioglycosides as potential anticancer agents. Z. Naturforsch 65(c):577–587
25. Elgemeie G, Abu-Zaied M, Azzam R (2016) Antimetabolites: a first synthesis of a new class of cytosine thioglycoside analogs. Nucleosides, Nucleotides and Nucleic Acids 1–12
26. FAAKE.-BOHG, Elgemeie GH (2014) Design, synthesis and cytotoxic evaluation of novel heterocyclic thioglycosides. Med Chem 4(12):814–820
27. El-Sayed WA, El-Sofany WI, Hussein HAR, Fathi NM (2017) Synthesis and anticancer activity of new [(Indolyl)pyrazolyl]-1,3,4-oxadiazole thioglycosides and acyclic nucleoside analogs. Nucleosides, Nucleotides Nucleic Acids 1–22
28. Elgemeie GH, Zaghary WA, Amin KM, Nasr TM (2009) First synthesis of thiophene thioglycosides. J Carbohydr Chem 28(3):161–178
29. MME.-EMI, Galal KA, Elgemeie H (2002) Nucleic acid components and their analogues: a novel and efficient method for the synthesis of a new class of bipyridyl and biheterocyclic-nitrogen thioglycosides from pyridine-2(1h)-thiones. Nucleosides, Nucleotides Nucleic Acids 21(6):477–493
30. Elgemeie GH, Zaghary WA, Nasr TM, Amin KM, Elgemeie H, Zaghary WA, Nasr TM, Amin KM (2008) First synthesis of thienopyrazole thioglycosides. J Carbohydr Chem 27(6):345–356
31. Gaidhane MK, Ghatole AM, Lanjewar KR, Hatzade KM (2020) Innovational combination of hetero-bifunctional N-PEG quinoline scaffolds derivatives with improved anticancer activity against breast and colon cancer cell lines and P-glycoprotein, cytochrome p450 enzyme activity prediction. Turk J Chem 44(6):1495–1514
32. Palak KP, Hiren MM, Dhrubo JS (2011) Synthesis and biological evaluation of 1,3,4-oxadiazole derivatives as potential antibacterial and antifungal agents. Int J Drug Dev Res 3(2):248–255
33. Shridhar AH, Keshavayya J, Joy HH (2012) Synthesis of 1,3,4-oxadiazole incorporated azo dye derivatives as a potent biological activity molecules. Int J Pharm and Pharm Sci 4:386–390

Microwave-Assisted Synthesis of Monohydroxy Pendant Polybenzobisthiazole Thermoplastic Polymer

Roli Verma, Kirti Srivastava, and R. S. Jagadish

1 Introduction

Rigid-rod-like aromatic heterocyclic polymers have attracted constant interest due to its exceptional combination of thermal stability and great resistance to adverse conditions. Rigid-rod ordered polymers are referred to poly[(benzo1,2-d:5,4-dibisoxazole-2,6-diyl)-1,4-phenylene] (PBO) and poly[benzo1,2-d:4,5-dibisthiazole-2,6-diyl)-1,4-phenylene (PBZT) [1–3]. Outstanding mechanical and thermal properties are displayed by these polymers [4–10]. PBO fibers have specific moduli and strength 9.4 and 9.0 times that of steel and tensile strengths and moduli as high as 5.6 and 352 GPA respectively [11, 12]. Both PBO and PBZT have high thermal stability and do not show softening behavior before thermal degradation, which takes place above 600 °C in air and nearly 700 °C in an inert atmosphere. Thermal degradation occurs above these temperatures [7, 13]. The polymer backbone is characterized by conjugated benzobisazole and phenyl rings, and it induces an extended delocalization of pi electrons. Conformational flexibility within the polymer arises solely from the rotation of bonds between phenylene and heterocyclic groups. These changes help clarify the relationship between the polymer structure and conformational flexibility.

Organic chemistry research increasingly relies on microwave-assisted synthesis. The key advantage of MW-assisted synthesis lies in its remarkable ability to accelerate reaction rates, particularly when conducted in unconventional reaction media. Water, known for its high dielectric constant, has garnered significant attention as a nonflammable and cost-efficient alternative to conventional organic solvents. Its high dielectric constant enables efficient absorption of microwave irradiation. This approach simplifies and expedites the synthetic process, making repeated

R. Verma (✉) · K. Srivastava · R. S. Jagadish
Department of Chemistry, JSS Academy of Technical Education, C-20/1, Sector 62, Noida, U.P., India
e-mail: roliverma@jssaten.ac.in

S. Rattan et al. (eds.), *Emerging Trends in Synthesis and Catalysis in Chemistry*,
Springer Proceedings in Materials 51, https://doi.org/10.1007/978-981-97-5169-3_21

experiments more manageable and faster with fewer steps. Compared to alternative methods, microwave-assisted synthesis offers several additional advantages, including faster reaction times, higher yields, and improved heating uniformity, while also minimizing the formation of byproducts. The development of heteroaromatic polymers dates back to the 1960s, driven by the increasing demand for heat-resistant plastics for military and space applications. These polymers, exemplified by rigid-rod structures like PBX, PBIS, PBZTS, and PBOS, distinguish themselves from the latest generation of structural polymeric materials by their ability to withstand high temperatures and harsh conditions. In this article, we present the synthesis of monohydroxy pendant PBZT using microwave-assisted methods. The result was a high-molecular weight polymer with excellent thermal properties, achieved in a remarkably short reaction time.

2 Experimental

2.1 Materials and Methods

The conventional synthesis of model compound and monohydroxy pendant PBZT polymer followed previously reported procedures [2]. Microwave-assisted reactions were conducted using a domestic microwave oven (Samsung M183DN) at power settings of 100, 200, and 300 W [14].

All solvents and reagents used for synthesis were of general reagent grade and were purchased from Lancaster. Elemental analysis for carbon (C), hydrogen (H), nitrogen (N), and sulfur (S) was performed using the Elementar Vario EL. Infrared spectra of model compounds and polymers were recorded using a Nicolet MX-1 FT-IR spectrometer. The ^{13}C solid-state NMR spectra were provided by SIF, IISc Bangalore, and were obtained using the Brucker-DSX 300 Solid State NMR instrument (carbon frequency: 75.45 MHz). Thermogravimetric analysis (TGA) and differential scanning calorimetry (DSC) were carried out on powdered polymer samples to determine their relative thermal stability. A Du Pont 1090 thermal analyzer with a 951 TG module was used for TGA, and thermal transitions were evaluated using a Du Pont 9900 thermal analyzer with a 910 DSC module. The heating rate was set to 200 °C/min, and the measurements were conducted in a nitrogen atmosphere.

2.2 Synthesis of Base Compound 1,3-Bis-(2-Benzothiazoly)-5-Hydroxy Phenylene (3)

The model compound was synthesized using the method reported by Roli et al. [15]. The reaction scheme is illustrated in Fig. 1. The yield obtained was 3.16 gm representing a quantitative yield. M.P. was determined to be—170 °C. The calculated

Fig. 1 Reaction setup for MW synthesis

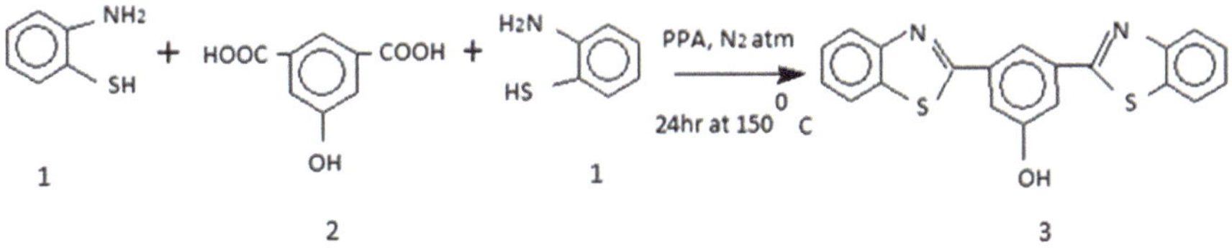

Fig. 2 Scheme of synthesis of 1,3-bis-(2-benzothiazoly)-5-hydroxy phenylene (3)

elemental composition for $C_{20}H_{12}N_2S_2$ is as follows: C: 69.75%, H: 3.47%, and N: 8.12%. The actual composition found through analysis was: C: 68.97%, H: 3.11%, and N: 6.68.

2.3 *Synthesis of Poly{(Benzo [1, 2-d: 4, 5-d']-Bisthiazole 2,6-Diyl) 1,3(5–Hydroxy Phenylene)} (5)*

2.3.1 Method A (Conventional Method)

The polymer was synthesized by the conventional method as reported by Roli et al. [15] The yield obtained was 3.50 g. The inherent viscosity was measured as 4.35 dL/g in MSA at room temperature. The reaction scheme is depicted in Fig. 2. The calculated elemental composition for $(C_{14}H_6N_2S_2)_n$ is as follows: C: 63.13%, H: 2.26%, and N: 10.53%. The actual composition found through analysis was: C: 62.17%, H: 2.22%, and N: 9.12.

2.3.2 Method B (MW-Assisted Synthesis)

The use of microwaves (MW) in organic mixtures is considered a green approach as it typically enhances reaction efficiency. This technique has also been extended

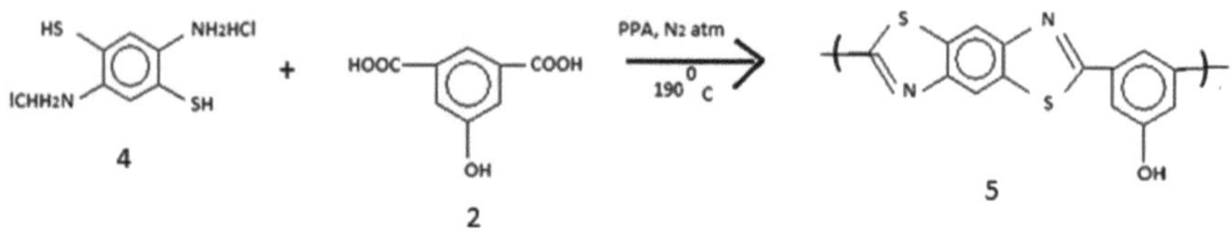

Fig. 3 Scheme of synthesis of poly{(benzo [1,2-d: 4,5-d′]-bisthiazole 2,6-diyl) 1,3(5–hydroxy phenylene)} (5) [15]

to the synthesis of polycondensation polymers. The synthesis of polybenzimidazoles (PBIs) using microwave-assisted solution polycondensation in polyphosphoric acid (PPA) has been previously reported [16, 17]. For the polycondensation reaction between 2,5-diamino-1,4-benzenedithiol dihydrochloride and 5-hydroxy isophthalic acid, the same molar quantities as in the conventional method were used [15] (Fig. 3). The reaction mixture was stirred for 24 h at room temperature in the presence of N_2, followed by an additional 24 h at temperatures ranging from 60 to 70 °C to dehydrate the amine monomer. After dehydrochlorination, the temperature was reduced to 50 °C, resulting in a clear and bubble-free solution. Under a nitrogen environment, 5-hydroxy isophthalic acid and more P_2O_5 were added to generate 82–84% polyphosphoric acid and compensate for H_2O formation during condensation polymerization. The reaction flask was then exposed to microwave radiation at three different power levels—100, 200, and 300W—to reach a temperature of 160 °C. The resulting polymers were separated and dried using the traditional procedure. Notably, chlorosulfonic acid and methanesulfonic acid only partially solubilized the polymer.

3 Result and Discussion

Monohydroxy PBZT was synthesized using two different methods: conventional and microwave-assisted syntheses, as illustrated in Fig. 2. The conventional method yielded a product with a 72% yield, while the microwave-assisted synthesis resulted in an 85% yield. Notably, the conventional synthesis required 75 h to complete, whereas the microwave-assisted synthesis finished in just 48 h and 15 minutes. This represents a significant reduction in synthesis time. However it is important to note that there were differences in solubility between the two synthesized polymers. The conventionally synthesized polymers exhibited solubility in chlorosulfonic acid and methanesulfonic acid, whereas the polymers synthesized using microwaves showed limited solubility in these solvents. The comparative parameters are given in Table 1.

Table 1 Comparison of various parameters of conventional and Microwave PBZT

Parameters	Conventional PBZT	Microwave PBZT
Total reaction time	75 h	48 h 15 min
% yield	72	85
Solubility	Methane sulfonic acid and chlosulfonic acid	Sparingly soluble in methane sulfonic acid and chlorosulfonic acid
Inherent viscosity(η_{inh} dL/g)	4.2	–

3.1 *Infrared Analysis*

The FT-IR spectrum of the model compound (3), 1,4-bis-(2-benzothiazolyl)-1,3(5-hydroxy) phenylene, exhibits significant changes. It shows the complete disappearance of NH_2 at 3320 cm^{-1}, SH at 2460 cm^{-1} (from 2-aminothiophenol), and C = O at 1700 cm^{-1} (from dicarboxylic acid). In addition, the spectrum displays various absorption bands at 3180, 3058, 1653, 1591, 1496, 1425, 1318, 1273, 1239, 1186, 1093, 978, 893, 857, 753, 722, and 674 cm^{-1}. The strong bands in the 1300–1500 cm^{-1} region suggest the creation of the benzdithiazole ring and are attributed to hetero ring stretching vibrations of varying intensities. The bands at 3180.3 and 3059 cm^{-1} are designated for (OH) stretching, confirming the formation of the model compound.

In contrast the FT-IR spectrum (Fig. 4b) of the polymer (5), poly{(benzo [1,2-d: 4,5-d']-bisthiazole 2,6-diyl) 1,3(5–hydroxy phenylene)}, synthesized by conventional methods, exhibits absorptions at 3432, 1652, 1560, 1513, 1458, 1392, 1288, 1212, 983, 878, 748, and 675 cm^{-1}. Notably, it shows an intense and broad band around 3400 cm^{-1}, indicating the presence of a normally molecularly bonded –OH group, confirming the hydrogen bonding in the monohydroxy PBZT molecule. Consequently, the spectra align well with the recommended structure of the polymer. The spectra (Fig. 4c) obtained for the MW-synthesized polymer are similar to those of the conventional polymer. The main difference is that the absorption bands are sharper, indicating that cyclization under MW conditions occurred very rapidly under these conditions.

3.2 *^{13}C Solid NMR*

Figure 5a and b presents the ^{13}C NMR spectra of polymers (3) and (5), respectively. Since the polymers were insoluble in common deuterated solvents, solid-state NMR analysis was employed. These spectra exhibit the characteristic chemical shifts of benzo-fused five-membered heterocycles with two heteroatoms bound to a side chain, as commonly found in benzothiazoles. The integrals of the bands in the spectra essentially reflect the corresponding carbon ratios in the repeat unit of the polymer chain.

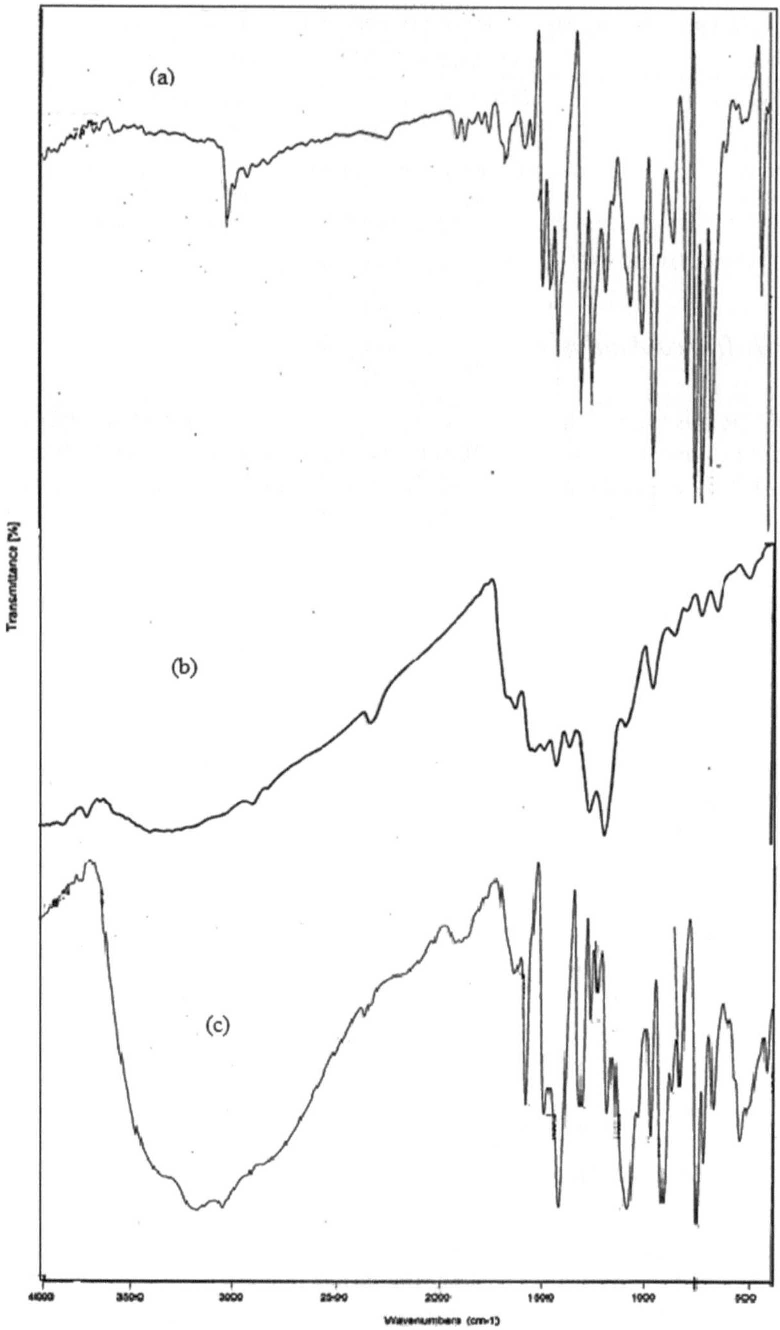

Fig. 4 Comparative IR of **a** model compound1,4-bis-(2-benzothiazolyl)–1,3 (5-hydroxy) phenylene (3), poly{(benzo [1,2-d: 4,5-d′]-bisthiazole 2,6-diyl) 1,3(5–hydroxy phenylene)}(5), **b** conventional method, and **c** MW-assisted

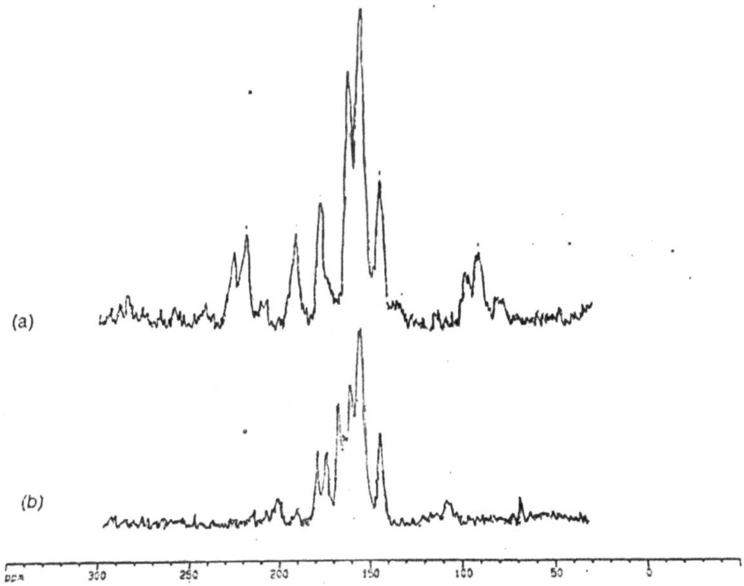

Fig. 5 NMR spectra of poly{(benzo [1, 2-d: 4, 5-d′]-bisthiazole 2, 6-diyl) 1, 3(5–hydroxy phenylene)} (5) **a** MW and **b** conventional

The conventionally synthesized polymer displays the typical ^{13}C chemical shifts for the benzothiazole carbons, consistent with previous reports [15]. In Fig. 5a, the monohydroxy PBZT NMR spectra, which correlate with the cyclized benzothiazole structure, are depicted. Four peaks are visible in the spectrum, appearing at 117.6, 135.06, 153.00, and 170.00 ppm, corresponding to the backbone carbons. Two additional peaks at 163.00 and 142.00 ppm are likely the result of a limited number of incomplete cyclized units. Overall, the ^{13}C signals align well with the suggested structure of the polymer.

3.3 *Thermal Characterization*

According to Differential Scanning Calorimetry (DSC) analysis, polymer did not displayed any expected softening or melting behavior before the breakdown point. This is thought to be a substantial result of the fused ring structure supporting the softening of the polymers based on benzobisthiazole (Fig. 6a, b). The thermogravimetric analysis (TGA) in nitrogen, both polymers, showed a one-step thermal breakdown profile. Fig. 7a, b shows the early weight loss from the ambient temperature to approx. 150 °C is the result of residual water and solvent. The degradation of polymer (5) synthesized by conventional method is around 400 °C and for MW-synthesized polymer. TGA graph (Fig. 7b) indicates that the polymer (5) is heat resistant with the onset of degradation temperature 455 °C in nitrogen. TGA result also confirms the degree of thermal stability of the polymer. The glass transition temperature (Tg) is not indicated by the DSC readings. These polymers' Tg being above their decomposition temperature, as previously suggested in the literature, is caused by the existence of a stiff backbone. Results on solubility provide additional evidence for this. These polymers could not be dissolved in common solvents as dimethylfuran, dimethylsulfoxide, dimethyacetamide, and N-methyl pyrrolidone. Conventionally synthesized polymer was soluble in CSA and MSA, but MW-synthesized polymer was found partially soluble in MSA and CSA.

According to the results from Differential Scanning Calorimetry (DSC) analysis, neither polymer exhibited any expected softening or melting behavior before reaching the breakdown point. This outcome is believed to be a significant result of the fused ring structure that provides support against softening in polymers based on benzobisthiazole, as depicted in Fig. 6a and b.

Thermogravimetric analysis (TGA) conducted in a nitrogen atmosphere revealed a one-step thermal breakdown profile for both polymers. In Fig. 7a and b, the initial weight loss from ambient temperature to approximately 150 °C can be attributed to residual water and solvents. The degradation temperature of the conventionally synthesized polymer (5) occurs around 400 °C. In contrast, the TGA graph (Fig. 7b)

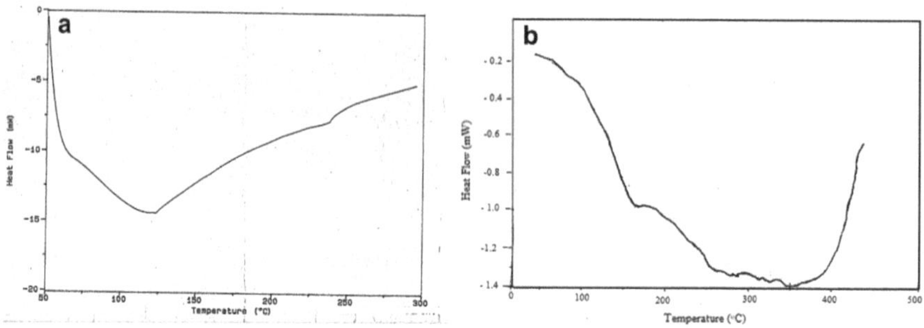

Fig. 6 **a** DSC of polymer 5 by conventional method. **b** DSC of polymer 5 by MW method

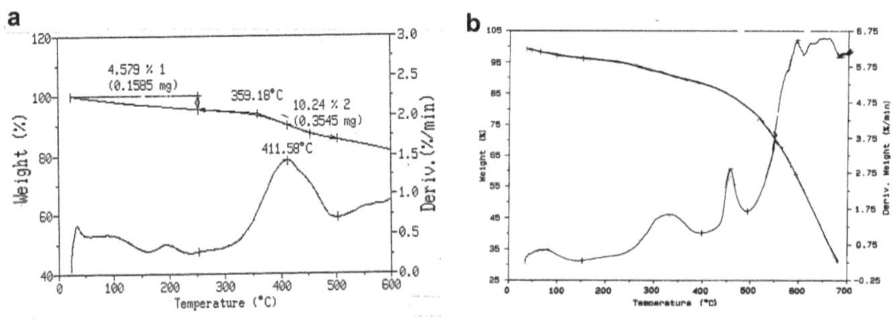

Fig. 7 **a** TGA of polymer 5 by conventional method. **b** TGA of polymer 5 by method

for the MW-synthesized polymer indicates greater heat resistance, with an onset degradation temperature of 455 °C in nitrogen. This TGA result confirms the high degree of thermal stability exhibited by the polymer.

Notably, the glass transition temperature (Tg) could not be detected by the DSC readings. This aligns with previous literature suggestions that the Tg of these polymers is above their decomposition temperature due to the presence of a stiff backbone. Solubility tests further support this finding, as these polymers could not be dissolved in common solvents such as dimethylfuran, dimethylsulfoxide, dimethylacetamide, and N-methyl pyrrolidone. The conventionally synthesized polymer was soluble in chlorosulfonic acid (CSA) and methanesulfonic acid (MSA), while the MW-synthesized polymer was only partially soluble in MSA and CSA.

4 Conclusion

We have discussed the microwave-assisted synthesis of 2,5-diamino-1,4-benzenedithiol dihydrochloride (DABDT.2HCl) with 5-hydroxyisophthalic acid in polyphosphoric acid (PPA) under a nitrogen (N_2) atmosphere. The results of our investigation have demonstrated that microwave-assisted synthesis yields higher quantities in significantly less time compared to conventional. Spectroscopic analysis has unveiled a fully planar structure, a consequence of strong intermolecular hydrogen interactions between the hydroxyl moieties and nitrogen atoms of adjacent benzothiazole units. These interactions contribute to the planarity of the structure. The resulting polymers exhibited limited solubility in both chlorosulfonic and methanesulfonic acid.

References

1. Wolfe J, Arnold FE (1981) Rigid-rod polymers. 1. Synthesis and thermal properties of para-aromatic polymers with 2,6-benzobisoxazole units in the main chain. Macromolecules 14(4):909–915
2. Wolfe JF, Loo BH, Arnold FE (1981) Rigid-rod polymers. 2. Synthesis and thermal properties of para-aromatic polymers with 2,6-benzobisthiazole units in the main chain. Macromolecules 14(4):915–920
3. Adams WW, Eby RK, Mclemore DE (1989) The materials science and engineering of rigid-rod polymers. In: Material resources of society symposium processing, vol 134. Eds., Pittsburgh, PA
4. Kumar S (1990) Lee SM (ed) International encyclopedia of composites, vol 4, VCH, New York, pp 51
5. Wolfe JF (1988) Encyclopedia of polymer science and engineering, vol 11. Wiley, New York pp 601
6. Kumar S (1996) Salamone JC (ed) The polymer materials encyclopedia, vol 10. CRC Press, pp 7512
7. Denny LR, Goldfarb IJ, Soloski EJ (1989) Mater Res Soc Symp Proc 134:395
8. Yang HH (1989) Aromatic high strength fibers. John Wiley, New York, pp 873

9. Hearle JWS (2001) High-performance fibers. CRC Woodhead Publishing Ltd, Cambridge, UK
10. Bourbigot S, Flambard X (2002) Heat resistance and flammability of high performance fibres. Rev Fire Mater 26, 155
11. Kitagawa T, Yabuki K, Young RJ (2001) An investigation into the relationship between processing, structure and properties for high-modulus PBO fibres. Part1. Raman band shifts and broadening in tension and compression Polymer, vol 42, pp 2101
12. So Y-H (2000) Rigid-rod polymers with enhanced lateral interactions. Prog Polym Sci 25:137
13. Kuroki T, Tanaka Y, Hokudoh T, Yabuki TK (1997) Heat resistance properties of poly(*p*-phenylene-2,6-benzobisoxazole) fiber. J Appl Polym Sci 65:1031
14. Olvera-Mancilla J, palacios-Alquisira J, Alexandrova L (2017) Eaton's reagent in polybenimidazole synthesis: the influence of temperature and microwave irradiation. High Perform Poly 306(6):699–709
15. Verma R, Srivastava K, Jagadish RS, Dubey R (2023) Synthesis and characterization of pseudo ladder mono hydroxy pendant polybenzobisthiazole aromatic hetero-cyclic thermoplastic polymer. Rasayan J Chem 16(2):964–971
16. Yang J, He R, Che Q, Gao X, Shi L (2010) A copolymer of poly[2,2′-(*m*-phenylene)-5,5′-bibenzimidazole] and poly(2,5-benzimidazole) for high-temperature proton-conducting membranes 59(12):1695–1700
17. Kang Y, Zou J, Sun Z, Wang F, Zhu H, Han K, Yang W, Song H, Meng Q (2013) Polybenzimidazole containing ether units as electrolyte for high temperature proton exchange membrane fuel cells. Int J Hydrogen Energy 38(15):6494–6650

Synthesis and Physicochemical Assessment of Anionic Polymeric Surfactant from Palm Oil for Improved Oil Recovery

Neha Saxena, Md. Merajul Islam, Prathibha Pillai, and Deepa Sharma

1 Introduction

As the demand for crude oil rises throughout the world and output falls from ageing fields, researchers have begun to focus more on improved oil recovery techniques. Polymers are commonly employed in EOR because water-soluble polymers can increase the volumetric sweep efficiency [1–4] by adjusting the water/oil mobility ratio. By decreasing the interfacial tension (IFT) between crude oil and water and altering the wettability of oil-wet rock to water-wet rock [5–8], surfactants also play a significant role in EOR. Polymeric surfactants, which combine polymer and surfactant characteristics, have received a lot of interest as of late.

Since increasing the viscosity of the surfactant solution and exhibiting favourable rheological behaviours is crucial to increasing the swept volume of displacement fluids in the EOR process [9–12], it is crucial that polymeric surfactants be water soluble [13–17]. This polymeric surfactant has the potential to replace more conventional surfactant polymer compositions since it has both IFT reduction and mobility control properties inside a single chemical component. Adsorption of polymeric surfactant in the oil reservoir's highly permeable zone lowers the relative permeability of the reservoir to oil.

In order to create polymeric surfactants with various types of substitution, several authors have worked on them and modified their structures [18–22]. In the presence of an initiator, the polymeric surfactants are synthesized via a reaction between the functional groups of various polymers and the hydrophobic groups of a surface-active monomer. Our previous research by Babu et al. [23] used castor oil to create a new polymeric surfactant with low IFT (102 mNm) and appropriate viscosity. Cao and

N. Saxena (✉) · Md. M. Islam · D. Sharma
School of Basic Sciences and Technology, IIMT University, Meerut, India
e-mail: gemini.neha1990@gmail.com

P. Pillai
Halliburton, Mumbai, India

S. Rattan et al. (eds.), *Emerging Trends in Synthesis and Catalysis in Chemistry*,
Springer Proceedings in Materials 51, https://doi.org/10.1007/978-981-97-5169-3_22

Li [24] used carboxy methyl cellulose and alkyl poly(etheroxy) acrylate to create a new class of polymeric surfactants. A novel polymeric surfactant was synthesized by grafting a sulphonated group onto the hydrophobic groups of the polymer backbone, as was described by Ye et al. [25]. Polymeric surfactant was created by Elraies et al. [18] by reaction of sodium methyl ester sulphonate with acrylamide monomer. With the goal of creating a material with superior interfacial characteristics, Xu et al. [26] synthesized an acrylamide-alkyl (phenyl) poly(oxyethylene) acrylate copolymer.

In this study, we used acrylamide as the monomer, potassium metabisulphite as the initiator, and water as the solvent for a free radical polymerization technique to create an anionic polymeric surfactant of methyl ester sulphonate derived from palm oil. FTIR and TGA analyses were used to characterize the synthesized polymeric surfactant. As a preliminary step in evaluating a synthetic polymeric surfactant for use in EOR, its interfacial physiochemical parameters, such as its interfacial free energy (IFT) and surface tension, were investigated.

2 Experimental Procedures

2.1 *Materials*

Palm oil was used to create methyl ester sulphonate (MES). Acetone, hydroquinone, sodium chloride, acetic acid, sodium metabisulphite, and acrylamide (99%) were bought from Merck India. The chemicals were utilized in their original form. The IFT crude oil was obtained from ONGC in Ahmedabad, India. At 30 °C, the oil has a viscosity of 0.247 Pa s and a gravity of 23.41°API. The total acid number is 0.058 mg KOH/g.

2.2 *Synthesis of Polymeric Methyl Ester Sulphonate*

Sulphonation of the methyl ester of palm oil was the first step in preparing methyl ester sulphonate for synthesis [27]. Palm oil mostly consists of the fatty acids oleic, stearic, and palmitic. Using ferric chloride and potassium metabisulphite as a redox pair, acrylamide was used in a free radical polymerization reaction to create a surfactant copolymer. Surfactant (MES) was dissolved in 40 mL of distilled water and placed in a three-necked flask fitted with a thermometer, inert gas (N_2) inlet tube, and mechanical stirrer. After adding 10 g of acrylamide, the mixture was heated to 20 °C. To the flask were added 0.5 mL of a 5% $FeCl_3$ aqueous solution and 3 mL of a 0.5% (w/v) K2S2O5 aqueous solution. The reaction was run for three to four hours at a temperature of 20 °C. Hydroquinone was used as a reaction terminator. The finished product was placed into the acetone solution (50 ml) to precipitate it. The precipitate was collected and washed many times in methanol. The last step was

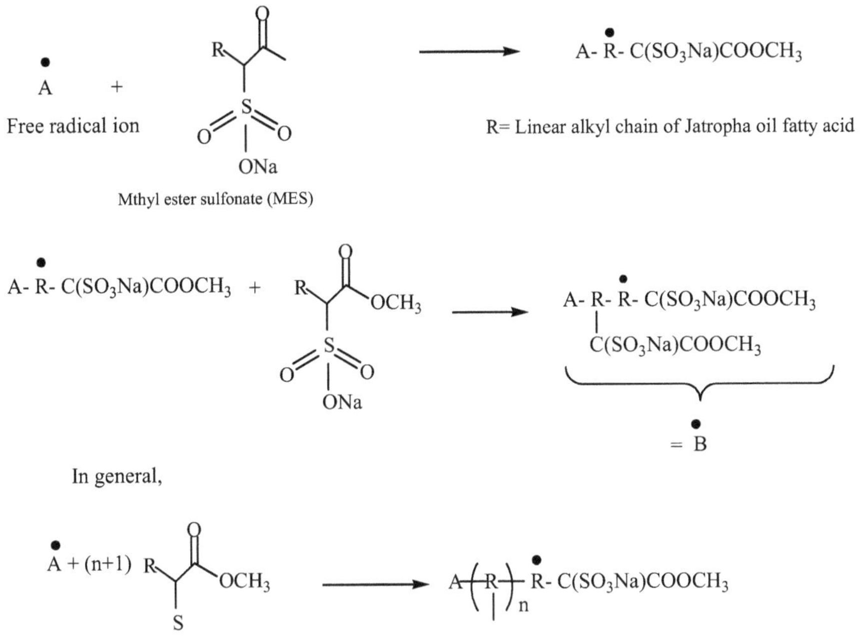

Fig. 1 Reaction scheme for synthesis of polymeric surfactant

drying the substance in a hoover drier for 24 h before characterizing it using FTIR spectroscopy. In Fig. 1 we have an illustration of the suggested chemical process for the production of polymeric surfactant.

2.3 *Analysis of Polymeric Surfactant Using FTIR*

The synthesized anionic polymeric surfactant was studied using a mid-IR spectrophotometer (FTIR; PerkinElmer, Spectrum-2) between 400 and 4000 cm^{-1}. The synthesized product was pelletized with potassium bromide (KBr) using a hydraulic press; the pellets thus prepared were used for FTIR analysis.

2.4 *Surface Properties of Synthesized Polymeric Surfactant*

2.4.1 Surface Tension Measurement

At 30 °C, the surface tension of the synthetic polymeric surfactant was determined using the Du Noüy ring technique in a tensiometer. Different quantities of synthesized polymeric surfactant aqueous solutions, from 2000 to 8000 ppm, were generated using distilled water. The surface tension of each solution was measured three times and averaged the results. A ring was cleaned in acetone and dried over flame before taking any measurements.

2.4.2 IFT Study

Using a spinning drop SVT 20 tensiometer (Dataphysics, Germany), authors determined the interfacial tension between aqueous solutions of the synthesized polymeric surfactant and crude oil (23.41 API gravity). At 30 °C, the IFT of both produced polymeric surfactant solutions (2000–10,000 ppm) and crude oil was determined. After filling the glass tube with the solution, 2 mL of crude oil was injected to determine the interfacial tension. After around a thousand seconds of spinning, the glass tube was inserted inside the tensiometer, and the equilibrium IFT value was recorded. To calculate the IFT, we used the following formula:

$$\sigma = \frac{\omega^2 R^3 \Delta\rho}{4} \tag{1}$$

σ is the interfacial tension, ω is the rotational speed, R radius of oil droplet, and $\Delta\rho$ is the density difference between crude oil and polymeric surfactant solution.

3 Results and Discussion

3.1 Characterization of Polymeric Surfactant by FTIR

An anionic polymeric surfactant, polymethyl ester sulphonate (PMES), was created from the anionic surfactant (MES) synthesized in a free polymerization procedure. FTIR analysis was used to characterize the product. The FTIR spectrum (Fig. 2) of a polymeric surfactant was measured at 30 °C. The stretching vibration at C = O is analysed with a peak at 1162 cm^{-1}, proving the presence of a sulphonate group. The prominent peak at 1450 cm^{-1} is due to the bending vibration of the methyl group (C–H), which is asymmetrical. As a result of N–H stretching, acrylamide (primary and secondary amides) is introduced into the surfactant, producing a strong peak at 3570 cm^{-1}. The presence of amide groups is indicated by a C = O stretching vibration in the infrared spectrum between 1685 and 1635 cm^{-1}.

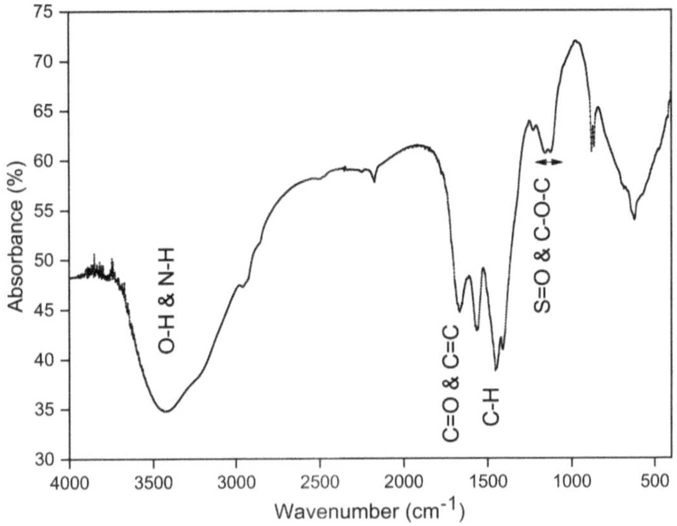

Fig. 2 FTIR spectrum of PMES surfactant at 30 °C

3.2 *Physiochemical Properties of Synthesized Polymeric Surfactant*

3.2.1 Surface Tension

At 30 °C Fig. 3 displays the results of a measurement of the surface tension of a diluted aqueous solution of an anionic polymeric surfactant synthesized in a laboratory. Increased adsorption of the surfactant at the surface causes the surface tension to drop as the PMES concentration rises, as seen in Fig. 3. The surface becomes saturated with the surfactant after a specific concentration, and subsequent additions of the compound have no effect on the surface tension. The term "critical micelles concentration" (CMC) is used to describe this value for the surfactant concentration. When a surfactant is dissolved in water, the hydrophilic moiety attracts water molecules while the hydrophobic group is repelled by the polar phase. Because polymeric surfactant molecules adsorb at the air–water contact, decreasing the repulsive force, the surface tension drops as the concentration does. Further increases in concentration had no effect on the surface tension because of the persistent repulsion force between water molecules and the negatively charged group (SO_3-Na +) in polymeric surfactant. A surfactant's CMC is proportional to the length of its hydrophobic chain. The CMC value is reached at a lower concentration for longer chains than it is for shorter chains [28]. It was shown in this research that the CMC of PMES is significantly lower than that of surfactant used as a monomer. The influence of salinity on the surface tension is also seen in Fig. 3. The lowest amount of surface tension, 32.5 m N/m, is reached with a salinity of 2.5 wt%. The electrolyte promotes the diffusion of surfactant from

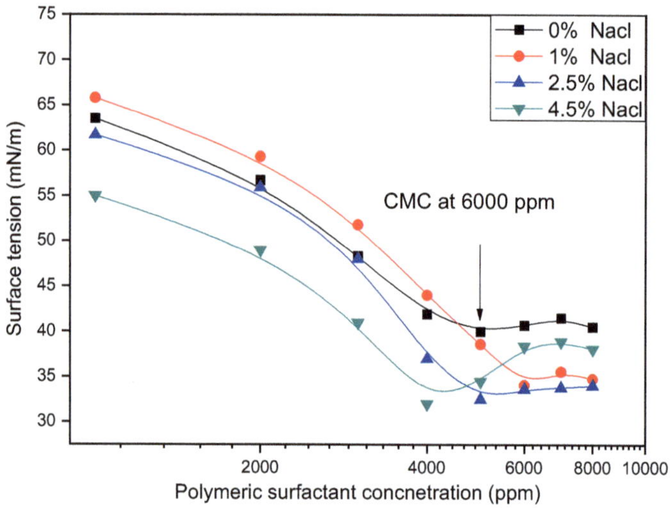

Fig. 3 Surface tension of the synthesized polymeric surfactant solution at different salinity and various concentration at 30 °C

the bulk of the solution to the interface; that is why surface tension decreases in its presence of salts [29].

3.2.2 Interfacial Tension (IFT) Behaviour

Traditional polymers, on the other hand, have been shown to have little to no influence on IFT reduction and, in certain cases, even raise the IFT value. Many polymeric surfactants have been shown to decrease IFT by anything from 2 to 7 mN/m [30]. Polymeric surfactants are not as effective as traditional surfactants for lowering ultra-low IFT. The equilibrium IFT shifts with PMES concentration are depicted in Fig. 4. The IFT value is reduced from 23 mN/m at 0 ppm to 2.69 mN/m at 6500 ppm of polymeric surfactant concentration, demonstrating the effectiveness of the synthetic anionic polymeric surfactant. Both surface tension and IFT are observed to be at their lowest at the CMC of PMES. As a result, we may deduce that a concentration of 6500 ppm for commercial use of polymeric surfactant in EOR process is optimal. As can be shown in Fig. 4 the equilibrium IFT falls with increasing PMES concentration due to adsorption of polymeric surfactant at the oil water interface, but this trend stops at the CMC. Micelles are formed in the majority of the solution after CMC when additional surfactant is added.

Saline water, which is often present in an oil reservoir, plays a critical role in the EOR process for residual oil extraction by mitigating IFT and wettability change caused by surfactant flooding. So, it is important to look at how salinity affects the efficiency with which synthetic polymeric surfactant inhibits IFT. At 30 °C, a spinning drop tensiometer was used to study how salinity affected the IFT decrease

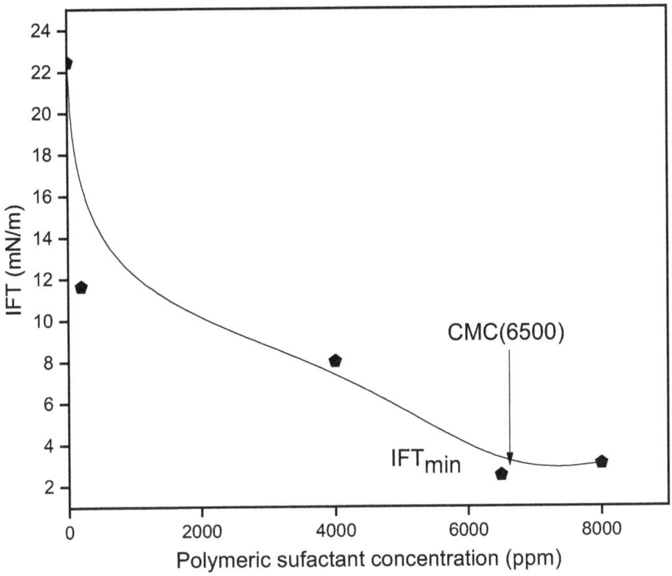

Fig. 4 Equilibrium IFT of PMES solution at various concentrations at 30 °C

of a synthetic polymeric surfactant. Figure 5 displays the influence of salinity on IFT suppression while using the optimum polymeric surfactant concentration of 6500 ppm. Figure 5 demonstrates that the IFT declines from 0 to 2.5 wt% salinity, with the least value occurring at 2.5 wt% salinity (0.35 mN/m). Surfactants have a propensity to adsorb at the interface when salts are present, which aids in the suppression of IFT [31]. Adding more salt to the solution has a negligible effect on the IFT. Crude oil characteristics, polymeric surfactant component (hydrophilic and hydrophobicity), and salt concentration are only a few of the factors that might lower or raise the interfacial tension (IFT) of a solution containing anionic polymeric surfactant. The presence of salts in a polymeric surfactant solution has been hypothesized to be changed by IFT through a number of different methods. While the presence of oil-in-water micro-emulsions with excess oil is a key mechanism in lowering IFT at lower salinities, the presence of water-in-oil micro-emulsions with excess water is a key mechanism in raising IFT at higher salinities beyond the optimum value (2.5 wt% in the present study). The polar component of the polymeric surfactant also has a significant impact on the oil-brine IFT change. The reduction in solubility of polymeric surfactant in water, which causes an increase in IFT of PMES solution at salinity above optimal salinity, may be caused by the attraction of Na+ ion and the negative charge of the polar moiety polymeric surfactant.

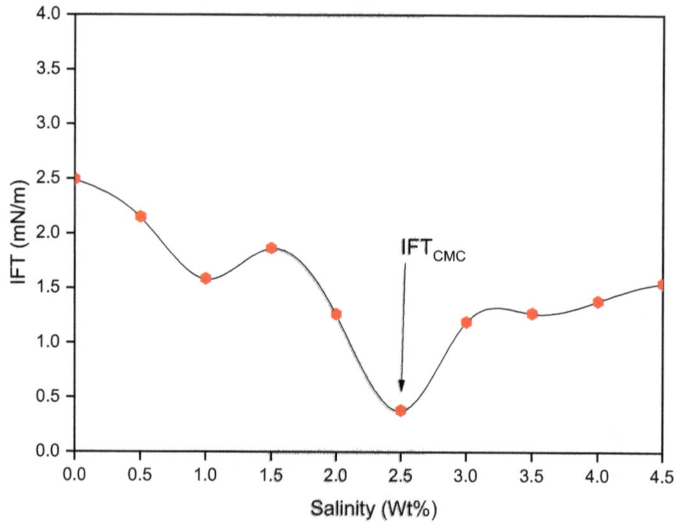

Fig. 5 Equilibrium IFT versus salt content at CMC of 6500 ppm PMES aqueous solution

4 Conclusion

In the current study a novel polymeric surfactant was synthesized from palm oil to improve the oil recovery. Chemical enhanced oil recovery benefits from the polymeric surfactant's unique combination of surfactant and polymer characteristics. Measurements of surface tension for CMC determination, IFT reduction, were used to assess the polymeric surfactant's viability for use in the EOR process. Good surface activity may be seen at low doses using the polymeric surfactant. At critical micelle concentration (CMC) of 6500 ppm PMES solution, the surface tension drops from 72 to 42 mN/m, and it drops even further to 32 mN/m in the presence of salt (2.5 wt%). In the absence of 2.5 wt% NaCl, the IFT value was 2.69 mN/m, but in its presence it was 0.35 mN/m at a 6500 ppm polymeric surfactant concentration. The measurement of surface tension and IFT are the screening criteria for the application of any surfactant or polymer in the oil recovery process. The newly synthesized polymer-based surfactant has potential in reducing the surface at oil-aqueous and oil-air interfaces and, hence, presents itself as a substitute to commercial used surfactant for enhanced oil recovery.

References

1. Zhang Z, Li J, Zhou J (2011) Microscopic roles of "viscoelasticity" in HPMA polymer flooding for EOR. Transp Porous Media 86(1):199

2. Chang HL (1978) Polymer flooding technology yesterday, today, and tomorrow. J Petrol Technol 30(8):1113
3. Sandiford BB (1964) Laboratory and field studies of water floods using polymer solutions to increase oil recoveries. J Petrol Technol 16(8):917
4. Urbissinova TS, Trivedi JJ, Kuru E (2010) Effect of elasticity during viscoelastic polymer flooding: a possible mechanism of increasing the sweep efficiency. J Can Pet Technol 49(12):49
5. Standnes DC, Austad T (2000) Wettability alteration in chalk 2. Mechanism for wettability alteration from oil-wet to water-wet using surfactants. J Petrol Sci Eng 28(3):123
6. Hou B, Wang Y, Huang Y (2015) Mechanistic study of wettability alteration of oil-wet sandstone surface using different surfactants. Appl Surf Sci 330:56
7. Gao B, Sharma MM (2013) A family of alkyl sulfate Gemini surfactants. 2. Water–oil interfacial tension reduction. J Colloid Interface Sci 407:375
8. Dehghan AA, Masihi M, Ayatollahi S (2015) Phase behavior and interfacial tension evaluation of a newly designed surfactant on heavy oil displacement efficiency; effects of salinity, wettability, and capillary pressure. Fluid Phase Equilib 396:20
9. Chen Q, Wang Y, Lu Z, Feng Y (2012) Thermoviscosifying polymer used for enhanced oil recovery: rheological behaviors and core flooding test. Polym Bull 70(2):391
10. Wei B, Romero-Zerón L, Rodrigue D (2014) Mechanical properties and flow behavior of polymers for enhanced oil recovery. J Macromol Sci Part B 53(4):625
11. Wever DAZ, Picchioni F, Broekhuis AA (2011) Polymers for enhanced oil recovery: a paradigm for structure–property relationship in aqueous solution. Prog Polym Sci 36(11):1558
12. Abidin AZ, Puspasari T, Nugroho WA (2012) Polymers for enhanced oil recovery technology. Procedia Chem 4:11
13. Sorbie KS (2013) Polymer-improved oil recovery. Springer Science & Business Media
14. Taylor KC, Nasr-El-Din HA (1998) Water-soluble hydrophobically associating polymers for improved oil recovery: a literature review. J Petrol Sci Eng 19(3):265
15. MacWilliams DC, Rogers JH, West TJ (1973) Water-soluble polymers in petroleum recovery. In: Water-soluble polymers. Springer, pp 105–126
16. Donaldson EC, Chilingarian GV, Yen TF (1989) Enhanced oil recovery, II: processes and operations, vol 17. Elsevier
17. Needham RB, Doe PH (1987) Polymer Flooding review. J Petrol Technol 39(12):1
18. Elraies KA, Tan IM (2003) The application of a new polymeric surfactant for chemical EOR
19. Akstinat MH (1980) Polymers for enhanced oil recovery in reservoirs of extremely high salinities and high temperatures. In: SPE oilfield and geothermal chemistry symposium; society of petroleum engineers
20. Chen CSH, Sheppard EW (1981) Oil recovery by waterflooding employing anionic polymeric surfactants. Google Patents
21. Depazbanez M (2000) Synthesis of novel cationic polymeric surfactants. Polymer 41(24):8501
22. Voronov S, Kohut A, Tarnavchyk I, Voronov A (2014) Advances in reactive polymeric surfactants for interface modification. Curr Opin Colloid Interface Sci 19(2):95
23. Babu K, Pal N, Saxena VK, Mandal A (2015) Synthesis and characterization of a new polymeric surfactant for chemical enhanced oil recovery. Korean J Chem Eng 32(4):1
24. Cao Y, Li H (2002) Interfacial activity of a novel family of polymeric surfactants. Eur Polymer J 38(7):1457
25. Ye L, Huang R, Wu J, Hoffmann H (2004) Synthesis and rheological behavior of poly [acrylamide–acrylic acid–n-(4-butyl) phenylacrylamide] hydrophobically modified polyelectrolytes. Colloid Polym Sci 282(4):305
26. Xu J, Shi L, Ye M (1997) Conformation of amphiphilic branch chain in copolymer surfactants on interface. J Polym Sci, Part B: Polym Phys 35(5):827
27. Elraies KA, Tan IM, Awang M, Saaid I (2010) The synthesis and performance of sodium methyl ester sulfonate for enhanced oil recovery. Pet Sci Technol 28(17):1799–1806
28. Han F, Deng Y, Wang P, Song J, Zhou Y, Xu B (2012) Synthesis and characterization of glucosamide surfactant. J Surfactants Deterg 16(2):155

29. Woolfrey SG, Banzon GM, Groves MJ (1986) The effect of sodium chloride on the dynamic surface tension of sodium dodecyl sulfate solutions. J Colloid Interface Sci 112(2):583
30. Bera A, Ojha K, Mandal A, Kumar T (2011) Interfacial tension and phase behavior of surfactant-brine–oil system. Colloids Surf, A 383(1–3):114
31. Ruckenstein E, Rao IV (1987) Interfacial tension of oil—brine systems in the presence of surfactant and cosurfactant. J Colloid Interface Sci 117(1):104

Synthesis of Metal Telluride Nanomaterials and Their Application in Catalysis

Nandita Chauhan, Nikhil Sharma, Prabhu Prasad Das, and Ashish Chalana

1 Introduction

The living organism suffers oxidative damage when reactive oxygen species (ROS) and hydroxyl radicals (·OH) are produced in excess during cellular metabolism [1, 2]. The human body contains reactive oxygen species, which are extremely harmful free radical ions that have been shown to oxidatively damage DNA, protein, and lipid [3]. Cardiovascular, neurological, and other illnesses are all influenced by this oxidative damage [4, 5]. It is widely known that when there are too many reactive oxygen species present, they combine with the transition metals in our bodies to produce extremely reactive free radicals [6, 7]. Equation 1 below illustrates Fenton and Fenton-type reactions, which produce the extremely poisonous ·OH radical that harms our biomolecules [8].

$$Fe^{2+} + Cu^{+} + H_2O_2 \rightarrow Fe^{3+} + Cu^{2+} + \cdot OH + OH^{-} \quad (1)$$

For all living things to carry out their typical physiological functions, copper and iron are essential redox active metals [9]. Metal homeostasis is often maintained by a unique mechanism in cells. However, overstress and improper management of the processes lead to an excessive concentration of unbound metal ions that either create the reactive oxygen species ·OH or poisonous molecules (H_2O_2). The term "oxidative stress" refers to this circumstance [10]. Increased levels of transition metal ions like copper and iron have also been linked to different diseases like Alzheimer's, Menke's, and Wilson's diseases as well as conditions like diabetes, cancer, and amyotrophic lateral sclerosis (ALS) [11].

N. Chauhan · N. Sharma · P. P. Das · A. Chalana (✉)
Department of Chemistry and Biochemistry, Sharda University, Greater Noida, U.P. 201306, India
e-mail: ashish.chalana@sharda.ac.in

A. Chalana
Center for Development of Biomaterial, Sharda University, Greater Noida, U.P. 201306, India

S. Rattan et al. (eds.), *Emerging Trends in Synthesis and Catalysis in Chemistry*, Springer Proceedings in Materials 51, https://doi.org/10.1007/978-981-97-5169-3_23

In dietary sources, chalcogens (an element of family of chalcogens: selenium) are largely present as organic selenomethionine and organic selenite [12]. Because there is ample clinical evidence supporting the value of selenium disease prevention, particularly chemo-preventions, researchers have been intensely interested in selenium supplements for the past several decades [13]. According to a trial report from Nutritional Prevention Cancer (NPC), supplementing with 200 g of selenium per day reduced cancer incidence by 37%, including prostate cancer [14].

Additionally, chalcogen (selenium) is now known to be incorporated into numerous human body proteins, including glutathione peroxidase, thioredoxin reductases, iodothyronine deiodinases, etc. [15]. According to reports, the elements selenium, selenomethionine, selenocysteine, and di-selenide derivatives can prevent oxidative damage caused by Cu^{+} and Fe^{2+} [16]. According to reports, selenium and its complexes can protect against ROS damage brought on by heavy metal poisoning [17]. Additionally, chalcogens (selenium) binding is considered to guard against lead-induced toxicity [18].

In addition, tellurides (another element of family of chalcogens) and its related metal compounds play a significant role in catalysis, particularly in heterogeneous catalysis [19]. The telluride composed with metal cation is commonly used as catalysts for hydrogenation and dehydrogenation reactions [20]. They can efficiently promote the addition and removal of hydrogen atoms from organic compounds. For example, palladium telluride (PdTe) is a well-known catalyst for hydrogenation reactions [21], while copper telluride (CuTe) [22] and nickel telluride (NiTe) are effective catalysts [23] for dehydrogenation reactions and CuSe in detoxification of mercury [24]. Metal tellurides, such as molybdenum telluride ($MoTe_2$), have shown excellent HDS activity [25].

Although numerous publications on the synthesis and application of M_nTe_m (where $M_nTe_m = Bi_2Te_3, Sb_2Te_3$, and MnTe, $Cu_{2\text{-}x}Se$ [26, 27]) have been published. Contrarily, there are surprisingly few papers on the synthesis of copper tellurides nanoparticles due to its complex phase diagram, which includes numerous stoichiometries and crystal morphologies and leads to a wide range of applications [26]. In this regard, Cu_2Te (hexagonal), CuTe (orthorhombic), Cu_3Te_2 (tetragonal), and Cu_7Te_4 (trigonal) are the most notable range of stoichiometries [28]. As a result, it is challenging to control the morphology, size, phase, and shape of the $Cu_{2\text{-}x}Te$ nanocrystal during synthesis because of their significant stoichiometry variations in the Cu-Te phase diagram, when compared to their analogues of sulfur and selenides [29]. The bulk of techniques for creating copper tellurides nanoparticles rely on pricy ingredients, intricate equipment, and unusual environments. As a result, a simple, affordable method for creating copper tellurides nanoparticles is widely required.

So, we report for the first time a simplest method to synthesis copper telluride nanoparticles in one step using copper (II) acetate and telluride powder as starting ingredients in the presence of oleic acid and oleyl amine (1:1) by using a solvothermal technique. Using a variety of methods, including TEM, SEM, and EDX XRD, the as-synthesized nanoparticles were comprehensively characterized. Herein we decided to explore the combination of both copper and chalcogen in the form of Cu_2Te nanoparticles for the application in catalyzing/neutralizing the free radical generated by copper/ascorbate system. We have used protein carbonyl assay to shown that Cu_2Te nanoparticle can catalyze/neutralize the production of free radical and protect the biomolecules such as protein from copper-induced oxidative damage.

2 Experimental Section

2.1 *Materials and Methods*

Bovine serum albumin, acetic acid, ascorbate, hydrogen peroxides, copper (II) acetate, telluride powder, oleic acid, and oleyl amine were acquired from Sigma Aldrich and used without additional purification. TEM was carried out using the TecnaiTM G2 20 apparatus. EDX on Zeiss EVO 50 and EVO combined with scanning electron was performed. Nanoparticle XRD data was collected using X'Pert PRO. For the XRD, 25 mg of the material was cast onto the glass slides. The glass slide was scanned in a range of 20–80° for two hours. The data was also matched using a JCPDS file. Prior to the EDS analysis, the sample underwent an hour-long ultrasonic treatment. The sample was cast onto carbon tape and given an hour to cure before analysis.

3 Synthetic Procedure

3.1 *Synthesis of Copper Tellurides Nanoparticles*

1 mmol of copper (II) acetate was dissolved in an equal mixture of oleic acid and oleyl amine in a flask with a three-necked circular bottom. The mixture was then expelled for 20 min at 30 °C. In 5 mL of oleic acid, 0.50 mmol of telluride powder was dissolved. The dispersed telluride powder solution was added drop by drop over a period of 60 min, and the temperature was raised to 180 °C. After injection, the reaction mixture immediately changed from green to a blacking-brown color. Over the course of two hours, the reaction was continually agitated. The reaction was stopped and the liquid was allowed to cool to room temperature. After being centrifuged three times with a toluene and ethanol solution at 8000 rpm, the as-nanoparticles were separated. The resulting nanocrystals were dispersed in acetone for further characterization.

3.2 The Procedure of Protection of Protein from Copper-Mediated Oxidative Damage

The protein carbonylation experiment was conducted using the protein known as bovine serum albumin (BSA) [30]. The procedure was followed as mentioned below by following the literature with some modification. BSA (1 mg/mL) was oxidize using 20 M $CuSO_4$ and 100 μM of ascorbate. About 300 μM of H_2O_2 was added to the experiment. The aforementioned solution was supplemented with various concentrations of copper tellurides nanoparticles, which was then incubated for a one hour at room temperature in phosphate buffer. Following an additional hour of incubation, 20 mM of DNPH (made in 2 N HCl) was added. The reaction mixture was once more incubated with 20% trichloroacetic acid for 20 min after 30 min. The sample was then centrifuged at a speed of 10,000 rpm. During centrifugation, the protein pellet precipitated out and was removed. The protein pellet was cleaned with a 5 mL solution of ethanol and ethyl acetate. The pellets that were produced were dissolved in 10 M guanidine at pH 2.3. UV–Vis spectroscopy was measured as the outcome of copper telluride nanoparticles' prevention of protein carbonylation or protein oxidation. Per milligram of protein, data was expressed as nanomoles of carbonyl group.

4 Result and Discussion

4.1 Synthesis and Characterization of Copper Telluride Nanoparticles

Copper telluride nanoparticles were synthesized using copper (II) acetate and telluride powder as a starting precursor. The schematic diagram of preparation of the copper telluride nanoparticles was shown in Fig. 1. The synthesis of copper telluride nanoparticles was carried out using solvothermal method, and the synthesis is just one single-step reaction. Oleyl amine and oleic acid were first mixed in a 1:1 ratio with copper acetate, and the mixture was then stirred for 30 min. The size and form of the nanoparticles were controlled using a combination of oleyl amine and oleic acid. After that, telluride powder was dissolved in oleic acid that was added drop by drop for 30 min. The reaction temperature was increased to 180 °C, the mixture was agitated for 2 h, and a brownish-black product was produced. This product was centrifuged and collected after drying in the oven. Here, we have used oleic acid (OAc) and oleylamine (OAm), a pair of chelating chemicals, that are the main focus in the regulated production of many nanoparticles. They are cheap ligands with a high boiling point (350 °C) that are liquid at room temperature and also both have almost same length and (almost 2 nm).

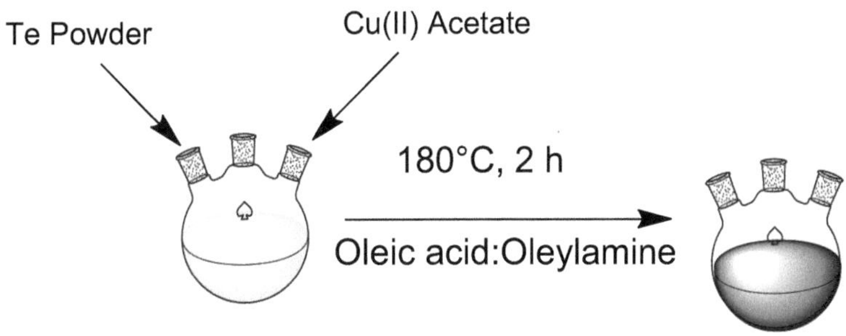

Fig. 1 Synthetic scheme for preparation of copper telluride nanoparticles using solvothermal method

4.2 Characterization of Copper Telluride Nanoparticles

The as-synthesized copper telluride nanoparticles were characterized using various characterization technique. After the synthesis, the first step is to confirm the elements and composition of the nanoparticles. We have performed EDS analysis. The acronym EDS stands for energy-dispersive X-ray spectroscopy. It is an analytical technique commonly used in materials science and chemistry to determine the elemental composition of a sample. EDS is typically performed in conjunction with SEM. The SEM produces a focused electron beam that interacts with the sample, causing the emission of characteristic X-rays from the elements present in the material. These emitted X-rays are then collected and analyzed by an energy-dispersive X-ray detector. By measuring the energy spectrum of the emitted X-rays, the elemental composition of the sample can be determined. The EDS technique provides qualitative and quantitative information about the elements present in the sample. Qualitative analysis involves identifying the elements based on their characteristic X-ray energies. Quantitative analysis involves determining the concentration or weight percentage of each element present in the sample. We have performed the EDS analysis of as-synthesized nanoparticles. The 20 mg/mL was dispersed in the acetone for 20 min and added on carbon tape using drop-casted method. The sample was dried in open air for a duration of 30 min. Once the sample is dried, it was loaded in the instrument for analysis. Figure 2a shows the EDS analysis of as-synthesized nanoparticles. The EDS clearly detects the copper and telluride nanoparticles. Table 1 shows the weight percentage and atomic percentage of the copper telluride nanoparticles. The atomic ratio of the copper to telluride nanoparticles was found to be 65.66:34.42 which clearly shows the copper telluride was found to be in Cu_2Te composition.

We have also carried out powder X-ray diffraction (XRD) research to further confirm the composition. An analytical method called powder X-ray diffraction (XRD) is used to identify the crystal structure and phase composition of a powdered material. A powdered sample is subjected to an X-ray beam in powder XRD. According to Bragg's rule, the sample's crystalline structure interacts with the X-rays, causing them to diffract at particular angles. To learn more about the sample's crystal

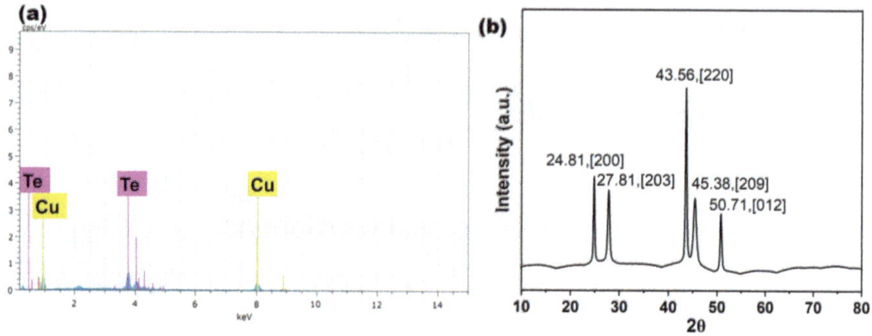

Fig. 2 EDS (**a**) and PXRD (**b**) analysis of copper telluride nanoparticles

Table 1 Weight and atomic (%) detected from EDS analysis

Element	Weight (%)	Atomic (%)
Copper	51.08	65.58
Telluride	48.92	34.42

structure and phase composition, the diffracted X-rays are collected, identified, and the resulting diffraction pattern is examined. In powder XRD, a series of peaks that correspond to the particular crystallographic planes in the sample form the diffraction pattern. The peak position (2θ angle) and intensity of these peaks provide valuable information about the arrangement of atoms within the crystal lattice and the type and abundance of different crystallographic phases present in the sample. As seen in Fig. 2b, we have assessed the various 2θ angle obtained from the PXRD instrument and correlated with the planes related to each angle by looking the literature reports [31]. The characteristic peak at 24.81°, 27.81°, 43.56°, 45.38°, and 50.71° can be indexed to the (200), (203), (220), (209), and (012) planes of Cu_2Te. Each peak that was assigned to it had a perfect match in the JCPDS file 04-019-0775 [31]. The results were confirmed by XRD analysis which shows the presence of Cu_2Te phase. The Cu_2Te was found to be in pure phase structure, meaning no other contaminants were identified. The XRD's sharp peaks further demonstrated the product is in the crystalline form. The calculated crystalline size from PXRD was 21.33 ± 15 nm.

Further, after getting the conformation of the phase and composition, it is important to know the size and morphology for the Cu_2Te nanoparticles. So, TEM and SEM are two powerful techniques used for imaging and analyzing materials at nanoscale, while both techniques utilize electron beams for imaging with different operational principles and applications. In TEM, a focused beam of electrons passes through a thin sample, transmitting through it. The sample needs to be very thin (typically less than 200 nm) to allow electrons to pass. TEM provides high-resolution images of the internal structure and morphology of the sample. Whereas in SEM, a focused beam of electrons scans the surface of a sample, causing various interactions with the sample surface. SEM provides high-resolution surface imaging of the sample, revealing details of the surface morphology, topography, and texture. So, the morphology and

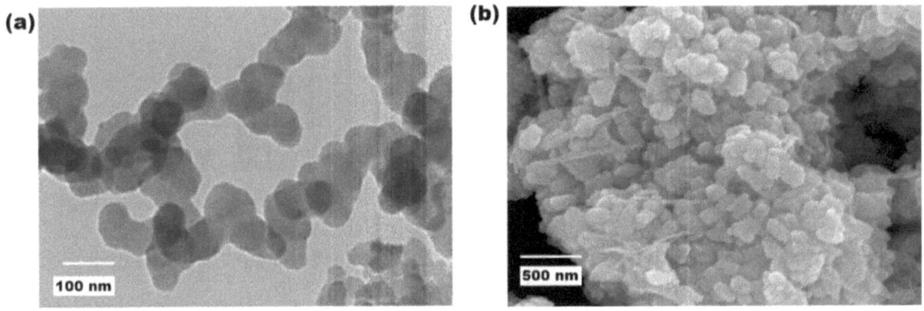

Fig. 3 TEM (**a**) and FE-SEM (**b**) analysis of copper telluride nanoparticles

size of Cu_2Te nanoparticles were assessed using TEM and FE-SEM. For the TEM and FE-SEM analysis, 1 mg/mL of sample was sonicated in acetone for 30 min. One or two drop was drop-casted at copper grid (in case of TEM) and carbon grid (in case of FE-SEM). After drying the sample in air, sample was loaded for the analysis. As shown in Fig. 3a, the nanoparticle was found to be spherical shape. In order to calculate the average of the nanoparticle, multiple images were taken and the size (average) was found to be 55 ± 15 nm. Similar way, FE-SEM was also well agreed with our TEM result. The shape of the nanoparticle was found to be in spherical shape. The FE-SEM image is shown in Fig. 3b. The particle was found to be uniformly distributed in TEM, whereas in FE-SEM, the particle was found to be in agglomeration form. However, both gave us the same conformation regarding the morphology of the Cu_2Te nanoparticles.

5 Application in Protecting Effect on Protein from Copper Catalyzed Oxidative Damage by Catalyzing/Neutralizing the Copper-Induced Free Radical

Since chalcogens-based compounds are known to reduce copper-mediated oxidative damage by reducing the free radical concentration [30], this motivated us to investigate the protective effect the copper telluride nanoparticles from copper-mediated oxidative damage by free radical to protein. The reactive oxygen species converts the $-NH_2$ (amino) group of protein into the oxidative carbonyl groups [31]. The conversion of the amino group to the carbonyl group from the copper-induced free radical oxidative damage reflects the extent of oxidation damage to protein and can be measured in a quantitative manner [32]. This data is known as protein carbonyl formation (PCF) and can be analyzed using UV–Vis spectrophotometry at an absorbance change at 370 nm [32]. The details were mentioned in the experimental section. Bovine serum albumin (BSA) protein was used to perform in this experiment. About 2 mg/mL of BSA was incubated for 60 h with copper (II) and an ascorbate system. This resulted in the formation of **47.88** nmol/mg protein carbonyls

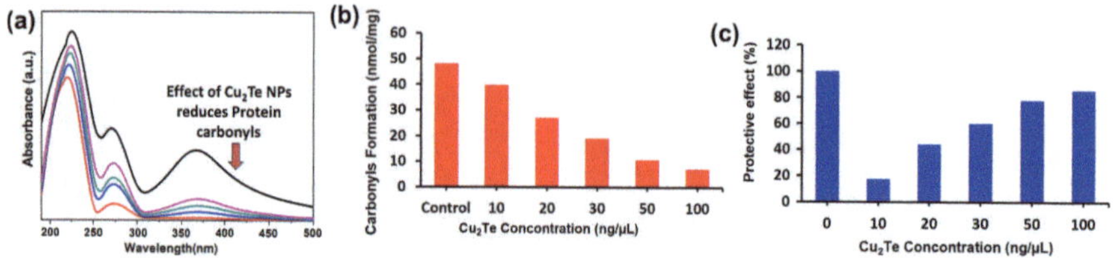

Fig. 4 **a** Reduction in carbonyl formation in the presence of Cu_2Te nanoparticles, and **b** protective effect on protein from copper-induced oxidative damage

Table 2 Effect of the copper telluride nanoparticles on the formation of protein carbonylation produced by copper (II) and ascorbate system

	Copper (II)/ascorbate/H_2O_2	
Cu_2Te concentration	Carbonyls (nmol/mg)	Percentage of inhibition (%)
Control	47.88	–
10 ng/μL	39.55	17.39
20 ng/μL	26.77	44.08
30 ng/μL	18.99	60.33
50 ng/μL	10.55	77.96
100 ng/μL	6.88	85.63

(Table 1). Whereas when Cu_2Te nanoparticle was used under the same physiological condition along with copper (II) and ascorbate/H_2O_2, it observed a significant reduction in the protein carbonylation percentage. At 10, 20, 30, 50, and 100 ng/μL concentrations, only 39.55, 26.77, 18.99, 10.55, and 6.88 nmol/mg carbonyls were found as shown in Fig. 4a. As shown in Fig. 4b and Table 2, the concentration of Cu_2Te increased the oxidative damage to protein that was significantly reduced. At the concentration of 100 ng/μL of Cu_2Te around 85.63% reduction was observed in protein carbonyls. This result concluded that Cu_2Te nanoparticles successfully neutralized the free radical produced by the copper/ascorbate/H_2O_2 system and inhibited oxidative damage to protein.

6 Conclusion

Copper tellurides nanoparticles were synthesized using the solvothermal method using the copper (II) acetate and tellurium powder as the starting precursors at 180 °C. A combination of oleylamine and oleic acid was used to control the size and shape of the Cu_2Te nanoparticles. The nanoparticles were assessed using various characterization techniques such as TEM, EDS, and PXRD. TEM showed that the nanoparticles were in a spherical shape with an average size of 55 nm. EDS showed that the nanoparticles the chemical composition of the nanoparticles were Cu_2Te. XRD also showed that the nanoparticle is in crystalline form. Cu_2Te nanoparticles were also assessed for the protection of biomolecules such as protein. It was shown that the nanoparticles were able to protect protein (BSA) from copper-induced oxidative damage. The Cu_2Te nanoparticles were found to neutralize the copper-induced free radical and inhibited the protein carbonyl formation by 17.39, 44.08, 60.33, 77.96, and 85.63% by using 10, 20, 30, 50, and 100 ng/μL. From these studies, it can be concluded that Cu_2Te nanoparticles can be an alternate source to protect biomolecules and cells from oxidative damage either induced by heavy metals such as copper or hydrogen peroxide.

Conflict of Interest On behalf of all authors, the corresponding author states that there is no conflict of interest.

References

1. Arora A, Sairam RK, Srivastava GC (2002) Oxidative stress and antioxidative system in plants. Curr Sci 1227–1238
2. Mohamed W, Sandhir R (eds) (2023) Trace elements in brain health and diseases. Springer Nature
3. Juan CA, Pérez de la Lastra JM, Plou FJ, Pérez-Lebeña E (2021) The chemistry of reactive oxygen species (ROS) revisited: outlining their role in biological macromolecules (DNA, lipids and proteins) and induced pathologies. Int J Mol Sci 22(9):4642
4. Krzemińska J, Wronka M, Młynarska E, Franczyk B, Rysz J (2022) Arterial hypertension—oxidative stress and inflammation. Antioxidants 11(1):172
5. Rotariu D, Babes EE, Tit DM, Moisi M, Bustea C, Stoicescu M, Bungau SG (2022) Oxidative stress–complex pathological issues concerning the hallmark of cardiovascular and metabolic disorders. Biomed Pharmacother 152:113238
6. Li Y, Wei J, Cui N, Li J, Xu M, Pan G, Li J (2023) Recent advance of atomically dispersed dual-metal sites carbocatalysts: properties, synthetic materials, catalytic mechanisms, and applications in persulfate-based advanced oxidation process. Adv Funct Mater 2301229
7. Halliwell B (1989) Free radicals, reactive oxygen species and human disease: a critical evaluation with special reference to atherosclerosis. Br J Exp Pathol 70(6):737
8. Gao H, Cao Z, Liu H, Chen L, Bai Y, Wu Q, Wang M (2023) Multifunctional nanomedicines-enabled chemodynamic-synergized multimodal tumor therapy via Fenton and Fenton-like reactions. Theranostics 13(6):1974
9. Jomova K, Baros S, Valko M (2012) Redox active metal-induced oxidative stress in biological systems. Transition Met Chem 37:127–134
10. Pisoschi AM, Pop A, Iordache F, Stanca L, Predoi G, Serban AI (2021) Oxidative stress mitigation by antioxidants-an overview on their chemistry and influences on health status. Eur J Med Chem 209:112891
11. Gaggelli E, Kozlowski H, Valensin D, Valensin G (2006) Copper homeostasis and neurodegenerative disorders (Alzheimer's, prion, and Parkinson's diseases and amyotrophic lateral sclerosis). Chem Rev 106(6):1995–2044
12. Nasim MJ, Zuraik MM, Abdin AY, Ney Y, Jacob C (2021) Selenomethionine: a pink Trojan redox horse with implications in aging and various age-related diseases. Antioxidants 10(6):882
13. Oldfield JE (1987) The two faces of selenium. J Nutr 117(12):2002–2008
14. Rayman MP (2005) Selenium in cancer prevention: a review of the evidence and mechanism of action. Proc Nutrition Soc 64(4):527–542

15. Mugesh G, du Mont WW, Sies H (2001) Chemistry of biologically important synthetic organoselenium compounds. Chem Rev 101(7):2125–2180
16. Kang D, Lee J, Wu C, Guo X, Lee BJ, Chun JS, Kim JH (2020) The role of selenium metabolism and selenoproteins in cartilage homeostasis and arthropathies. Exp Mol Med 52(8):1198–1208
17. Ercal N, Gurer-Orhan H, Aykin-Burns N (2001) Toxic metals and oxidative stress part I: mechanisms involved in metal-induced oxidative damage. Curr Top Med Chem 1(6):529–539
18. Sharma VK, McDonald TJ, Sohn M, Anquandah GA, Pettine M, Zboril R (2017) Assessment of toxicity of selenium and cadmium selenium quantum dots: a review. Chemosphere 188:403–413
19. Bouroushian M (2010) Electrochemistry of metal chalcogenides. Springer Science & Business Media
20. Trofimov BA (2002) Acetylene and its derivatives in reactions with nucleophiles: recent advances and current trends. Curr Org Chem 6(13):1121–1162
21. Arora A, Oswal P, Rao GK, Kumar S, Singh AK, Kumar A (2021) Catalytically active nano-sized Pd 9 Te 4 (telluropalladinite) and PdTe (kotulskite) alloys: first precursor-architecture controlled synthesis using palladium complexes of organotellurium compounds as single source precursors. RSC Adv 11(13):7214–7224
22. Pathan HM, Lokhande CD, Amalnerkar DP, Seth T (2003) Preparation and characterization of copper telluride thin films by modified chemical bath deposition (M-CBD) method. Appl Surf Sci 218(1–4):291–297
23. Tang YJ, Zou Y, Zhu D (2022) Efficient water oxidation using an Fe-doped nickel telluride–nickel phosphide electrocatalyst by partial phosphating. J Mater Chem A 10(23):12438–12446
24. Chalana A, Karri R, Mandal SC, Pathak B, Roy G (2019) Chemical degradation of mercury alkyls mediated by copper selenide nanosheets. Chem Asian J 14(24):4582–4587
25. Genin HS, Ibers JA (1998) Solid state transition metal sulphides: some aspects of structure and physical properties. Trans Met Sulphides: Chem Catal 1–35
26. Guo M, Gu S, Xu S, Lu J, Wang Y, Zhou G (2022) Design, synthesis and application of two-dimensional metal tellurides as high-performance electrode materials. Front Chem 10:1023003
27. Chalana A, Karri R, Das R, Kumar B, Rai RK, Saxena H, Gupta A, Banerjee M, Jha KK, Roy G (2019) Copper-driven deselenization: a strategy for selective conversion of copper ion to nanozyme and its implication for copper-related disorders. ACS Appl Mater Interfaces 11(5):4766–4776
28. Buathet S, Kodchakorn S, Pakpoom R, Veeramol V, Pichanan T, Duanghatai R, Nareerat K, Auttasit T (2020) Electrochemical performance of Bi_2Te_3 heterostructure thin film and Cu_7Te_4 nanocrystals on undoped and In 3+-doped WO3 films for energy storage applications. Electrochim Acta 341:136049
29. Pashinkin AS, Fedorov VA (2003) Phase equilibria in the Cu–Te system. Inorg Mater 39:539–554
30. Chalana A, Rai RK, Karri R, Jha KK, Kumar B, Roy G (2022) Interplay of the intermolecular and intramolecular interactions in stabilizing the thione-based copper (I) complexes and their significance in protecting the biomolecules against metal-mediated oxidative damage. Polyhedron 215:115647
31. Mishra M, Patil NN, Zubair M, Kapuria N, Lebedev V, Adegoke TE, Ryan KM, Singh S (2022) Phosphine free synthesis of copper telluride nanocrystals in 1D and 2D shapes using Dipehylditelluride (DPDTe) as an air-stable source. Nanotechnology 33(30):305602
32. Suzuki D, Miyata T, Saotome N, Horie K, Inagi R, Yasuda Y, Uchida K, Izuhara Y, Yagame M, Sakai H, Kurokawa K (1999) Immunohistochemical evidence for an increased oxidative stress and carbonyl modification of proteins in diabetic glomerular lesions. J Am Soc Nephrol 10(4):822–832

MIX
Papier aus verantwortungsvollen Quellen
Paper from responsible sources
FSC® C105338

FSC
www.fsc.org

If you have any concerns about our products,
you can contact us on
ProductSafety@springernature.com

In case Publisher is established outside the EU,
the EU authorized representative is:
Springer Nature Customer Service Center GmbH
Europaplatz 3, 69115 Heidelberg, Germany

Printed by Libri Plureos GmbH
in Hamburg, Germany